What Is Justice?

What Is Justice?

Classic and Contemporary Readings

Edited by

Robert C. Solomon

Mark C. Murphy

NEW YORK · OXFORD
Oxford University Press
1990

Oxford University Press

Oxford   New York   Toronto
Delhi   Bombay   Calcutta   Madras   Karachi
Petaling Jaya   Singapore   Hong Kong   Tokyo
Nairobi   Dar es Salaam   Cape Town
Melbourne   Auckland

and associated companies in
Berlin   Ibadan

Copyright © 1990 by Oxford University Press, Inc.

Published by Oxford University Press, Inc.,
200 Madison Avenue, New York, New York 10016

Oxford is a registered trademark of Oxford University Press

Library of Congress Cataloging-in-Publication Data
What Is Justice? : classic and contemporary readings / edited by
Robert C. Solomon, Mark C. Murphy.
p.   cm.
Includes bibliographical references.
ISBN 0-19-506544-1
ISBN 0-19-506050-4 (pbk)
1. Justice.   I. Solomon, Robert C.   II. Murphy, Mark C.
JC578.W47   1990   89-77183
320.01'1—dc20

9 8 7 6 5 4

Printed in the United States of America
on acid-free paper

for
Thomas and Ann Murphy
and
Eugene and Kathryn Higgins

ACKNOWLEDGMENTS

Our thanks to

Jon Solomon, Victor Caston, Don Becker, Douglas Kellner,

Kenneth Riley, Paul Saladino, and our editor at Oxford,

Cynthia Read.

CONTENTS

PART IV

PART V

What Is Justice?

INTRODUCTION

"What is justice?" asked Socrates in Plato's *Republic*, and ever since, it has been one of the leading questions of philosophy and all social thinking. For Plato and for Aristotle after him, justice, in its most general sense, was the essential virtue, the summary virtue, the virtue most important for the "social animals" that we are, living together in ever-larger communities, cities, and nation-states. But even in the *Republic*, the answer to that question—the definition of "justice" (or rather, *dikaiosune*)—is at best controversial and provides no clear criteria for making the choices—just or unjust—that we make in everyday life. Socrates dispatches his various interlocutors and their proposed answers, such as "justice is giving and getting one's due," but it is not at all evident that, after rejecting such context-bound and merely conventional replies, Socrates is able to supply a single definition and provide a singular criterion that accounts for justice in its various contexts. He says in the *Republic* Book IV that justice is "doing one's own"—every person's performing his or her proper role in the community—but it is by no means obvious how we should translate this into concrete decisions and policies.

We do become clear, however, about what justice is not. In one of the classic early exchanges in the *Republic*, the philosophical thug Thrasymachus argues the ultracynical line that justice always serves the interests of the rulers of the society, "the advantage of the stronger." If you are an ordinary person, you are only hurting yourself by trying to live in accordance with justice. This shocking thesis is refuted by Socrates, right is properly distinguished from mere might, and Thrasymachus walks off in a huff. In Thrasymachus' abandoned place, Glaucon suggests a more modest hypothesis, that justice is ultimately just a matter of self-interest, and people adhere to its conventions only to avoid punishment. Socrates takes this suggestion much more seriously, but he ultimately insists that justice is not merely a matter of convention and, in the vulgar sense intended, it is not a matter of self-interest either. But then Socrates spends the rest of the *Republic* taking us through a whirlwind of philosophical considerations as he speculates about metaphysics and human nature, praises (some of) the political ideals of ancient Greece, and introduces his own rather radical republicanism to show that justice must be counted as desirable for its own sake, that justice is harmony in the soul as it is harmony in the state, that justice is the rule of reason and, finally, that justice even "pays off" in the end, for the just man can ultimately suffer no harm (to his soul, at least). But what we do not get— what we thought we would get—is anything like an adequate criterion concerning what sorts of considerations we should use in evaluating this or that social arrangement or rule. Plato does tell us that responsibility should be delegated in accordance with ability and "place," but what about the distribution of wealth in society? As a spokesman for the aristocracy, Plato was disdainful of money and markets and said very little about them. How should the goods of society get distributed? What should we do about the poor? How much should a doctor or a lawyer or a soldier or a tradesman be paid? What Socrates promises us is a standard of justice; what we get is an elaborate metaphor. But so it has been ever since, with the great discussions of justice.

On the one hand, the question "what is justice?" is an invitation to the most abstract sort of philosophical speculation. What is the good (the best) society? What makes a government legitimate? What kind of creatures did (or does) God (or Nature) intend us to be? What is our essential relationship to our fellow beings, and what obligations do we have to one another? Where do these obligations come from? On the other hand, the question of justice focuses our attention on the concrete problems of our times: Is it just for there to be poor people living virtually next to people who have more money than they could ever possibly spend? Is it fair that hard-working people of considerable talent go unrewarded, while others, smiled-upon by fortune and raised with wealth and power, are constantly "rewarded" in return for no work and no contribution to society whatever? Should the rich be taxed to help the poor? Should "unearned income" be taxed the same as "earned" income? Should men and women receive the same wages for the same jobs, without regard to need? Should people be paid or should students be graded on the basis of their efforts or their results? Do people whose ancestors were treated unfairly deserve compensation for what their grandfathers suffered? What should a society do with those who break the law? Should we execute criminals for the most heinous crimes? A theory of justice has the extremely difficult task of bridging this abyss between the abstract and the eminently practical. No theory of justice can long remain on the luxurious level of philosophical speculation without diving down into the particular-ities of social life, but no attempt to solve the problems of daily politics can long sustain itself without reaching up to the heights of philosophy, struggling as Socrates struggled to come to grips with the definition of justice, with its essential nature and justification.

What is justice? The original meaning of the term, for instance in the Old Testament and in the *Iliad*, is very much bound up with punishment, or more accurately, with retribution and revenge. But even by the time of Socrates and in the teachings of Jesus, revenge had been clearly separated from justice and was viewed as a vice instead of a virtue. The problem of punishment would remain a central concern of every imperfect society, but justice became more a matter of social harmony (in Plato) and mercy (in Christian ethics). In Plato, the ideal of justice was wholly bound up with the ideal community, while already in the Torah, as later in the Gospels and the Koran, the idea of justice was bound up with the belief in and obedience to a merciful (if also retributive) God. But even in ancient times, the concept of justice had particular application to the details of ordinary household life, to the question of fair wages and the distribution of rewards and honors. The particular concept of justice developed by Aristotle, for example, has much to do with fair exchange and "equality" (that is, more accurately, proportion) or what we would call "desert," what a person deserves or has earned. In modern times, the focus of justice has come to be more and more concentrated on such questions of distribution and exchange and, in particular, questions of private property and individual liberty. But here, too, the apparent focus reveals a blur of competing images: How are we to reconcile the supposedly inalienable right of individuals to own private property with the tragic existence of misery and poverty in the same society, misery and poverty which is often not unrelated to precisely the same

activities through which the wealthy obtained their property? How are we to reconcile the rights of individuals to hold onto property that they did not in any sense earn
(e.g., through inheritance or a lucky bet in the state lottery) with our insistence that
people should earn and thus deserve what they've got? Indeed, insofar as justice is
the legal and moral right to hold onto what one (or one's family or community)
already has, is it too far-fetched to suggest that, once again, we find ourselves faced,
in neo-Thrasymachian language, with the rights of the stronger, the entitlement of
the powerful and the privileged to hold onto what they already have? What are the
rights, in such a framework, of the poor and unpropertied? Or are we putting too
much emphasis on property rights as the focus of justice? Are other rights and
liberties not ultimately more important than the right to unlimited ownership? And
is the overall public good not more important than the rights and liberties of any
particular individual? Or should our focus be somewhere else as well, on the
conception of universal equality, according to which it is any *difference* in the
distribution of wealth that needs justification instead of the alternative presumption
that established differences are by that very fact legitimate, and it is any redistribution (e.g., through taxes) that needs justification.

Such are the dimensions of the question, "what is justice?" The very general
issues that faced Socrates in the *Republic* are still with us now. Are our standards of
justice, ultimately, in the interests of the stronger, the more established and most
powerful citizens? Is faithfully following the principles of justice in any sense in one's
own interest and to one's own advantage? Is our concept of justice really just a social
convention, perhaps a matter of agreement among the members of this society but
possibly quite different in others? Indeed, given the enormous amount of disagreement among us concerning the right conception of justice, can we suppose with any
confidence that there is, ultimately, some single standard or "definition" to be found?
But in addition, there are all of those other, more specific and more urgent practical
questions that need resolution, both ancient and very modern. How should we think
of punishment, as retaliation or as retribution or as mere public revenge against
those who have criminally assaulted us or violated our laws, or as a more future-
oriented attempt to deter future crime and reform wayward citizens? How should we
understand that very modern sense in which everyone is "created equal"? How do we
(or can we) justify—and from what perspective should we view—the often enormous
disparities between the rich and the poor? How do we weigh the importance of
individual rights and liberties against the public good? And how much trust should
we put in that peculiarly modern social institution—the free market—as a vehicle for
assuring justice? Or is that the wrong way to look at the role of the marketplace in
society? Perhaps the market defines its own conception of justice and should not be
treated as a mere means to other, possibly archaic conceptions. Perhaps the market
replaces or is incompatible with the concept of justice. All of this presumes, of
course, that there is an adequate answer to our original question, "what is justice?" Is
it fairness? Is it equal treatment? Is it desert? Is it "getting one's due"? And does this
concept of justice depend on a particular context, a particular set of social goals and
conventions, or is it something bigger and more universal than that, perhaps provided by God or in any case an intrinsic part of human nature?

Despite its obvious importance and eminent practicality, the question of justice
seems to come and go as a central topic of concern, and by the beginning of this
century it got eclipsed somewhere along the line and fell out of favor in contempo-

rary philosophy. Perhaps the rather expansive question of justice was put aside to make room for other social and moral questions that seemed to have more precision. Perhaps it was because social and political philosophy had been eclipsed for such a long time by metaphysics and theology and, more recently, dismissed as tangential to "mainstream" interests in epistemology and the philosophy of language. Perhaps it was because the question of justice came to seem too practical to philosophers concerned with much more abstract questions and too immense to philosophers focused on seemingly much more manageable issues. Perhaps it was because the interest in human nature fell under the anthropologist's or the existentialist's ax or simply shifted into the social science departments. Perhaps it became too apparent that seemingly abstract theories of justice had long stood as façades for other more immediate concerns—to legitimize a revolution or a dictatorship, or to defend the status quo and the sanctity of private property, for example. But it is safe to say that the subject of justice, perhaps in somewhat altered form or as part of some related concern, perhaps in the guise of a shadow or even in its very absence, has been at the core of social thinking ever since Plato and Aristotle?

Today, however, it is clear that the question of justice has returned to center stage in Anglo-American philosophy. In 1971 Harvard philosopher John Rawls published his epochal book, *A Theory of Justice*, and the old Socratic question has never been more alive. Only 3 years later, Rawls's younger colleague, Robert Nozick, published his own theory of justice, a very different sort of theory indeed and something of a rejoinder to Rawls's. Between the two of them, quite a dialectic has been established (though the two of them have rarely responded in public to one another), and the ferocity of the debate has even spilled over into popular press (e.g., *Esquire* magazine (March 1983) "Robert Nozick vs. John Rawls"). The difference between the two might be (and is often) characterized as the difference between a "liberal" and a "libertarian" theory, but such politically loaded designations do little to help philosophical understanding. In a nutshell, Rawls tries to find a proper ordering between equality and liberty with particular concern for the needs of the "least advantaged" in society; Nozick is anxious to defend a particularly strong notion of "entitlement," such that a just world would be one in which everyone had just what they were entitled to, without reference to needs or inequalities. But the dispute is, for all of its current interest and importance, only the most recent and rather narrow manifestation of a 3,000- (and more) year-old debate. Rawls's concern for universal equality and individual liberty would not have been intelligible to Plato and Aristotle, and Nozick's exclusive insistence on private property rights and virtually total neglect of any concept of "community" would have horrified them (as it still does most cultures around the world). But what is particularly revealing is that neither Rawls nor Nozick adequately acknowledge what the ancients and many moderns would consider the heart of justice, and that is the concept of desert. Moreover, both of them are concerned only tangentially with questions of punishment and with questions of social status and honors that cannot be "cashed out" in economic terms. To be sure, the question of whether the state has the right to tax some people in order to help others in need—one of the central questions that arise from the Rawls and Nozick debate—is a lively and emotional issue in these days of neo-Reaganomics, mass homelessness, billion-dollar weapons systems, and trillion-dollar deficits, but the question of justice and the range of human concerns that it expresses are much more expansive than that, and we have tried to capture some of that expanse in this book.

What is justice? What are the origins of justice? What are the different types of justice and how do they relate to one another? Can we submit our concern for the poor, our insistence that people earn what they have, and our need to punish those who break the law under a single, all-embracing criterion? Is there a single overriding concept of justice that encompasses all of our activities and crosses the borders of all cultures, no matter how socially different? Is justice, as Rawls and Cicero long before him argued, the leading virtue of societies—or are there other virtues more central, more important even than justice? Could justice be, in certain circumstances, dispensible, or perhaps irrelevant? What is the relationship between justice and the law— is there or could there be any justice apart from legal institutions? Is there some "natural" (or Divine) law, from which our civil laws (at least, those that are right) are derived? What or where are the sources of justice, in God or human nature, in our rational capacities to reason dispassionately and abstractly, in the history of (particular) human practices, in our search for social harmony and utility or in the recently discovered supposedly rock-bottom fact of individual freedom? Is justice conventional or "natural"? Does it consist primarily in respect for authority, respect for principles, or respect for persons? Can we conceive of a unifying answer that does justice to all of these, or is justice—as Socrates' interlocutors unsuccessfully insisted—whatever rules or rulers dominate a particular community at a particular time?

What are the definitive ingredients in justice? Is it need (as the Marxists say, "to each according to [his] needs")? Is it merit (in Aristotle, for example, who one is and what one has done)? What is the role of punishment in justice—central or peripheral? Is it equality (and in what sense is it true that "all [people] are created equal")? How much weight should tradition and the status quo have in determinations of justice, or are they simply irrelevant to what is just and fair? How much consideration should the history of injustices place on our current disputes over justice, or are they too "just history" (as an American president recently said) and irrelevant to what is just and fair now? What special pressures does our "free enterprise" system put on justice, or does it embody a special theory of justice itself? Does the free enterprise system reward merit (e.g., good ideas and hard work) as is often suggested by its promoters, or is it much more a matter of luck? What are we to make of the concept of private property, which played such an enormous role in the development of modern British social philosophy but surprisingly little elsewhere? (Rousseau suggested, in his "Discourse on the Origins of Inequality," that the invention of private property was tantamount to a crime against humanity, and Pierre Proudhon flatly asserted that "property is theft.") What are we to make of the notion of "rights" that has become so central to our own thinking about justice—and so prone to abuse? (Do ordinary citizens really have a "right" to possess and carry military assault weapons?) How do we weigh these ingredients, need against merit against rights in an adequate conception of justice? This is no abstract question, a mere plaything for the philosophers. Our public policy depends on (or simply assumes) an answer, and everyone who pays even a dollar in taxes (to pick out one particularly tangible consequence) has a stake in it.

And what are we to make of punishment? Is it "natural" to want to punish someone who has wronged you, to hurt them as they've hurt you, or is retribution an invention of society and thus, perhaps, properly and strictly a function of the law? Should we actually *punish* people at all, or should we rather restrict our punitive efforts to deterring crime and reforming criminals? Should we instead turn our attention to rooting out the conditions of crime, placing much more emphasis on rewards than punishments, or is this unrealistic, and if it is unrealistic, is it therefore any less just? Such are the questions of justice. Our current debates are but the most recent manifestations of a set of concerns that are as old as "civilized" society itself. Indeed, one might hypothesize that it is just that set of concerns that defines a society as "civilized." It is with this in mind that we present the following selections, not as an attempt to answer Socrates' question, but in an effort to get a new generation engaged in one of the most important and practical philosophical activities of our time.

This book is divided into five parts, although not according to any single dimension or set of categories. Our aim has been to provide a wide variety of topics and selections of both historical and current interest ranging from the presuppositions of our own conceptions of justice and society (to be found, for example, in the United States Constitution and Bill of Rights) to the nature of justice as such (if there is such a thing), from the most basic obligations one citizen or one human being owes another to the relative roles of deterrence and retribution in various theories of punishment.

Part I provides a number of ancient discussions of justice, from which our contemporary conceptions and debates are inevitably derived. It includes excerpts from the Bible (both Old and New Testaments) and the Koran, an excerpt from Homer's *Iliad* in which the primitive conception of justice as vengeance is uncompromisingly expressed, and a sizable selection from Plato's *Republic*. We have included most of Aristotle's discussion of justice in his *Nicomachean Ethics* and Saint Thomas Aquinas's later reflections on the wisdom of "the Philosopher" (Aristotle). To avoid provincialism (but not only for that reason) we have also included an expression of the Confucian view of justice in ancient China, according to Confucius' leading disciple, Mencius.

Part II concerns what is today the single most prevalent metaphor in the discussion of the conceptual and/or historical origins of justice, the idea of a "social contract." Whether this is supposed to be a historically real or merely imaginary and hypothetical event, the social contract is an agreement on the basis of which society was (or should have been) founded by rational, autonomous human beings who saw their way to compromising their various and often divisive self-interests in order to establish a just society in which both their own and others' self-interests would be maximized. Prior to such an agreement, so the harshest version of the story (true or merely imagined) goes, there was an undisciplined "state of nature" in which, whether individuals were hostile or "indifferent" to one another, they did not cooperate and, consequently, had none of the benefits or securities made possible by civilized society. In particular, there was no justice, for justice, according to the promoters of this pervasive and popular idea, is only an "artificial" virtue, a matter of social convention or invention and not a "natural" virtue at all. Social contract theory came to all but define social and political philosophy by the eighteenth century, particularly from the works of Thomas Hobbes, John Locke, and Jean-Jacques Rousseau, and as embodied in our own Declaration of Independence and

Constitution. Today, the dual imagery of the social contract and the state of nature remains at the very core of our debates about justice, for example, in the more abstract form of what John Rawls calls "the original position."

Fascination with the idea of presocietal humanity goes back many centuries before modern philosophy, however, and there is even some semblance of a social contract theory in Plato. The selections in Part II, accordingly, begin once again with the ancient philosophers of Greece, and Plato in particular, and then trace various formulations of the social contract and the supposedly preceding state of nature through modern and contemporary social philosophy. Not all of the authors included in Part II actually endorse the idea of the social contract, or the state of nature, or both. Robert Nozick, for instance, accepts the argument from the state of nature but rejects Rawls's reliance on any form of social contract, however hypothetical or merely speculative. Michael Sandel rejects the very idea of a presocial humanity as well as the idea of the formation of society by way of the collective agreement by autonomous presocial individuals. His argument is not just that there was *in fact* no state of nature or social agreement (virtually everyone would concede that point), but rather that such an idea, even imaginary, is unintelligible. But Sandel's argument is hardly novel, and its ancestry is to be found, in particular, in one author who is too rarely included in these discussions. The German philosopher G. W. F. Hegel attempted his own allegorical representation of presocial beings, in conscientious contrast with Hobbes and Rousseau in particular, in the famous "master and slave" section of his book, *The Phenomenology of Spirit*. But Hegel's point, like Sandel's, is that the very idea of autonomous individuals in a presocial state of nature, even as a hypothetical alternative to society, makes no real sense and, in any case, is a disastrous presupposition for social philosophy.

Part III moves beyond the discussion of the origins of justice into the more contemporary question of the role of justice and its justification in modern society. Quite apart from the origins of justice, what function does it play in our culture—or in any culture? Is it, as Hobbes insisted, an attempt to guarantee security and safety? Is it, as John Locke insists, primarily a matter of "convenience" and protection for our natural right to private property? The place and justification of the notion of private property, accordingly, is a central concern of Part III, and along with the right to private property the notion of rights in general. So, too, the subject of the public good, the primary concern of that all-important moral philosophy called "utilitarianism," is much at issue here. Is social "utility" in fact the primary goal of justice? Is it possibly the only goal? Or is it, as the critics of utilitarianism have maintained, incompatible and antithetical to justice?

The selections in Part III begin with our own Declaration of Independence and the Bill of Rights (plus selected amendments) to our Constitution. We have then provided readings from John Locke, David Hume, and Adam Smith (his *Theory of the Moral Sentiments* as well as his better known and more influential *Wealth of Nations*) expressing the central concern with private property that characterized traditional modern British social philosophy, and a short excerpt from Rousseau's second *Discourse* expressing a more pessimistic view of the role of private property in social history. We have included an important selection from Kant's *Metaphysical Elements of Justice*, another excerpt from Hegel (this time from his *Philosophy of Right*), and a pair of excerpts from Karl Marx's co-author and colleague Friedrich Engels (who was much influenced by Hegel) that are, predictably, deeply critical of the institution of private property and the injustices it engenders. We have included a substantial selection from the rightly famous and controversial fifth chapter of John Stuart Mill's *Utilitarianism*. As we mentioned earlier, utilitarianism was often and still is charged with being incompatible with justice (for example, by John Rawls in *A Theory of Justice*) on the grounds that social utility may sometimes sanction or indeed require violation of the rights of one or a few individuals. Mill here argues (as Hume did before him) that there is no real incompatibility between justice properly understood and utility, and the former serves the latter.

One of the key presuppositions of utilitarianism, but of most modern social philosophy as well, is that all people are in some essential sense equal. But this insistence on equality, as we have already warned, is not clearly compatible with several other crucial ingredients of justice. In particular, it is not obviously compatible with the inevitable inequalities that result from the hustle and bustle of the free market and it is clearly incompatible with the idea of individual merit, in which it is our differences that define justice rather than our shared similarities. And so to round out Part III, we have included an excerpt from the German liberal (now we would call him "libertarian") philosopher Friedrich von Hayek, who argues that the liberties basic to the free market preclude efforts to assure equality or recognize merit. In contrast, we have included an excerpt from Cambridge-Berkeley philosopher Bernard Williams on "Equality," defending that notion. Finally, to put all of this in perspective, we have included a cross-cultural analysis of justice by British philosopher/sociologist David Miller—with the suggestion that there may be no one conception of justice that is valid in very different societies.

Part IV involves a shifting of gears into the study of retributive rather than distributive justice—the logic of punishment. Current debates tend to focus on the effectiveness of punishment (and capital punishment in particular) as a *deterrent* and as a *reformative* agent, but a glance at the literature shows that the debates about punishment go back long before the reform movements of the eighteenth century (on which these debates are based) to the days when punishment was still, more than anything else, a matter of retribution and revenge. The famous passage from Exodus, "an eye for an eye, a tooth for a tooth . . ." is the culmination of a long ancient debate, in which the mark of justice was made out to be the "measure" of retribution, the proper payment of a "debt" to society or to the wronged person or party. The main goal was neither deterrence nor reform but rather to find the punishment that would "fit" the crime and "pay off" the criminal's debt. But in the development of our increasingly civilized views of justice, this ancient sense of retribution—and with it the even more dangerous emotion of vengefulness—fell by the wayside, so much so that vengeance is hardly even mentioned in most contemporary debates, except to emphatically cast it out of the conversation. In Part IV, we have attempted to correct this bit of historical repression, and in addition to a number of readings from the current debates about deterrence and capital punishment we have included an equal number of reminders of the more brutal aspects of punishment. For example, we have included an infamous section from Nietzsche's *Genealogy of Morals* in which he discusses our apparent need for cruelty, a selection from a recent book on vengeance that gives us some details about Mediterranean codes of "honor" in which revenge plays a critical part and Susan Jacoby's recent study of "wild justice" as the most dangerous source of repression of our time. We have also included both the majority and minority opinions of the U.S. Supreme Court on one of the most controversial death penalty cases of recent years and a number of philosophical arguments and reflections on both sides of this unpleasant and brutal but essential aspect of justice.

Part V, finally, includes a number of selections from and reactions to John Rawls's monumental *Theory of Justice.* We begin with an earlier essay of Rawls, in which some of the basic positions of the book are clearly laid out and follow with his development of two principles of justice that define the structure of much of his later work. Critical responses include those of his colleagues Robert Nozick and Michael Sandel, two very different approaches to justice by Michael Walzer and Elizabeth Wolgast, a reply to both Rawls and Nozick together by Alasdair MacIntyre, and a reply to MacIntyre by Charles Taylor.

The volume is arranged so that readers and instructors will have considerable freedom and flexibility in ordering selections. We have tried as much as possible to keep our own biases and opinions out of both the text and its structure and to avoid current political labels and other devices that would tend to prejudice the reader one way or another. Needless to say, such even-handedness in editing as in justice is, however desirable, an ideal all but impossible to carry out in practice. But we hope that the result is as fair-minded and representative of the field—and as readable and enjoyable—as possible.

PART I

Our concept of justice has a long, convoluted history, steeped in philosophy and theology and in ancient practices that no doubt preceded recorded history altogether. Every human society, no doubt dating back far into prehistory, had its conceptions of right and wrong, the forbidden, the permissible and the commendable, the way things ought to be and the way they ought not to be. The great religious texts of the West, the Old and New Testaments and the Koran, are about justice—both human and divine—as well as more spiritual and transcendent matters. The great Homeric epics were not just chauvinist adventure stories but morality tales and lessons in justice. And it is no accident that the question of justice as posed by Socrates in the *Republic* marks something of an official beginning to Western philosophy.

In ancient Greece, there were at least two different terms that might be translated into English as "justice," one ("*dikaiosune*") usually translated as "justice," the other ("*ison*") better translated as "equality" (Vlastos, "Justice and Equality" in Brandt, R., ed. *Social Justice*). The former is graphically represented in Homer's *Iliad*. What Menelaus and Agamemnon mean in the enclosed short excerpt is justice as vengeance, indeed justice as legitimate genocide! So, too, the Old Testament punishments visited by God, for instance, on the sad cities of Sodom and Gomorrah underscore the original meaning of divine justice. It is against this brutal background that we should read Plato's *Republic* and Socrates' inspired defense of what was in fact a radically new conception of justice, justice as social and psychic harmony. But it is against the war-bound background of the ancient world, too, that we should read and understand the similarly radical instructions in the Bible. The well-known Old Testament injunction, "an eye for an eye, a tooth for a tooth" is not incitement to brutality but, quite the contrary, an attempt to limit vengeance and punishment to "measure for measure," harming just the culprit instead of, as in the *Iliad*, destroying his or her entire family, tribe, or city. But it is against this "eye for an eye" conception of justice that the renewed emphasis on mercy in the teachings of Jesus should strike us, much like the contrast between the Homeric and Socratic conceptions of justice. The Koran, too, displays a close kinship between justice and retribution but always tempered with mercy. But if this emphasis on retribution strikes us as uncivilized, in juxtaposition with our own strictly civilian emphasis on distributive justice, we should remind ourselves that the circumstances in which these ancient codes of justice were formulated were anything but peaceful. That, indeed, makes the antiretributive shift in justice in the great Greek philosophers and in the New Testament all the more remarkable. It also makes the almost wholesale neglect of retribution (and vengeance) in contemporary discussions of justice lamentable.

Plato's conception of justice as republican community and harmony requires getting rid of the more warlike and cynical images of justice, and this
explains the particular importance of the rather belligerent debate with Thrasymachus in Book I. By the time that Aristotle picks up the discussion a generation
later, in his *Nicomachean Ethics*, the centrality of civil justice and justice as a
personal virtue, quite apart from questions of power and vengeance, has been
established. Artistotle's conception of justice as "equality" may be easily mis-
understood, for while Aristotle was no doubt more of a democrat than his
teacher(s) Plato (and Socrates), he certainly did not have anything in mind like
the egalitarian distribution defended by modern liberal and leftist philosophers.
Indeed, Aristotle (like Socrates and Plato) was not concerned with what we call
"distributive" (or redistributive) justice at all, but rather "commutative" justice,
that is, the notion of a fair exchange. The central notion there is "proportion,"
and what Aristotle has in mind by "equality" is not that people should be treated
equally but rather the near-tautology that like cases should be treated alike, and
different cases treated differently. But with Aristotle, justice clearly completes
the shift from retribution and vengeance to the domestic virtues. We can see his
lasting influence over a thousand years later, when Saint Thomas Aquinas,
reviewing, defending and revising Aristotle from a Christian point of view,
invokes the long tradition of Aristotelian "natural law," a conception of justice
as built into the "nature" of human beings.

It is not just in the West that this struggle between warlike retribution and
civil virtues is taking place. It is, perhaps, with a due amount of Western humility
that we should read Confucius or his disciple Mencius, who encouraged a sense
of compassion and understanding when such sentiments were still struggling for
recognition in "the West," about the same time as Plato and Aristotle. But the
Confucian conception of justice embodies another struggle, too, that makes
itself evident in the ancient writings of the Mediterranean and Middle East. That
is the struggle for the recognition of a singular ultimate authority, whether it be
political or divine. In warlike worlds of scattered and mutually antagonistic
tribes and city-states the hunger for such a central authority had to be overwhelming. And so we should not be surprised by the authoritarianism of Plato's
Republic, any more than we are shocked by the emphasis on divine authority
and obedience to God in the Bible or the Koran. In ancient China, too, the
appeal to justice was also an appeal to authority, whether this appeal was to be
found in the state (as in Plato and Confucius) or in God (as in the Bible and the
Koran).

Homer, from the *Iliad*

Homer probably lived in Asia Minor about 800 B.C., at least 400 years after the bitter war in Troy that he so vividly described in the *Iliad*. There is considerable controversy about his life, his dates, his authorship, and, indeed, his very existence, for it has often been argued that there was no single poet named "Homer" who was the author of the *Iliad* and the *Odyssey*. The concept of justice as vengeance threads throughout the *Iliad* as part of its essential plot, from the opening offense of Agamemnon (kidnapping Apollo's priestess, Chryseis) for which Apollo takes his revenge, and Paris's taking of Helen (whom he had justly won as a prize from Aphrodite), to the wholesale act of revenge visited by the combined Greek forces on Paris's home-city of Troy. The action of the *Iliad* is largely defined by such acts of injustice and retribution, and in the brief passage that follows, Agamemnon summarizes a brutal ancient notion of justice (*"dike"*). The excerpt is taken from Book Six, in the midst of a ferocious battle between the Achaians and the Trojans.

     Now Menelaos of the great war cry captured Adrestos
alive; for his two horses bolting over the level land
got entangled in a tamarisk growth, and shattered the curving
chariot at the tip of the pole; so they broken free went
on toward the city, where many beside stampeded in terror.
So Adrestos was whirled beside the wheel from the chariot
headlong into the dust on his face; and the son of Atreus,
Menelaos, with the far-shadowed spear in his hand, stood over him.
But Adrestos, catching him by the knees, supplicated:
'Take me alive, son of Atreus, and take appropriate ransom.
In my rich father's house the treasures lie piled in abundance;
bronze is there, and gold, and difficultly wrought iron,
and my father would make you glad with abundant repayment
were he to hear that I am alive by the ships of the Achaians.'
     So he spoke, and moved the spirit inside Menelaos.
And now he was on the point of handing him to a henchman
to lead back to the fast Achaian ships; but Agamemnon
came on the run to join him and spoke his word of argument:
'Dear brother, o Menelaos, are you concerned so tenderly
with these people? Did you in your house get the best of treatment
from the Trojans? No, let not one of them go free of sudden
death and our hands; not the young man child that the mother carries

still in her body, not even he, but let all of Ilion's
people perish, utterly blotted out and unmourned for.'

The hero spoke like this, and bent the heart of his brother
since he urged justice. Menelaos shoved with his hand Adrestos
the warrior back from him, and powerful Agamemnon
stabbed him in the side.

The Bible, (Old and New Testament)

The book(s) that we know as The Bible were probably written over a substantial period of time, by a great many writers, from the Five Books of Moses (The Pentateuch or Torah), which is now thought to have been written well before 1000 B.C. (on the basis of stories that possibly dated back to 1800 B.C.) to the last book of the New Testament, The Book of Revelations, which was probably completed during the reign of the Emperor Domitian, between 81 and 96 A.D. There are volumes of literature describing the links and discontinuities between the Old and New Testaments (and the "intertestamental" Apocrypha between them), but it is enough for us here that we note one special difference in emphasis, at least, from the notion of divine retribution in the Old Testament to the insistence on mercy and forgiveness in the New. The following excerpts are taken from Genesis 18:20–33; 19:1–28, where the destruction of Sodom and Gomorrah is described and justified, from Leviticus 24:17–22 (Exodus 21:23) where the "Law of the Talon," "an eye for an eye" principle is announced, and the Sermon on the Mount according to Matthew 5:1–12, 38–42, where Jesus repudiates that doctrine. Then follows a selection from Isaiah (59:1–21), which details God's role as judge of human injustice, and then Job's curse and God's reply (from the Book of Job, 3:1–26; 8:1–22; 38:1–18) concerning what would seem to be the most flagrant case of divine *in*justice in the Old Testament. We end with the story of the Prodigal Son from Luke (15:11–32) to underscore the New Testament stress on the importance of repentance and redemption of the wicked and unjust, even in comparison with the already just.

Genesis 18:20–33; 19:1–28

20 And Jehovah said, 'Because the cry of Sodom and Gomorrah is great, and because their sin is very grievous; 21 I will go down now, and see whether they have done altogether according to the cry of it, which is come unto me; and if not, I will know.

22 And the men turned from thence, and went toward Sodom: but Abraham stood yet before Jehovah. 23 And Abraham drew near, and said, Wilt thou consume the righteous with the wicked? 24 Peradventure there are fifty righteous within the city: wilt thou consume and not spare the place for the fifty righteous that are therein? 25 That be far from thee to do after this manner, to slay the righteous with the wicked, that so the righteous should be as the wicked; that be far from thee: shall not the Judge of all the earth do right? 26 And Jehovah said, If I find in Sodom fifty righteous within the city, then I will spare all the place for their sake. 27 And

Abraham answered and said, Behold now, I have taken upon me to speak unto the Lord, who am but dust and ashes: 28 peradventure there shall lack five of the fifty righteous: wilt thou destroy all the city for lack of five? And he said, I will not destroy it, if I find there forty and five. 29 And he spake unto him yet again, and said, Peradventure there shall be forty found there. And he said, I will not do it for the forty's sake. 30 And he said, Oh let not the Lord be angry, and I will speak: peradventure there shall thirty be found there. And he said, I will not do it, if I find thirty there. 31 And he said, Behold now, I have taken upon me to speak unto the Lord: peradventure there shall be twenty found there. And he said, I will not destroy it for the twenty's sake. 32 And he said, Oh let not the Lord be angry, and I will speak yet but this once: peradventure ten shall be found there. And he said, I will not destroy it for the ten's sake. 33 And Jehovah went his way, as soon as he had left off communing with Abraham: and Abraham returned unto his place.

19 And the two angels came to Sodom at even: and Lot sat in the gate of Sodom: and Lot saw them, and rose up to meet them; and he bowed himself with his face to the earth; 2 and he said, Behold now, my lords, turn aside, I pray you, into your servant's house, and tarry all night, and wash your feet, and ye shall rise up early, and go on your way. And they said, Nay; but we will abide in the street all night. 3 And he urged them greatly; and they turned in unto him, and entered into his house; and he made them a feast, and did bake unleavened bread, and they did eat. 4 But before they lay down, the men of the city, *even* the men of Sodom, compassed the house round, both young and old, all the people from every quarter; 5 and they called unto Lot, and said unto him, Where are the men that came in to thee this night? bring them out unto us, that we may know them. 6 And Lot went out unto them to the door, and shut the door after him. 7 And he said, I pray you, my brethren, do not so wickedly. 8 Behold now, I have two daughters that have not known man; let me, I pray you, bring them out unto you, and do ye to them as is good in your eyes; only unto these men do nothing, forasmuch as they are come under the shadow of my roof. 9 And they said, Stand back. And they said, This one fellow came in to sojourn, and he will needs be a judge: now will we deal worse with thee, than with them. And they pressed sore upon the man, even Lot, and drew near to break the door. 10 But the men put forth their hand, and brought Lot into the house to them, and shut to the door. 11 And they smote the men that were at the door of the house with blindness, both small and great, so that they wearied themselves to find the door.

12 And the men said unto Lot, Hast thou here any besides? son-in-law, and thy sons, and thy daughters, and whomsoever thou hast in the city, bring them out of the place: 13 for we will destroy this place, because the cry of them is waxed great before Jehovah; and Jehovah hath sent us to destroy it. 14 And Lot went out, and spake unto his sons-in-law, who married his daughters, and said, Up, get you out of this place; for Jehovah will destroy the city. But he seemed unto his sons-in-law as one that mocked. 15 And when the morning arose, then the angels hastened Lot, saying, Arise, take thy wife, and thy two daughters that are here, lest thou be consumed in the iniquity of the city. 16 But he lingered; and the men laid hold upon his hand, and upon the hand of his wife, and upon the hand of his two daughters, Jehovah being merciful unto him: and they brought him forth, and set him without the city. 17 And it came to pass, when they had brought them forth abroad, that he said, Escape for thy life; look not behind thee, neither stay thou in all the Plain; escape to the mountain, lest thou be consumed. 18 And Lot said unto them, Oh, not so, my lord:

19 behold now, thy servant hath found favor in thy sight, and thou hast magnified thy loving kindness, which thou hast showed unto me in saving my life; and I cannot escape to the mountain, lest evil overtake me, and I die: 20 behold now, this city is near to flee unto, and it is a little one. Oh let me escape thither (is it not a little one?), and my soul shall live. 21 And he said unto him, See, I have accepted thee concerning this thing also, that I will not overthrow the city of which thou hast spoken. 22 Haste thee, escape thither; for I cannot do anything till thou be come thither. Therefore the name of the city was called Zoar.

23 The sun was risen upon the earth when Lot came unto Zoar. 24 Then Jehovah rained upon Sodom and upon Gomorrah brimstone and fire from Jehovah out of heaven; 25 and he overthrew those cities, and all the Plain, and all the inhabitants of the cities, and that which grew upon the ground. 26 But his wife looked back from behind him, and she became a pillar of salt. 27 And Abraham gat up early in the morning to the place where he had stood before Jehovah: 28 and he looked toward Sodom and Gomorrah, and toward all the land of the Plain, and beheld, and, lo, the smoke of the land went up as the smoke of a furnace.

Isaiah 59:1–21

59 Behold, Jehovah's hand is not shortened, that it cannot save; neither his ear heavy, that it cannot hear: 2 but your iniquities have separated between you and your God, and your sins have hid his face from you, so that he will not hear. 3 For your hands are defiled with blood, and your fingers with iniquity; your lips have spoken lies, your tongue muttereth wickedness. 4 None sueth in righteousness, and none pleadeth in truth: they trust in vanity, and speak lies; they conceive mischief, and bring forth iniquity. 5 They hatch adders' eggs, and weave the spider's web: he that eateth of their eggs dieth; and that which is crushed breaketh out into a viper. 6 Their webs shall not become garments, neither shall they cover themselves with their works: their works are works of iniquity, and the act of violence is in their hands. 7 Their feet run to evil, and they make haste to shed innocent blood: their thoughts are thoughts of iniquity; desolation and destruction are in their paths. 8 The way of peace they know not; and there is no justice in their goings: they have made them crooked paths; whosoever goeth therein doth not know peace.

9 Therefore is justice far from us, neither doth righteousness overtake us: we look for light, but, behold, darkness; for brightness, but we walk in obscurity. 10 We grope for the wall like the blind; yea, we grope as they that have no eyes: we stumble at noonday as in the twilight; among them that are lusty we are as dead men. 11 We roar all like bears, and moan sore like doves: we look for justice, but there is none; for salvation, but it is far off from us. 12 For our transgressions are multiplied before thee, and our sins testify against us; for our transgressions are with us, and as for our iniquities, we know them: 13 transgressing and denying Jehovah, and turning away from following our God, speaking oppression and revolt, conceiving and uttering from the heart words of falsehood. 14 And justice is turned away backward, and righteousness standeth afar off; for truth is fallen in the street, and uprightness cannot enter. 15 Yea, truth is lacking; and he that departeth from evil maketh himself a prey.

And Jehovah saw it, and it displeased him that there was no justice. 16 And he saw that there was no man, and wondered that there was no intercessor: therefore his own arm brought salvation unto him; and his righteousness, it upheld him. 17 And he put on righteousness as a breastplate; and a helmet of salvation upon his head; and he put on garments of vengeance for clothing, and was clad with zeal as a mantle. 18 According to their deeds, accordingly he will repay, wrath to his adversaries, recompense to his enemies; to the islands he will repay recompense. 19 So shall they fear the name of Jehovah from the west, and his glory from the rising of the sun; for he will come as a rushing stream, which the breath of Jehovah driveth. 20 And a Redeemer will come to Zion, and unto them that turn from transgression in Jacob, saith Jehovah. 21 And as for me, this is my covenant with them, saith Jehovah: my Spirit that is upon thee, and my words which I have put in thy mouth, shall not depart out of thy mouth, nor out of the mouth of thy seed, nor out of the mouth of thy seed's seed, saith Jehovah, from henceforth and for ever.

Leviticus 24:17–22

17 And he that smiteth any man mortally shall surely be put to death. 18 And he that smiteth a beast mortally shall make it good, life for life. 19 And if a man cause a blemish in his neighbor; as he hath done, so shall it be done to him: 20 breach for breach, eye for eye, tooth for tooth; as he hath caused a blemish in a man, so shall it be rendered unto him. 21 And he that killeth a beast shall make it good: and he that killeth a man shall be put to death. 22 Ye shall have one manner of law, as well for the sojourner, as for the home-born; for I am Jehovah your God.

Matthew 5:1–12

5 And seeing the multitudes, he went up into the mountain: and when he had sat down, his disciples came unto him: 2 and he opened his mouth and taught them, saying,

3 Blessed are the poor in spirit: for theirs is the kingdom of heaven.

4 Blessed are they that mourn: for they shall be comforted.

5 Blessed are the meek: for they shall inherit the earth.

6 Blessed are they that hunger and thirst after righteousness: for they shall be filled.

7 Blessed are the merciful: for they shall obtain mercy.

8 Blessed are the pure in heart: for they shall see God.

9 Blessed are the peacemakers: for they shall be called sons of God.

10 Blessed are they that have been persecuted for righteousness' sake: for theirs is the kingdom of heaven. 11 Blessed are ye when *men* shall reproach you, and persecute you, and say all manner of evil against you falsely, for my sake. 12 Rejoice, and be exceeding glad: for great is your reward in heaven: for so persecuted they the prophets that were before you.

Matthew 5:38–42

38 Ye have heard that it was said, An eye for an eye, and a tooth for a tooth: 39 but I say unto you, Resist not him that is evil: but whosoever smiteth thee on thy right

cheek, turn to him the other also. 40 And if any man would go to law with thee, and take away thy coat, let him have thy cloak also. 41 And whosoever shall compel thee to go one mile, go with him two. 42 Give to him that asketh thee, and from him that would borrow of thee turn not thou away.

Job

3 After this opened Job his mouth, and cursed his day.

2 And Job answered and said:

3 Let the day perish wherein I was born. And the night which said, There is a man-child conceived.

4 Let that day be darkness; Let not God from above seek for it, Neither let the light shine upon it.

19 The small and the great are there: And the servant is free from his master. 20 Wherefore is light given to him that is in misery, And life unto the bitter in soul; 21 Who long for death, but it cometh not, And dig for it more than for hid treasures; 22 Who rejoice exceedingly, And are glad, when they can find the grave? 23 *Why is light given* to a man whose way is hid, And whom God hath hedged in? 24 For my sighing cometh before I eat, And my groanings are poured out like water. 25 For the thing which I fear cometh upon me, And that which I am afraid of cometh unto me. 26 I am not at ease, neither am I quiet, neither have I rest; But trouble cometh.

8 Then answered Bildad the Shuhite, and said, 2 How long wilt thou speak these things? And *how long* shall the words of thy mouth be *like* a mighty wind? 3 Doth God pervert justice? Or doth the Almighty pervert righteousness? 4 If thy children have sinned against him, And he hath delivered them into the hand of their transgression; 5 If thou wouldest seek diligently unto God, And make thy supplication to the Almighty; 6 If thou wert pure and upright: Surely now he would awake for thee, And make the habitation of thy righteousness prosperous. 7 And though thy beginning was small, Yet thy latter end would greatly increase.

20 Behold, God will not cast away a perfect man, Neither will he uphold the evildoers. 21 He will yet fill thy mouth with laughter, And thy lips with shouting. 22 They that hate thee shall be clothed with shame; And the tent of the wicked shall be no more.

38 Then Jehovah answered Job out of the whirlwind, and said, 2 Who is this that darkeneth counsel By words without knowledge? 3 Gird up now thy loins like a man; For I will demand of thee, and declare thou unto me. 4 Where wast thou when I laid the foundations of the earth? Declare, if thou hast understanding. 5 Who determined the measures thereof, if thou knowest? Or who stretched the line upon it? 6 Whereupon were the foundations thereof fastened? Or who laid the corner-stone thereof, 7 When the morning stars sang together, And all the sons of God shouted for joy?

8 Or *who* shut up the sea with doors, When it brake forth, *as if* it had issued out of the womb; 9 When I made clouds the garment thereof, And thick darkness a swaddling-band for it, 10 And marked out for it my bound, And set bars and doors,

11 And said, Hitherto shalt thou come, but no further; And here shall thy proud waves be stayed?

12 Hast thou commanded the morning since thy days *began, And* caused the dayspring to know its place; 13 That it might take hold of the ends of the earth, And the wicked be shaken out of it? 14 It is changed as clay under the seal; And *all things* stand forth as a garment: 15 And from the wicked their light is withholden, And the high arm is broken.

16 Hast thou entered into the springs of the sea? Or hast thou walked in the recesses of the deep? 17 Have the gates of death been revealed unto thee? Or hast thou seen the gates of the shadow of death? 18 Hast thou comprehended the earth in its breadth? Declare, if thou knowest it all.

Luke 15:11–32

11 And he said, A certain man had two sons: 12 and the younger of them said to his father, Father, give me the portion of *thy* substance that falleth to me. And he divided unto them his living. 13 And not many days after, the younger son gathered all together and took his journey into a far country; and there he wasted his substance with riotous living. 14 And when he had spent all, there arose a mighty famine in that country; and he began to be in want. 15 And he went and joined himself to one of the citizens of that country; and he sent him into his fields to feed swine. 16 And he would fain have filled his belly with the husks that the swine did eat: and no man gave unto him. 17 But when he came to himself he said, How many hired servants of my father's have bread enough and to spare, and I perish here with hunger! 18 I will arise and go to my father, and will say unto him, Father, I have sinned against heaven, and in thy sight: 19 I am no more worthy to be called thy son: make me as one of thy hired servants. 20 And he arose, and came to his father. But while he was yet afar off, his father saw him, and was moved with compassion, and ran, and fell on his neck, and kissed him. 21 And the son said unto him, Father, I have sinned against heaven, and in thy sight: I am no more worthy to be called thy son. 22 But the father said to his servants, Bring forth quickly the best robe, and put it on him; and put a ring on his hand, and shoes on his feet: 23 and bring the fatted calf, *and* kill it, and let us eat, and make merry: 24 for this my son was dead, and is alive again; he was lost, and is found. And they began to be merry. 25 Now his elder son was in the field: and as he came and drew nigh to the house, he heard music and dancing. 26 And he called to him one of the servants, and inquired what these things might be. 27 And he said unto him, Thy brother is come; and thy father hath killed the fatted calf, because he hath received him safe and sound. 28 But he was angry, and would not go in: and his father came out, and entreated him. 29 But he answered and said to his father, Lo, these many years do I serve thee, and I never transgressed a commandment of thine; and *yet* thou never gavest me a kid, that I might make merry with my friends: 30 but when this thy son came, who hath devoured thy living with harlots, thou killedst for him the fatted calf. 31 And he said unto him, Son, thou art ever with me, and all that is mine is thine. 32 But it was meet to make merry and be glad: for this thy brother was dead, and is alive *again*; and *was* lost, and is found.

Plato, from the *Republic* (Books I, II, IV) (427 B.C.–347 B.C.)

Plato probably wrote the *Republic* around 380 B.C. and, in it, canonized the question "What is justice?" as one of the leading questions of philosophy. In casual conversation with some of his colleagues in the Athenian marketplace, Socrates evokes the question, and probes it extensively and critically—much to the frustration of some of his interlocutors. The conversation with Cephalus, which begins the discussion, is short and inconclusive. Cephalus suggests that justice is paying your debts and giving each his due, and Socrates quickly shows him that it would not be just to pay a debt—say, of arms—to a wicked or a crazed man. Cephalus, a grand and kindly old man, makes it quite clear that he is not quite up to the rigors of a philosophical argument, and politely excuses himself from the discussion. The next excerpt is the famous exchange with Thrasymachus, who argues that justice is nothing but the interests of the strong. Socrates catches him in a contradiction, and Thrasymachus, his conversation punctuated by insults already, leaves with a final rebuff. The subsequent dialogue with Glaucon, which in fact continues for the remainder of the book, concerns the question, "why should one be just?" Glaucon suggests, using the fabled "Ring of Gyges" as his example, that people would not be just if they could in fact get away with being unjust. Socrates is determined to argue, against Glaucon's sincere skepticism, that it is good to be just, for its own sake, and good for the person who is just as well. Finally, we have included a short excerpt from Book IV, in which Plato gives us (as close as he gets to) his own theory of justice, the twin ideal of the various parts of the republic and the various parts of the individual soul each doing their own assigned task and working together in harmony.

We should note a problem of language here, which, in fact, introduces an enormous problem for the study of justice, not only through the ages and in translation from different far-flung cultures, but even within a single society (and within the same language) as well. There are at least two different words employed by the Greeks as "justice," and neither of them translates easily into an English equivalent. The problem is as much political as linguistic. On the one hand, there is *to eson* or *isotes*, which mean "equality," but, according to the dean of Plato scholars Gregory Vlastos is also the most common word for justice. Then there is *dikaiosune*—the word that both Plato and Aristotle use for justice, which more properly means righteousness. Thus many of the odder things that Plato and Aristotle worry about, such as whether justice is complete virtue or only one of the virtues, and Plato's central claim that justice is "performing the function(s) for which one's nature is best fitted," makes much more sense of "righteousness" than of justice. But, then, one might well wonder whether we are talking about what we think of as *justice* at all. Moreover, insofar as justice means "equality," it is more than odd that both Plato and Aristotle notoriously defend *inegalitarian* views of justice. For example, Plato sums up his dismissal of democracy in a single line: "distributing an odd sort of equality to equals and unequals" (Rep 558c). Sir Karl Popper was thus pressed to exasperation when he wrote, "why did Plato claim that justice meant inequality if, in general usage, it meant equality?" and he answered, "to make propaganda for his totalitarian state" [*The Open Society and its Enemies* (London 1949, p. 79)]. The meaning of "equality" in consideration of justice will be one of the continuing concerns of this volume, and we will discuss its particular difficulties in Greek philosophy when we come to Aristotle, next. But the divergence of meaning and the problem of translation should be kept in mind when reading Plato, too, and any comparisons with what we would ordinarily say about justice should thus be cautiously considered. Indeed, even within the same culture and the same language, it is clear that the Greeks do not agree on the proper usage or interpretation of these sensitive terms, nor, in our own culture and language, do we.

[Cephalus describes the benefits and burdens of old age to Socrates] if it is moderate and contented, then old age too is but moderately burdensome; if it is not, then both old age and youth are hard to bear.

[Socrates] I wondered at his saying this and I wanted him to say more, so I urged him on by saying: Cephalus, when you say this, I don't think most people would agree with you; they think you endure old age easily not because of your manner of life but because you are wealthy, for the wealthy, they say, have many things to encourage them.

What you say is true, he said. They would not agree. And there is something in what they say, but not as much as they think. What Themistocles said is quite right: when a man from Seriphus was insulting him by saying that his high reputation was due to his city and not to himself, he replied that, had he been a Seriphian, he would not be famous, but neither would the other had he been an Athenian. The same can be applied to those who are not rich and find old age hard to bear—namely that a good man would not very easily bear old age in poverty, nor would a bad man, even if wealthy, be at peace with himself.

Did you inherit most of your wealth, Cephalus, I asked, or did you acquire it?

How much did I acquire, Socrates? As a moneymaker I stand between my grandfather and my father. My grandfather and namesake inherited about the same amount of wealth which I possess but multiplied it many times. My father, Lysanias, however, diminished that amount to even less than I have now. As for me, I am satisfied to leave to my sons here no less but a little more than I inherited.

The reason I asked, said I, is that you did not seem to me to be overfond of money, and this is generally the case with those who have not made it themselves. Those who have acquired it by their own efforts are twice as fond of it as other men. Just as poets love their own poems and fathers love their children, so those who have made their money are attached to it as something they have made themselves, besides using it as other men do. This makes them poor company, for they are unwilling to give their approval to anything but money.

What you say is true, he said.

It surely is, said I. Now tell me this much more: What is the greatest benefit you have received from the enjoyment of wealth?

I would probably not convince many people in saying this, Socrates, he said, but you must realize that when a man approaches the time when he thinks he will die, he becomes fearful and concerned about things which he did not fear before. It is then that the stories we are told about the underworld, which he ridiculed before—that the man who has sinned here will pay the penalty there—torture his mind lest they be true. Whether because of the weakness of old age, or because he is now closer to what happens there and has a clearer view, the man himself is filled with suspicion and fear, and he now takes account and examines whether he has wronged anyone. If he finds many sins in his own life, he awakes from sleep in terror, as children do, and he lives with the expectation of evil. However, the man who knows he has not sinned has a sweet and good hope as his constant companion, a nurse to his old age, as Pindar too puts it. The poet has expressed this charmingly, Socrates, that whoever lives a just and pious life

> Sweet is the hope that nurtures his heart,
> companion and nurse to his old age,
> a hope which governs the rapidly changing thoughts of mortals.

This is wonderfully well said. It is in this connection that I would say that wealth has its greatest value, not for everyone but for a good and well-balanced man. Not to have lied to or deceived anyone even unwillingly, not to depart yonder in fear, owing either sacrifices to a god or money to a man: to this wealth makes a great contribution. It has many other uses, but benefit for benefit I would say that its greatest usefulness lies in this for an intelligent man, Socrates.

Beautifully spoken, Cephalus, said I, but are we to say that justice or right is simply to speak the truth and to pay back any debt one may have contracted? Or are these same actions sometimes right and sometimes wrong? I mean this sort of thing, for example: everyone would surely agree that if a friend has deposited weapons with you when he was sane, and he asks for them when he is out of his mind, you should not return them. The man who returns them is not doing right, nor is one who is willing to tell the whole truth to a man in such a state.

What you say is correct, he answered.

This then is not a definition of right or justice, namely to tell the truth and pay one's debts.

It certainly is, said Polemarchus interrupting, if we are to put any trust in Simonides.

And now, said Cephalus, I leave the argument to you, for I must go back and look after the sacrifice.

Do I then inherit your role? asked Polemarchus.

You certainly do, said Cephalus laughing, and as he said it he went off to sacrifice.

Then do tell us, Polemarchus, said I, as the heir to the argument, what it is that Simonides stated about justice which you consider to be correct.

He stated, said he, that it is just to give to each what is owed to him, and I think he was right to say so.

Well now, I said, it is hard not to believe Simonides, for he is a wise and inspired man, but what does he mean? Perhaps you understand him, but I do not. Clearly he does not mean what we were saying just now, that anything he has deposited must be returned to a man who is not in his right mind; yet anything he has deposited is owing to him. Is that not so?—Yes.

But it is not to be returned to him at all if he is out of his mind when he asks for it?—That's true.

Certainly Simonides meant something different from this when he says that to return what is owed is just.

He did indeed mean something different by Zeus, said he. He believes that one owes it to one's friends to do good to them, and not harm.

I understand, said I, that one does not give what is owed or due if one gives back gold to a depositor, when giving back and receiving are harmful, and the two are friends. Is that not what you say Simonides meant?—Quite.

Well then, should one give what is due to one's enemies?

By all means, said he, what is in fact due to them, and I believe that is what is properly due from an enemy to an enemy, namely something harmful.

It seems, I said, that Simonides was suggesting the nature of the just poetically and in riddles. For he thought this to be just, to give to each man what is proper to him, and he called this what is due.—Surely.

* * *

When you say friends, do you mean those whom a man believes to be helpful to him, or those who are helpful even if they do not appear to be so, and so with enemies?

Probably, he said, one is fond of those whom one thinks to be good and helpful to one, and one hates those whom one considers bad and harmful.

Surely people make mistakes about this, and consider many to be helpful when they are not, and often make the opposite mistake about enemies?—They do.

Then good men are their enemies, and bad people their friends?—Quite so.

And so it is just and right for these mistaken people to benefit the bad and harm the good?—It seems so.

But the good are just and able to do no wrong?—True.

But according to your argument it is just to harm those who do no wrong.

Never, Socrates, he said. It is the argument that is wrong.

It is just to harm the wrongdoers and to benefit the just?

That statement, Socrates, seems much more attractive than the other.

Then, Polemarchus, for many who are mistaken in their judgment it follows that it is just to harm their friends, for these are bad, and to benefit their enemies, who are good, and so we come to a conclusion which is the opposite of what we said was the meaning of Simonides.

That certainly follows, he said, but let us change our assumption; we have probably not defined the friend and the enemy correctly.

Where were we mistaken, Polemarchus?

—When we said that a friend was one who was thought to be elpful.

How shall we change this now? I asked.

Let us state, he said, that a friend is one who is both thought to be helpful and also is; one who is thought to be, but is not, helpful is thought to be a friend but is not. And so also with the enemy.

According to this argument then, the good man will be a friend, and the bad man an enemy.—Yes.

You want us to add to what we said before about the just, namely that it is just to benefit one's friend and harm one's enemy; to this you want us to make an addition and say that it is just to benefit the friend who is good and to harm the enemy who is bad?

Quite so, he said. This seems to me to be well said.

But, I said, is it the part of the just man to harm anyone at all?

Why certainly, he said, those who are bad and one's enemies.

Do horses become better or worse when they are harmed?—Worse.

* * *

Shall we not say so about men too, that when they are harmed they deteriorate in their human excellence?—Quite so.

And is not justice a human excellence?—Of course.

Then men who are harmed, my friend, necessarily become more unjust.—So it appears.

* * *

Well then, can the just, by the practice of justice, make men unjust? Or, in a word, can good men, by the practice of their virtue, make men bad?—They cannot.

* * *

It is not then the function of the just man, Polemarchus, to do harm to a friend or anyone else, but it is that of his opposite, the unjust man?—I think that you are entirely right, Socrates.

If, then, anyone tells us that it is just to give everyone his due, and he means by this that from the just man harm is due to his enemies and benefit due to his friends—the man who says that is not wise, for it is not true. We have shown that it is never just to harm anyone.—I agree.

* * *

While we were speaking Thrasymachus often started to interrupt, but he was restrained by those who were sitting by him, for they wanted to hear the argument to the end. But when we paused after these last words of mine he could no longer keep quiet. He gathered himself together like a wild beast about to spring, and he came at us as if to tear us to pieces.

Polemarchus and I were afraid and flustered as he roared into the middle of our company: What nonsense have you two been talking, Socrates? Why do you play the fool in thus giving way to each other? If you really want to know what justice is, don't only ask questions and then score off anyone who answers, and refute him. You know very well that it is much easier to ask questions than to answer them. Give an answer yourself and tell us what you say justice is. And don't tell me that it is the needful, or the advantageous, or the beneficial, or the gainful, or the useful, but tell me clearly and precisely what you mean, for I will not accept it if you utter such rubbish.

His words startled me, and glancing at him I was afraid. I think if I had not looked at him before he looked at me, I should have been speechless. As it was I had glanced at him first when our discussion began to exasperate him, so I was able to answer him and I said, trembling: do not be hard on us, Thrasymachus, if we have erred in our investigation, he and I; be sure that we err unwillingly. You surely do not believe that if we were searching for gold we would be unwilling to give way to each other and thus destroy our chance of finding it, but that when searching for justice, a thing more precious than much gold, we mindlessly give way to one another, and that we are not thoroughly in earnest about finding it. You must believe that, my friend, for I think we could not do it. So it is much more seemly that you clever people should pity us than that you should be angry with us.

When he heard that he gave a loud and bitter laugh and said: By Heracles, that is just Socrates' usual irony. I knew this, and I warned these men here before that you would not be willing to answer any questions but would pretend ignorance, and that you would do anything rather than give an answer, if anyone questioned you.

* * *

Listen then, said he. I say that the just is nothing else than the advantage of the stronger. Well, why don't you praise me? But you will not want to.

I must first understand your meaning, said I, for I do not know it yet. You say that the advantage of the stronger is just. What do you mean, Thrasymachus? Surely you do not mean such a thing as this: Poulydamas, the pancratist athlete, is stronger than we are; it is to his advantage to eat beef to build up his physical strength. Do you mean that this food is also advantageous and just for us who are weaker than he is?

You disgust me, Socrates, he said. Your trick is always to take up the argument at the point where you can damage it most.

Not at all, my dear sir, I said, but tell us more clearly what you mean.

Do you not know, he said, that some cities are ruled by a despot, others by the people, and others again by the aristocracy?—Of course.

And this element has the power and rules in every city?—Certainly.

Yes, and each government makes laws to its own advantage: democracy makes democratic laws, a despotism makes despotic laws, and so with the others, and when they have made these laws they declare this to be just for their subjects, that is, their own advantage, and they punish him who transgresses the laws as lawless and unjust. This then, my good man, is what I say justice is, the same in all cities, the advantage of the established government, and correct reasoning will conclude that the just is the same everywhere, the advantage of the stronger.

Now I see what you mean, I said. Whether it is true or not I will try to find out. But you too, Thrasymachus, have given as an answer that the just is the advantageous whereas you forbade that answer to me. True, you have added the words "of the stronger."

Perhaps, he said, you consider that an insignificant addition!

It is not clear yet whether or not it is significant. Obviously, we must investigate whether what you say is true. I agree that the just is some kind of advantage, but you

add that it is the advantage of the stronger. I do not know. We must look into this.—
Go on looking, he said.

We will do so, said I. Tell me, do you also say that obedience to the rulers is just?—I do.

And are the rulers in all cities infallible, or are they liable to error?—No doubt they are liable to error.

When they undertake to make laws, therefore, they make some correctly and make others incorrectly?—I think so.

"Correctly" means that they make laws to their own advantage, and "incorrectly" not to their own advantage. Or how would you put it?—As you do.

And whatever laws they make must be obeyed by their subjects, and this is just?—Of course.

Then, according to your argument, it is just to do not only what is to the advantage of the stronger, but also the opposite, what is not to their advantage.

What is that you are saying? he asked.

The same as you, I think, but let us examine it more fully. Have we not agreed that, in giving orders to their subjects, the rulers are sometimes in error as to what is best for themselves, yet it is just for their subjects to do whatever their rulers order. Is that much agreed?—I think so.

Think then also, said I, that you have agreed that it is just to do what is to the disadvantage of the rulers and the stronger whenever they unintentionally give orders which are bad for themselves, and you say it is just for the others to obey their given orders. Does it not of necessity follow, my wise Thrasymachus, that it is just to do the opposite of what you said? The weaker are then ordered to do what is to the disadvantage of the stronger.

Yes by Zeus, Socrates, said Polemarchus, that is quite clear.

Yes, if you bear witness for him, interrupted Cleitophon.

What need of a witness? said Polemarchus. Thrasymachus himself agrees that the rulers sometimes give orders that are bad for themselves, and that it is just to obey them.

Thrasymachus maintained that it is just to obey the orders of the rulers, Polemarchus.

He also said that the just was the advantage of the stronger, Cleitophon. Having established those two points he went on to agree that the stronger sometimes ordered the weaker, their subjects, to do what was disadvantageous to themselves. From these agreed premises it follows that what is of advantage to the stronger is no more just than what is not.

But, Cleitophon replied, he said that the advantage of the stronger is what the stronger believes to be of advantage to him. This the weaker must do, and that is what he defined the just to be.

When we reached this point in our argument and it was clear to all that the definition of justice had turned into its opposite, Thrasymachus, instead of answering, said: Tell me, Socrates, do you have a nanny?

What's this? said I. Had you not better answer than ask such questions?

Because, he said, she is letting you go around with a snotty nose and does not wipe it when she needs to, if she leaves you without any knowledge of sheep or shepherds.

What is the particular point of that remark? I asked.

You think, he said, that shepherds and cowherds seek the good of their sheep or cattle, whereas their sole purpose in fattening them and looking after them is their own good and that of their master. Moreover, you believe that rulers in the cities, true rulers that is, have a different attitude towards their subjects than one has towards sheep, and that they think of anything else, night and day, than their own advantage. You are so far from understanding the nature of justice and the just, of injustice and the unjust, that you do not realize that the just is really another's good, the advantage of the stronger and the ruler, but for the inferior who obeys it is a personal injury. Injustice on the other hand exercises its power over those who are truly naive and just, and those over whom it rules do what is of advantage to the other, the stronger, and, by obeying him, they make him happy, but themselves not in the least.

You must look at it in this way, my naive Socrates: the just is everywhere at a disadvantage compared with the unjust. First, in their contracts with one another: wherever two such men are associated you will never find, when the partnership ends, the just man to have more than the unjust, but less. Then, in their relation to the city: when taxes are to be paid, from the same income the just man pays more, the other less; but, when benefits are to be received, the one gets nothing while the other profits much; whenever each of them holds a public office, the just man, even if he is not penalized in other ways, finds that his private affairs deteriorate through neglect while he gets nothing from the public purse because he is just; moreover, he is disliked by his household and his acquaintances whenever he refuses them an unjust favour. The opposite is true of the unjust man in every respect. I repeat what I said before: the man of great power gets the better deal. Consider him if you want to decide how much more it benefits him privately to be unjust rather than just. You will see this most easily if you turn your thoughts to the most complete form of injustice which brings the greatest happiness to the wrongdoer, while it makes those whom he wronged, and who are not willing to do wrong, most wretched. This most complete form is despotism; it does not appropriate other people's property little by little, whether secretly or by force, whether public or private, whether sacred objects or temple property, but appropriates it all at once.

When a wrongdoer is discovered in petty cases, he is punished and faces great opprobrium, for the perpetrators of these petty crimes are called temple robbers, kidnappers, housebreakers, robbers, and thieves, but when a man, besides appropriating the possessions of the citizens, manages to enslave the owners as well, then, instead of those ugly names he is called happy and blessed, not only by his fellow-citizens but by all others who learn that he has run through the whole gamut of injustice. Those who give injustice a bad name do so because they are afraid, not of practising but of suffering injustice.

And so, Socrates, injustice, if it is on a large enough scale, is a stronger, freer, and more powerful thing than justice and, as I said from the first, the just is what is advantageous to the stronger, while the unjust is to one's own advantage and benefit.

Having said this and poured this mass of close-packed words into our ears as a bathman might a flood of water, Thrasymachus intended to leave, but those present did not let him, and made him stay for a discussion of his views. I too begged him to stay and I said: My dear Thrasymachus, after throwing such a speech at us, you want to leave before adequately instructing us or finding out whether you are right or not? Or do you think it a small thing to decide on a whole way of living, which, if each of us adopted it, would make him live the most profitable life?

Come then, Thrasymachus, I said, answer us from the beginning. You say that complete injustice is more profitable than complete justice?

I certainly do say that, he said, and I have told you why.

Well then, what about this: you call one of the two a virtue and the other a vice?—Of course.

That is, you call justice a virtue, and injustice a vice?

Is that likely, my good man, said he, since I say that injustice is profitable, and justice is not?

What then?—The opposite.

Do you call being just a vice?—No, but certainly high-minded foolishness.

And you call being unjust low-minded?—No, I call it good judgment.

You consider the unjust then, Thrasymachus, to be good and knowledgeable?

Yes, he said, those who are able to carry injustice through to the end, who can bring cities and communities of men under their power. Perhaps you think I mean purse-snatchers? Not that those actions too are not profitable, if they are not found out, but they are not worth mentioning in comparison with what I am talking about.

I am not unaware of what you mean, I said, but this point astonishes me: do you include injustice under virtue and wisdom, and justice among their opposites?—I certainly do.

That makes it harder, my friend, and it is not easy now to know what to say. If you had declared that injustice was more profitable, but agreed that it was a vice or shameful as some others do, we could have discussed it along the lines of general opinion. Now, obviously, you will say that it is fine and strong, and apply to it all the attributes which we used to apply to justice, since you have been so bold as to include it under virtue and wisdom.—Your guess, he said, is quite right.

We must not, however, shrink from pursuing our argument and looking into this, so long as I am sure that you mean what you say. For I do not think you are joking now, Thrasymachus, but are saying what you believe to be true.

What difference, said he, does it make to you whether I believe it or not? Is it not my argument you are refuting?

No difference, said I, but try to answer this further question: do you think that the just man wants to get the better of the just?

Never, said he, for he would not then be well mannered and simple, as he is now.

Does he want to overreach a just action?

Not a just action either, he said.

Would he want to get the better of an unjust man, and would he deem that just or not?

He would want to, he said, and he would deem it right, but he would not be able to.

That was not my question, said I, but whether the just man wants and deems it right to outdo not a just man, but an unjust one?—That is so.

What about the unjust man? Would he deem it right to outdo the just man and the just action?

Of course he does, he said, since he deems it right to get the better of everybody.

So the unjust man will get the better of another unjust man or an unjust action and he will strive to get all he can from everyone?—That is so.

Let us put it this way, I said. The just man does not try to get the better of one like him but of one unlike him, whereas the unjust man overreaches the like and the unlike?—Very well put.

The unjust man, I said, is knowledgeable and good, and the just man is neither?—That is well said too.

It follows, I said, that the unjust man is like the knowledgeable and the good, while the just man is unlike them?

Of course that will be so, he said, being such a man he will be like such men, while the other is not like them.

Now Thrasymachus, I said, we found that the unjust man tries to get the better of both those like and those unlike him. Did you not say so?—I did.

Yes, and the just man will not get the better of his like, but of one unlike him?—Yes.

The just man then, I said, resembles the wise and good, while the unjust resembles the bad and ignorant?—It may be so.

Further, we agreed that each will be such as the man he resembles?—We did so agree.

So we find that the just man has turned out to be good and wise, and the unjust man ignorant and bad.

Thrasymachus agreed to all this, not easily as I am telling it, but reluctantly and after being pushed. It was summer and he was perspiring profusely. And then I saw something I had never seen before: Thrasymachus blushing. After we had agreed that justice was virtue and wisdom, and injustice vice and ignorance, I said: Very well, let us consider this as established,

* * *

Come now, consider this point next: There is a function of the soul which you could not fulfill by means of any other thing, as for example: to take care of things, to rule, to deliberate, and other things of the kind; could we entrust these things to any other agent than the soul and say that they belong to it?—To no other.

What of living? Is that not a function of the soul?—It most certainly is.

So there is also an excellence of the soul?—We say so.

And, Thrasymachus, will the soul ever fulfill its function well if it is deprived of its own particular excellence, or is this impossible?—Impossible.

It is therefore inevitable that the bad soul rules and looks after things badly and that the good soul does all these things well.—Inevitable.

Now we have agreed that justice is excellence of the soul, and that injustice is vice of soul?—We have so agreed.

The just soul and the just man, then, will live well, and the unjust man will live badly.—So it seems, according to your argument.

Surely the one who lives well is blessed and happy, and the one who does not is the opposite.—Of course.

So the just man is happy, and the unjust one is wretched.—So be it.

It profits no one to be wretched, but to be happy.—Of course.

And so, my good Thrasymachus, injustice is never more profitable than justice.

I, before finding the answer to our first enquiry into the nature of justice, let that go and turned to investigate whether it was vice and ignorance or wisdom and virtue. Another argument came up after, that injustice was more profitable than justice, and I could not refrain from following this up and abandoning the previous one so that the result of our discussion for me is that I know nothing; for, when I do not know what justice is, I shall hardly know whether it is a kind of virtue or not, or whether the just man is unhappy or happy.

When I had said this I thought I had done with the discussion, but evidently this was only a prelude. Glaucon on this occasion too showed that boldness which is characteristic of him, and refused to accept Thrasymachus' abandoning the argument. He said: Do you, Socrates, want to appear to have persuaded us, or do you want truly to convince us that it is better in every way to be just than unjust?

I would certainly wish to convince you truly, I said, if I could.

They say that to do wrong is naturally good, to be wronged is bad, but the suffering of injury so far exceeds in badness the good of inflicting it that when men have done wrong to each other and suffered it, and have had a taste of both, those who are unable to avoid the latter and practise the former decide that it is profitable to come to an agreement with each other neither to inflict injury nor to suffer it. As a result they begin to make laws and covenants, and the law's command they call lawful and just. This, they say, is the origin and essence of justice; it stands between the best and the worst, the best being to do wrong without paying the penalty and the worst to be wronged without the power of revenge. The just then is a mean between two extremes; it is welcomed and honoured because of men's lack of the power to do wrong. The man who has that power, the real man, would not make a compact with anyone not to inflict injury or suffer it. For him that would be madness. This then, Socrates, is, according to their argument, the nature and origin of justice.

Even those who practise justice do so against their will because they lack the power to do wrong. This we could realize very clearly if we imagined ourselves granting to both the just and the unjust the freedom to do whatever they liked. We could then follow both of them and observe where their desires led them, and we would catch the just man redhanded travelling the same road as the unjust. The reason is the desire for undue gain which every organism by nature pursues as a good, but the law forcibly sidetracks him to honour equality. The freedom I just mentioned would most easily occur if these men had the power which they say the ancestor of the Lydian Gyges possessed. The story is that he was a shepherd in the service of the ruler of Lydia. There was a violent rainstorm and an earthquake which broke open the ground and created a chasm at the place where he was tending sheep. Seeing this and marvelling, he went down into it. He saw, besides many other wonders of which we are told, a hollow bronze horse. There were window-like openings in it; he climbed through them and caught sight of a corpse which seemed of more than human stature, wearing nothing but a ring of gold on its finger. This ring the shepherd put on and came out. He arrived at the usual monthly meeting which reported to the king on the state of the flocks, wearing the ring. As he was sitting among the others he happened to twist the hoop of the ring towards himself, to the inside of his hand, and as he did this he became invisible to those sitting near him and they went on talking as if he had gone. He marvelled at this and, fingering the ring, he turned the hoop outward again and became visible. Perceiving this he tested whether the ring had this power and so it happened: if he turned the hoop inwards he became invisible, but was visible when he turned it outwards. When he realized this, he at once arranged to become one of the messengers to the king. He went, committed adultery with the king's wife, attacked the king with her help, killed him, and took over the kingdom.

Now if there were two such rings, one worn by the just man, the other by the unjust, no one, as these people think, would be so incorruptible that he would stay on the path of justice or bring himself to keep away from other people's property and not touch it, when he could with impunity take whatever he wanted from the market, go into houses and have sexual relations with anyone he wanted, kill anyone, free all those he wished from prison, and do the other things which would make him like a god among men. His actions would be in no way different from those of the other and they would both follow the same path. This, some would say, is a great proof that no one is just willingly but under compulsion, so that justice is not one's private good, since wherever either thought he could do wrong with impunity he would do so. Every man believes that injustice is much more profitable to himself than justice, and any exponent of this argument will say that he is right. The man who did not wish to do wrong with that opportunity, and did not touch other people's property, would be thought by those who knew it to be very foolish and miserable. They would praise him in public, thus deceiving one another, for fear of being wronged. So much for my second topic.

As for the choice between the lives we are discussing, we shall be able to make a correct judgment about it only if we put the most just man and the most unjust man face to face; otherwise we cannot do so. By face to face I mean this: let us grant to the unjust the fullest degree of injustice and to the just the fullest justice, each being perfect in his own pursuit. First, the unjust man will act as clever craftsmen do—a top navigator for example or physician distinguishes what his craft can do and what it cannot; the former he will undertake, the latter he will pass by, and when he slips he can put things right. So the unjust man's correct attempts at wrongdoing must remain secret; the one who is caught must be considered a poor performer, for the extreme of injustice is to have a reputation for justice, and our perfectly unjust man must be granted perfection in injustice. We must not take this from him, but we must allow that, while committing the greatest crimes, he has provided himself with the greatest reputation for justice; if he makes a slip he must be able to put it right; he must be a sufficiently persuasive speaker if some wrongdoing of his is made public; he must be able to use force, where force is needed, with the help of his courage, his strength, and the friends and wealth with which he has provided himself.

Having described such a man, let us now in our argument put beside him the just man, simple as he is and noble, who, as Aeschylus put it, does not wish to appear just but to be so. We must take away his reputation, for a reputation for justice would bring him honour and rewards, and it would then not be clear whether he is what he is for justice's sake or for the sake of rewards and honour. We must strip him of everything except justice and make him the complete opposite of the other. Though he does no wrong, he must have the greatest reputation for wrongdoing so

that he may be tested for justice by not weakening under ill repute and its consequences. Let him go his incorruptible way until death with a reputation for injustice
throughout his life, just though he is, so that our two men may reach the extremes,
one of justice, the other of injustice, and let them be judged as to which of the two is
the happier.

Whew! My dear Glaucon, I said, what a mighty scouring you have given those
two characters, as if they were statues in a competition.

Besides this, Socrates, look at another kind of argument which is spoken in private,
and also by the poets, concerning justice and injustice. All go on repeating with one
voice that justice and moderation are beautiful, but certainly difficult and burdensome, while incontinence and injustice are sweet and easy, and shameful only by
repute and by law. They add that unjust deeds are for the most part more profitable
than just ones. They freely declare, both in private and in public, that the wicked
who have wealth and other forms of power are happy. They honour them but pay
neither honour nor attention to the weak and the poor, though they agree that these
are better men than the others.

What men say about the gods and virtue is the most amazing of all, namely that
the gods too inflict misfortunes and a miserable life upon many good men, and the
opposite fate upon their opposites.

[from Book IV, where Socrates defines justice as "doing one's own"]
. . . everyone must pursue one occupation of those in the city, that for which his
nature best fitted him.—Yes, we kept saying that.

Further, we have heard many people say, and have often said ourselves, that
justice is to perform one's own task and not to meddle with that of others.—We have
said that.

This then, my friend, I said, when it happens, is in some way justice, to do one's
own job. And do you know what I take to be a proof of this?—No, tell me.

Look at it this way and see whether you agree: you will order your rulers to act
as judges in the courts of the city?—Surely.

And will their exclusive aim in delivering judgment not be that no citizen should
have what belongs to another or be deprived of what is his own?—That would be
their aim.

That being just?—Yes.

In some way then possession of one's own and the performance of one's own
task could be agreed to be justice.—That is so.

Consider then whether you agree with me in this: if a carpenter attempts to do the
work of a cobbler, or a cobbler that of a carpenter, and they exchange their tools and
the esteem that goes with the job, or the same man tries to do both, and all the other
exchanges are made, do you think that this does any great harm to the city?—No.

But I think that when one who is by nature a worker or some other kind of
moneymaker is puffed up by wealth, or by the mob, or by his own strength, or some
other such thing, and attempts to enter the warrior class, or one of the soldiers tries
to enter the group of counsellors and guardians, though he is unworthy of it, and
these exchange their tools and the public esteem, or when the same man tries to
perform all these jobs together, then I think you will agree that these exchanges and
this meddling bring the city to ruin.—They certainly do.

The meddling and exchange between the three established orders does very great harm to the city and would most correctly be called wickedness.—Very definitely.

And you would call the greatest wickedness worked against one's own city injustice?—Of course.

That then is injustice. And let us repeat that the doing of one's own job by the moneymaking, auxiliary, and guardian groups, when each group is performing its own task in the city, is the opposite, it is justice and makes the city just.—I agree with you that this is so.

Do not let us, I said, take this as quite final yet. If we find that this quality, when existing in each individual man, is agreed there too to be justice, then we can assent to this—for what can we say?—but if not, we must look for something else. For the present, let us complete that examination which we thought we should make, that if we tried to observe justice in something larger which contains it, this would make it easier to observe it in the individual. We thought that this larger thing was a city, and so we established the best city we could, knowing well that justice would be present in the good city. It has now appeared to us there, so let us now transfer it to the individual and, if it corresponds, all will be well. But if it is seen to be something different in the individual, then we must go back to the city and examine this new notion of justice. By thus comparing and testing the two, we might make justice light up like fire from the rubbing of firesticks, and when it has become clear, we shall fix it firmly in our own minds.—You are following the path we set, and we must do so.

Well now, when you apply the same name to a thing whether it is big or small, are these two instances of it like or unlike with regard to that to which the same name applies?—They are alike in that, he said.

So the just man and the just city will be no different but alike as regards the very form of justice.—Yes, they will be.

Now the city was thought to be just when the three kinds of men within it each performed their own task, and it was moderate and brave and wise because of some other qualities and attitudes of the same groups.—True.

And we shall therefore deem it right, my friend, that the individual have the same parts in his own soul, and through the same qualities in those parts will correctly be given the same names.—That must be so.

* * *

Well, then, I said, we are surely compelled to agree that each of us has within himself the same parts and characteristics as the city? Where else would they come from? It would be ridiculous for anyone to think that spiritedness has not come to be in the city from individuals who are held to possess it, like the inhabitants of Thrace and Scythia and others who live to the north of us, or that the same is not true of the love of learning which one would attribute most to our part of the world, or the love of money which one might say is conspicuously displayed by the Phoenicians and the Egyptians.—Certainly, he said.

This then, is the case, I said, and it is not hard to understand.—No indeed.

But this is: whether we do everything with the same part of our soul, or one thing with one of the three parts, and another with another. Do we learn with one part of ourselves, get angry with another, and with some third part desire the pleasures of food and procreation and other things closely akin to them, or, when we set out after something, do we act with the whole of our soul in each case? This will be hard to determine satisfactorily.—I think so too.

The position of the spirited part seems the opposite of what we thought a short time ago. Then we thought of it as something appetitive, but now we say it is far from being that; in the civil war of the soul it aligns itself far more with the reasonable part.—Very much so.

Is it different from that also, or is it some part of reason, so that there are two parts of the soul instead of three, the reasonable and the appetitive? Or, as we had three separate parts holding our city together, the money-making, the auxiliary and the deliberative, so in the soul the spirited is a third part, by nature the helper of reason, if it has not been corrupted by a bad upbringing?—It must be a third part.

Yes, I said, if it now appears to be different from the reasonable part, as earlier from the appetitive part.

It is not difficult, he said, to show that it is different. One can see this in children; they are full of spirit from birth, whereas a few of them seem to me never to acquire a share of reason, while the majority do not do so until late.

By Zeus, I said, that is very well put. One can see this also in animals. Besides, our earlier quotation from Homer bears witness to it, where he says:

> Striking his chest, he addressed his heart,

for clearly Homer represents the part which reasons about the better and the worse course, and which strikes his chest, as different from that which is angry without reasoning.—You are definitely right.

We have now made our difficult way through a sea of argument to reach this point, and we have fairly agreed that the same kinds of parts, and the same number of parts, exist in the soul of each individual as in our city.—That is so.

It necessarily follows that the individual is wise in the same way, and in the same part of himself, as the city.—Quite so.

And the part which makes the individual brave is the same as that which makes the city brave, and in the same manner, and everything which makes for virtue is the same in both?—That necessarily follows.

Moreover, Glaucon, I think we shall say that a man is just in the same way as the city is just.—That too is inevitable.

We have surely not forgotten that the city was just because each of the three classes in it was fulfilling its own task.—I do not think, he said, that we have forgotten that.

We must remember then that each one of us within whom each part is fulfilling its own task will himself be just and do his own work.—We must certainly remember this.

Therefore it is fitting that the reasonable part should rule, it being wise and exercising foresight on behalf of the whole soul, and for the spirited part to obey it and be its ally.—Quite so.

Aristotle, from *Nicomachean Ethics* (Book V) (364 B.C.–322 B.C.)

Aristotle was born in Macedonia but spent most of his life in Athens, where he was a student of Plato and, later, the teacher of Alexander the Great. In the *Ethics*, Aristotle gives us a complicated account of a complex concept of justice. He begins by dividing justice into two categories, a general concept of justice as "the lawful" (though this does not necessarily mean obedience to the laws of any particular state), and a particular concept as "the fair and equal." Once again, we should keep in mind the more literal meaning of *dikaiosyne* as "righteousness," which makes more sense of Aristotle's claim that "this form of justice . . . is complete virtue, . . . not absolutely, but in relation to our neighbor" (p. 41). He then divides "particular justice" into "distributive" and "rectificatory" justice, and now it is clear that he is analyzing a notion much closer to our own notion(s) of justice as fairness. One of Aristotle's key terms here is the notion of "grasping" (*pleonexia*), which is something along the lines of what we mean by "greedy," clearly a vice in the Greek way of thinking. Aristotle also makes much of the opposition between justice and injustice, for it is often through particular acts of injustice that we become clear about justice. Distributive justice, for Aristotle, is primarily concerned with what people *deserve*. Aristotle is also particularly concerned with the justice of *transactions*—a form of rectificatory (or commutative) justice, whether such "voluntary" matters as buying, selling, and lending, or such "involuntary" matters as being the victim of an insult, a theft, or an assassination. It is in the discussion of distributive justice that Aristotle's peculiar and perhaps perverse notion of "equality" enters into the argument. What Aristotle means by "equality," he makes quite clear to us, is really *proportion*, or what he specifically calls "geometrical proportion." What he means is that equals deserve equal but unequals deserve unequal, in proportion to their merit. The better warrior deserves a larger proportion of the spoils. The unjust is whatever violates the proportion. Thus Vlastos describes the "acrobatic linguistic posture" of Aristotle and Plato, too, in trying to say that an "equal" distribution, to be just, is almost always an *unequal* one ("Justice and Equality," in R. Brandt, ed., *Social Justice* (Prentice Hall, 1962, p. 32). Emphasizing again the importance of the concept of equality in Greek thinking, Vlastos adds, "the meritarian view of justice paid reluctant homage to the equalitarian one by using the vocabulary of equality to assert the justice of inequality" (ibid).

Central to Aristotle's overall argument (and the context into which Book Five fits into the *Ethics*) is the idea of justice as a state of character, a cultivated set of dispositions, attitudes, and good habits. It is not an abstract scheme or principle. In its particular manifestation, justice is concerned with good judgment and a sense of fairness. It is a virtue of particular importance, naturally, to those who rule, and those who judge. In rectificatory justice, the justice of exchanges, such judgment again involves a sense of equality, here not as proportion but as a straightforward equivalence. (Aristotle misleadingly calls it "arithmetical" as opposed to "geometrical.") It involves equality in the sense that two men who are before the court for the same crime are to be considered as equals "before the law," even if they are very different otherwise. It is this sense of equality, too, that enters into considerations of punishment. Aristotle says, for example, that a man who has wounded or slain another man should receive a penalty, by which means "the judge tries to equalize things . . . taking away from the gain of the assailant." But notice that there is no clear indication of whether the penalty in such cases is to be conceived as retribution or retaliation, or as some other form of redress or equalization. Indeed, the very notion of "rectificatory" as opposed to "retributive" justice seems to obscure any such query. Finally, Aristotle briefly discusses his famous doctrine that justice—like all virtues—is a "mean between the extremes."

5. Justice

5.1 The Definition of Justice

5.11 JUSTICE AS A STATE OF CHARACTER

The questions we must examine about justice and injustice are these: What sorts of actions are they concerned with? What sort of mean is justice? What are the extremes between which justice is intermediate? Let us examine them by the same type of investigation that we used in the topics discussed before.

We see that the state everyone means in speaking of justice is the state that makes us doers of just actions, that makes us do justice and wish what is just. In the same way they mean by injustice the state that makes us do injustice and wish what is unjust. Let us also, then, [follow the common beliefs and] begin by assuming this in outline.

Since justice is a state, its relation to just actions is different from the relation of a capacity to its activities
For what is true of sciences and capacities is not true of states. For while one and the same capacity or science seems to have contrary activities, a state that is a contrary has no contrary activities. Health, e.g., only makes us do healthy actions, not their contraries; for we say we are walking in a healthy way if [and only if] we are walking in the way a healthy person would.

States may be studied by reference to their contraries
Often one of a pair of contrary states is recognized from the other contrary; and often the states are recognized from their subjects. For if, e.g., the good state is evident, the bad state becomes evident too; and moreover the good state becomes

evident from the things that have it, and the things from the state. For if, e.g., the good state is thickness of flesh, then the bad state will necessarily be thinness of flesh, and the thing that produces the good state will be what produces thickness of flesh.

It follows, usually, that if one of a pair of contraries is spoken of in more ways than one, so is the other; if, e.g., what is just is spoken of in more ways than one, so is what is unjust.

5.12 THE TWO TYPES OF JUSTICE AND INJUSTICE

Now it would seem that justice and injustice are both spoken of in more ways than one, but since the different ways are closely related, their homonymy is unnoticed, and is less clear than it is with distant homonyms where the distance in appearance is wide (e.g., the bone below an animal's neck and what we lock doors with are called keys homonymously).

5.13 JUSTICE AS LAWFULNESS AND JUSTICE AS FAIRNESS

Let us, then, find the number of ways an unjust person is spoken of. Both the lawless person and the greedy and unfair person seem to be unjust; and so, clearly, both the lawful and the fair person will be just. Hence what is just will be both what is lawful and what is fair, and what is unjust will be both what is lawless and what is unfair.

5.14 THE GOODS AND EVILS RELEVANT TO FAIRNESS

Since the unjust person is greedy, he will be concerned with goods—not with all goods, but only with those involved in good and bad fortune, goods which are, [considered] unconditionally, always good, but for this or that person not always good. Though human beings pray for these and pursue them, they are wrong; the right thing is to pray that what is good unconditionally will also be good for us, but to choose [only] what is good for us.

Now the unjust person [who chooses these goods] does not choose more in every case; in the case of what is bad unconditionally he actually chooses less. But since what is less bad also seems to be good in a way, and greed aims at more of what is good, he seems to be greedy. In fact he is unfair; for unfairness includes [all these actions], and is a common feature [of his choice of the greater good and of the lesser evil].

5.2 General Justice

5.21 IT REQUIRES OBSERVANCE OF LAW

Since, as we saw, the lawless person is unjust and the lawful person is just, it clearly follows that whatever is lawful is in some way just; for the provisions of legislative science are lawful, and we say that each of them is just. Now in every matter they deal with the laws aim either at the common benefit of all, or at the benefit of those in control, whose control rests on virtue or on some other such basis. And so in one way what we call just is whatever produces and maintains happiness and its parts for a political community.

5.22 THE SCOPE OF LAW EXTENDS TO ALL THE VIRTUES . . .

Now the law instructs us to do the actions of a brave person—not to leave the battle-line, e.g., or to flee, or to throw away our weapons; of a temperate person—not to commit adultery or wanton aggression; of a mild person—not to strike or revile another; and similarly requires actions that express the other virtues, and prohibits those that express the vices. The correctly established law does this correctly, and the less carefully framed one does this worse.

5.23 . . . AND HENCE GENERAL JUSTICE IS COMPLETE VIRTUE

This type of justice, then, is complete virtue, not complete virtue unconditionally, but complete virtue in relation to another. And this is why justice often seems to be supreme among the virtues, and 'neither the evening star nor the morning star is so marvellous', and the proverb says 'And in justice all virtue is summed up.'

Moreover, justice is complete virtue to the highest degree because it is the complete exercise of complete virtue. And it is the complete exercise because the person who has justice is able to exercise virtue in relation to another, not only in what concerns himself; for many are able to exercise virtue in their own concerns but unable in what relates to another.

And hence Bias seems to have been correct in saying that ruling will reveal the man, since a ruler is automatically related to another, and in a community. And for the same reason justice is the only virtue that seems to be another person's good, because it is related to another; for it does what benefits another, either the ruler or the fellow-member of the community.

The worst person, therefore, is the one who exercises his vice towards himself and his friends as well [as towards others]. And the best person is not the one who exercises virtue [only] towards himself, but the one who [also] exercises it in relation to another, since this is a difficult task.

This type of justice, then, is the whole, not a part, of virtue, and the injustice contrary to it is the whole, not a part, of vice.

At the same time our discussion makes clear the difference between virtue and this type of justice. For virtue is the same as justice, but what it is to be virtue is not the same as what it is to be justice. Rather, in so far as virtue is related to another, it is justice, and in so far as it is a certain sort of state unconditionally it is virtue.

5.3 Special Justice Contrasted With General

5.31 SPECIAL JUSTICE MUST BE A VIRTUE DISTINCT FROM GENERAL JUSTICE

But we are looking for the type of justice, since we say there is one, that consists in a part of virtue, and correspondingly for the type of injustice that is a part [of vice].

Here is evidence that there is this type of justice and injustice:

First, if someone's activities express the other vices—if, e.g., cowardice made him throw away his shield, or irritability made him revile someone, or ungenerosity made him fail to help someone with money—what he does is unjust, but not greedy. But when one acts from greed, in many cases his action expresses none of these vices—certainly not all of them; but it still expresses some type of wickedness, since we blame him, and [in particular] it expresses injustice. Hence there is another type of injustice that is a part of the whole, and a way for a thing to be unjust that is a part of the whole that is contrary to law.

Moreover, if A commits adultery for profit and makes a profit, while B commits adultery because of his appetite, and spends money on it to his own loss, B seems intemperate rather than greedy, while A seems unjust, not intemperate. Clearly, then, this is because A acts to make a profit.

Further, we can refer every other unjust action to some vice—to intemperance if he committed adultery, to cowardice if he deserted his comrade in the battle-line, to anger if he struck someone. But if he made an [unjust] profit, we can refer it to no other vice except injustice.

Hence evidently (a) there is another type of injustice, special injustice, besides the whole of injustice; and (b) it is synonymous with the whole, since the definition is in the same genus. For (b) both have their area of competence in relation to another. But (a) special injustice is concerned with honour or wealth or safety, or whatever single name will include all these, and aims at the pleasure that results from making a profit; but the concern of injustice as a whole is whatever concerns the excellent person.

Clearly, then, there is more than one type of justice, and there is another type besides [the type that is] the whole of virtue; but we must still grasp what it is, and what sort of thing it is.

5.32 THE DISTINCTION REFLECTS THE DISTINCTION BETWEEN WHAT IS LAWLESS AND WHAT IS UNFAIR

What is unjust is divided into what is lawless and what is unfair, and what is just into what is lawful and what is fair. The [general] injustice previously described, then, is concerned with what is lawless. But what is unfair is not the same as what is lawless, but related to it as part to whole, since whatever is unfair is lawless, but not everything lawless is unfair. Hence also the type of injustice and the way for a thing to be unjust [that expresses unfairness] are not the same as the type [that expresses lawlessness], but differ as parts from wholes. For this injustice [as unfairness] is a part of the whole of injustice, and similarly justice [as fairness] is a part of the whole of justice.

Hence we must describe special [as well as general] justice and injustice, and equally this way for a thing to be just or unjust.

5.33 A DETAILED DESCRIPTION OF GENERAL JUSTICE IS UNNECESSARY, SINCE IT IS SIMPLY THE WHOLE OF VIRTUE

Let us, then, set to one side the type of justice and injustice that corresponds to the whole of virtue, justice being the exercise of the whole of virtue, and injustice of the whole of vice, in relation to another.

And it is evident how we must distinguish the way for a thing to be just or unjust that expresses this type of justice and injustice; for the majority of lawful actions, we might say, are the actions resulting from virtue as a whole. For the law instructs us to express each virtue, and forbids us to express each vice, in how we live. Moreover, the actions producing the whole of virtue are the lawful actions that the laws prescribe for education promoting the common good.

We must wait till later, however, to determine whether the education that makes an individual an unconditionally good man is a task for political science or for another science; for, presumably, being a good man is not the same as being every sort of good citizen.

5.34 BUT A DETAILED DESCRIPTION OF SPECIAL JUSTICE IS NEEDED

Special justice, however, and the corresponding way for something to be just [must be divided].

One species is found in the distribution of honours or wealth or anything else that can be divided among members of a community who share in a political system; for here it is possible for one member to have a share equal or unequal to another's.

Another species concerns rectification in transactions. This species has two parts, since one sort of transaction is voluntary, and one involuntary. Voluntary transactions include selling, buying, lending, pledging, renting, depositing, hiring out—these are called voluntary because the origin of these transactions is voluntary. Some involuntary ones are secret, e.g. theft, adultery, poisoning, pimping, slave-deception, murder by treachery, false witness; others are forcible, e.g. assault, imprisonment, murder, plunder, mutilation, slander, insult.

5.4 Justice in Distribution

5.41 JUSTICE, FAIRNESS AND EQUALITY

Since the unjust person is unfair, and what is unjust is unfair, there is clearly an intermediate between the unfair [extremes], and this is what is fair; for in any action where too much and too little are possible, the fair [amount] is also possible. And so if what is unjust is unfair, what is just is fair (*ison*), as seems true to everyone even without argument.

And since what is equal (*ison*) [and fair] is intermediate, what is just is some sort of intermediate. And since what is equal involves at least two things [equal to each other], it follows that what is just must be intermediate and equal, and related to some people. In so far as it is intermediate, it must be between too much and too little; in so far as it is equal, it involves two things; and in so far as it is just, it is just for some people. Hence what is just requires four things at least; the people for whom it is just are two, and the [equal] things that are involved are two.

5.42 HOW EQUALITY IS DETERMINED

Equality for the people involved will be the same as for the things involved, since [in a just arrangement] the relation between the people will be the same as the relation between the things involved. For if the people involved are not equal, they will not [justly] receive equal shares; indeed, whenever equals receive unequal shares, or unequals equal shares, in a distribution, that is the source of quarrels and accusations.

This is also clear from considering what fits a person's worth. For everyone agrees that what is just in distributions must fit some sort of worth, but what they call worth is not the same; supporters of democracy say it is free citizenship, some supporters of oligarchy say it is wealth, others good birth, while supporters of aristocracy say it is virtue.

5.43 JUSTICE IS PROPORTIONATE EQUALITY

Hence what is just [since it requires equal shares for equal people] is in some way proportionate. For proportion is special to number as a whole, not only to numbers consisting of [abstract] units, since it is equality of ratios and requires at least four terms.

This is the sort of proportion that mathematicians call geometrical, since in geometrical proportion the relation of whole to whole is the same as the relation of each [part] to each [part]. But this proportion [involved in justice] is not continuous, since there is no single term for both the person and the item.

What is just, then, is what is proportionate, and what is unjust is what is counter-proportionate. Hence [in an unjust action] one term becomes more and the other less; and this is indeed how it turns out in practice, since the one doing injustice has more of the good, and the victim less. With an evil the ratio is reversed, since the lesser evil, compared to the greater, counts as a good; for the lesser evil is more choiceworthy than the greater, what is choiceworthy is good, and what is more choiceworthy is a greater good.

This, then, is the first species of what is just

5.5 Justice in Rectification

5.51 IT IS DISTINCT FROM DISTRIBUTIVE JUSTICE

The other way of being just is the rectificatory, found in transactions both voluntary and involuntary; and this way of being just belongs to a different species from the first.

For what is just in distribution of common assets will always fit the proportion mentioned above, since distribution from common funds will also fit the ratio to one another of different people's deposits. Similarly, the way of being unjust that is opposed to this way of being just is what is counter-proportionate. On the other hand, what is just in transactions is certainly equal in a way, and what is unjust is unequal; but still it fits numerical proportion, not the [geometrical] proportion of the other species.

5.52 IT INVOLVES NUMERICAL PROPORTION AND EQUALITY

For here it does not matter if a decent person has taken from a base person, or a base person from a decent person, or if a decent or a base person has committed adultery. Rather, the law looks only at differences in the harm [inflicted], and treats the people involved as equals, when one does injustice while the other suffers it, and one has done the harm while the other has suffered it. Hence the judge tries to restore this unjust situation to equality, since it is unequal.

These apply to other wrongs besides theft

For [not only both when one steals from another but also] and when one is wounded and the other wounds him, or one kills and the other is killed, the action and the suffering are unequally divided [with profit for the offender and loss for the victim]; and the judge tries to restore the [profit and] loss to a position of equality, by subtraction from [the offender's] profit. For in such cases, stating it without qualification, we speak of profit for, e.g., the attacker who wounded his victim, even if that is not the proper word for some cases, and of loss for the victim who suffers the wound. At any rate, when what was suffered has been measured, one part is called the [victim's] loss, and the other the [offender's] profit.

In fact, however, these names 'loss' and 'profit' are derived from voluntary exchange. For having more than one's own share is called making a profit, and having less than what one had at the beginning is called suffering a loss, e.g. in buying and selling and in other transactions permitted by law. And when people get neither more nor less, but precisely what belongs to them, they say they have their own share, and make neither a loss nor a profit.

5.53 HENCE RECTIFICATORY JUSTICE IS A MEAN

Hence what is equal is intermediate between more and less; profit and loss are more and less in contrary ways, since more good and less evil is profit, and the contrary is loss; and the intermediate area between [profit and loss], we have found, is what is equal, which we say is just. Hence what is just in rectification is what is intermediate between loss and profit.

This is confirmed by the judge's rectificatory role in justice

Hence parties to a dispute resort to a judge, and an appeal to a judge is an appeal to what is just; for the judge is intended to be a sort of living embodiment of what is just. Moreover, they seek the judge as an intermediary, and in some cities they actually call judges mediators, assuming that if they are awarded an intermediate amount, the award will be just. If, then, the judge is an intermediary, what is just is in some way intermediate.

5.54 RECTIFICATORY JUSTICE IS ADMINISTERED
BY RESTORING NUMERICAL EQUALITY

The judge restores equality, as though a line [AB] had been cut into unequal parts [AC and CB], and he removed from the larger part [AC] the amount [DC] by which it exceeds the half [AD] of the line [AB], and added this amount [DC] to the smaller part [CB]. And when the whole [AB] has been halved [into AD and DB], then they say that each person has what is properly his own, when he has got an equal share. This is also why it is called just (*dikaion*), because it is a bisection (*dicha*), as though we said bisected (*dichaion*), and the judge (*dikastes*) is a bisector (*dichastes*).

What is equal [in this case] is intermediate, by numerical proportion, between the larger [AC] and the smaller line [CB]. For when [the same amount] is subtracted from one of two equal things and added to the other, then the one part exceeds the other by the two parts; for if a part had been subtracted from the one, but not added to the other, the larger part would have exceeded the smaller by just one part. Hence the larger part exceeds the intermediate by one part, and the intermediate from which [a part] was subtracted [exceeds the smaller] by one part.

In this way, then, we will recognize what we must subtract from the one who has more and add to the one who has less [to restore equality]; for to the one who has less we must add the amount by which the intermediate exceeds what he has, and from the greatest amount [which the one who has more has] we must subtract the amount by which it exceeds the intermediate.

5.64 MONEY IS DESIGNED TO SECURE PROPORTIONATE RECIPROCITY, BY FACILITATING EXCHANGE

This is why all items for exchange must be comparable in some way. Currency came along to do exactly this, and in a way it becomes an intermediate, since it measures everything, and so measures excess and deficiency—how many shoes are equal to a house.

Hence, as builder is to shoemaker, so must the number of shoes be to a house; for if this does not happen, there will be no exchange and no association, and the proportionate equality will not be reached unless they are equal in some way. Everything, then, must be measured by some one measure, as we said before.

Exchange rests on need, and hence is facilitated by money

In reality, this measure is need, which holds everything together; for it people required nothing, or needed things to different extents, there would be either no exchange or not the same exchange.

* * *

5.7 Political Justice

5.71 CONDITIONS FOR POLITICAL JUSTICE

We have previously described, then, the relation of reciprocity to what is just. We must now notice that we are looking not only for what is just unconditionally but also for what is just in a political association. This is found among associates in a life aiming at self-sufficiency, who are free and either proportionately or numerically equal.

Hence those who lack these features have nothing politically just in their relations, though they have something just in so far as it is similar [to what is politically just].

For what is just is found among those who have law in their relations. Where there is law, there is injustice, since the judicial process is judgement that distinguishes what is just from what is unjust. Where there is injustice there is also doing injustice, though where there is doing injustice there need not also be injustice. And doing injustice is awarding to oneself too many of the things that, [considered] unconditionally, are good, and too few of the things that, [considered] unconditionally, are bad.

5.72 THE NATURE OF POLITICAL JUSTICE EXPLAINS WHY INDIVIDUALS
ARE TEMPTED TO DO INJUSTICE

This is why we allow only reason, not a human being, to be ruler; for a human being awards himself too many goods and becomes a tyrant, but a ruler is a guardian of what is just and hence of what is equal [and so must not award himself too many goods].

If a ruler is just, he seems to profit nothing by it. For since he does not award himself more of what, [considered] unconditionally, is good if it is not proportionate to him, he seems to labour for another's benefit; that is why justice is said, as we also remarked before, to be another person's good. Hence some payment [for ruling] should be given; this is honour and privilege, and the people who are unsatisfied by these are the ones who become tyrants.

5.73 FORMS SIMILAR TO POLITICAL JUSTICE

What is just for a master and a father is similar to this, not the same. For there is no unconditional injustice in relation to what is one's own; one's own possession, or one's child until it is old enough and separated, is as though it were a part of oneself, and no one decides to harm himself. Hence there is no injustice in relation to them, and so nothing politically unjust or just either. For we found that what is politically just must conform to law, and apply to those who are naturally suited for law, hence to those who have equality in ruling and being ruled. [Approximation to this equality] explains why relations with a wife more than with children or possessions allow something to count as just—for that is what is just in households; still, this too is different from what is politically just.

5.74 JUSTICE BY NATURE AND BY LAW

One part of what is politically just is natural, and the other part legal. What is natural is what has the same validity everywhere alike, independent of its seeming so or not. What is legal is what originally makes no difference [whether it is done] one way or another, but makes a difference whenever people have laid down the rule— e.g. that a mina is the price of a ransom, or that a goat rather than two sheep should be sacrificed; and also laws passed for particular cases, e.g. that sacrifices should be offered to Brasidas; and enactments by decree.

Variations may seem to show there is no natural justice
     Now it seems to some people that everything just is merely legal, since what is natural is unchangeable and equally valid everywhere—fire, e.g., burns both here and in Persia—while they see that what is just changes [from city to city].

However, variations are consistent with the existence of natural justice
     This is not so, though in a way it is so. With us, though presumably not at all with the gods, there is such a thing as what is natural, but still all is changeable; despite the change there is such a thing as what is natural and what is not.
     What sort of thing that [is changeable and hence] admits of being otherwise is natural, and what sort is not natural, but legal and conventional, if both natural and

legal are changeable? It is clear in other cases also, and the same distinction [between the natural and the unchangeable] will apply; for the right hand, e.g., is naturally superior, even though it is possible for everyone to become ambidextrous.

The sorts of things that are just by convention and expediency are like measures. For measures for wine and for corn are not of equal size everywhere, but in wholesale markets they are bigger, and in retail smaller. Similarly, the things that are just by human [enactment] and not by nature differ from place to place, since political systems also differ; still, only one system is by nature the best everywhere.

The Koran

The Qur'an or Koran, the Holy Book of half a billion Muslims, is said to be an exact transcription of a tablet that exists for all eternity in Heaven, revealed to Mohammed over a period of some 20 years in the seventh century. Technically, the text is untranslatable from the original Arabic, so what we offer the reader here is a piece of a "paraphrase" into English. The Holy Book of Islam has a great deal to say about justice, and the key to this justice is "submission" (or "*Islam*") and "the fear of God (Allah)." Allah is described as "terrible in his retribution" but yet "all-forgiving and compassionate." The ultimate standard of justice is belief itself, but with belief and obedience go a system of divine demands concerning the details of daily life. A central theme in the Koran, both as a matter of practical concern and as a metaphor, is the "debt," for ancient as modern Arab and Middle Eastern culture was rich in commerce and trading. The metaphor of debt is also central to the conception of justice; for example, we are told that thieves are to have their hands cut off "as a recompense for what they have earned." The concepts of vengeance and retribution are prevalent, but they are tempered throughout by mercy. The Biblical injunction of "a life for a life, an eye for an eye" is repeated, but with the qualification that "whosoever forgoes it [retaliation] . . . that shall be for him an expiation."

———

And fear the Day
When ye shall be
Brought back to God.
Then shall every soul
Be paid what it earned,
And none shall be
Dealt with unjustly.

O ye who believe!
When ye deal with each other,
In transactions involving
Future obligations
In a fixed period of time,
Reduce them to writing
Let a scribe write down
Faithfully as between
The parties: let not the scribe
Refuse to write: as God

Has taught him,
So let him write.
Let him who incurs
The liability dictate,
But let him fear
His Lord God,
And not diminish
Aught of what he owes.
If the party liable
Is mentally deficient,
Or weak, or unable
Himself to dictate,
Let his guardian
Dictate faithfully.
And get two witnesses,
Out of your own men,
And if there are not two men,
Then a man and two women,
Such as ye choose,
For witnesses,
So that if one of them errs,
The other can remind her.
The witnesses
Should not refuse
When they are called on
(For evidence).
Disdain not to reduce
To writing (your contract)
For a future period,
Whether it be small
Or big: it is juster
In the sight of God,
More suitable as evidence,
And more convenient
To prevent doubts
Among yourselves
But if it be a transaction
Which ye carry out
On the spot among yourselves,
There is no blame on you
If ye reduce it not
To writing.
But take witnesses
Whenever ye make
A commercial contract;
And let neither scribe
Nor witness suffer harm.

If ye do (such harm),
It would be wickedness
In you. So fear God;
For it is God
That teaches you.
And God is well acquainted
With all things.

* * *

O ye who believe!
Stand out firmly
For God, as witnesses
To fair dealing, and let not
The hatred of others
To you make you swerve
To wrong and depart from
Justice. Be just: that is
Next to Piety: and fear God.
For God is well-acquainted
With all that ye do.

To those who believe
And do deeds of righteousness
Hath God promised forgiveness
And a great reward.
Those who reject faith
And deny Our Signs
Will be Companions
Of Hell-fire.

O ye who believe!
Call in remembrance
The favour of God
Unto you when
Certain men formed the design
To stretch out
Their hands against you,
But (God) held back
Their hands from you:
So fear God. And on God
Let Believers put
(All) their trust.

* * *

As to the thief,
Male or female,
Cut off his or her hands:
A punishment by way

Of example, from God,
For their crime:
And God is Exalted in Power.

But if the thief repent
After his crime,
And amend his conduct,
God turneth to him
In forgiveness; for God
Is Oft-forgiving, Most Merciful.

Knowest thou not
That to God (alone)
Belongeth the dominion
Of the heavens and the earth?
He punisheth whom He pleaseth,
And He forgiveth whom He pleaseth:
And God hath power
Over all things.

* * *

Therefore fear not men,
But fear Me, and sell not
My Signs for a miserable price.
If any do fail to judge
By (the light of) what God
Hath revealed, they are
(No better than) Unbelievers.

We ordained therein for them:
"Life for life, eye for eye,
Nose for nose, ear for ear,
Tooth for tooth, and wounds
Equal for equal." But if
Any one remits the retaliation
By way of charity, it is
An act of atonement for himself.
And if any fail to judge
By (the light of) what God
Hath revealed, they are
(No better than) wrong-doers.

* * *

O ye who believe!
God doth but make a trial of you
In a little matter
Of game well within reach
Of your hands and your lances,

That He may test
Who feareth Him unseen:
Any who transgress
Thereafter, will have
A grievous penalty.

O ye who believe!
Kill not game
While in the Sacred
Precincts or in pilgrim garb.
If any of you doth so
Intentionally, the compensation
Is an offering, brought
To the Ka'ba, of a domestic animal
Equivalent to the one he killed,

As adjudged by two just men
Among you; or by way
Of atonement, the feeding
Of the indigent; or its
Equivalent in fasts: that he
May taste of the penalty
Of his deed. God
Forgives what is past:
For repetition God will
Exact from him the penalty.
For God is Exalted,
And Lord of Retribution.

Lawful to you is the pursuit
Of water-game and its use
For food,—for the benefit
Of yourselves and those who
Travel; but forbidden
Is the pursuit of land-game;—
As long as ye are
In the Sacred Precincts
Or in pilgrim garb.
And fear God, to Whom
Ye shall be gathered back.

* * *

Know ye that God
Is strict in punishment

And that God is
Oft-forgiving, Most Merciful.

Aquinas, from *Summa Theologica* (1224–1274)

Saint Thomas synthesized the Christianity of the Church with Aristotle's ethics and metaphysics, and the result was a theology that has ruled the Catholic Church for many centuries since. In the following passages, Saint Thomas examines and defends "the Philosopher"—in every case Aristotle—on his analysis of justice, against the prevalent objections of the day. He considers, in particular, Aristotle's defense of desert in distributive justice, whether each person should get "his right," Aristotle's insistence that justice (as "righteousness") is essentially a virtue directed at other people, whether there is such a general virtue and a "particular" virtue as well, whether justice is or is not about the passions and pleasure and pain, and whether justice is, indeed, the "mean between the extremes." Finally, Aquinas considers the Aristotelian view of whether (distributive) justice is primarily concerned with desert, "to each his own," or if it is more properly conceived of in terms of need and benevolence, as Saint Augustine had argued with regard to our obligation to care for the needy, and Cicero had suggested regarding charity and generosity ("liberality" as an essential part of justice).

―――――――――

Of Justice

We must now consider justice. Under this head there are twelve points of inquiry: (1) What is justice? (2) Whether justice is always towards another? (3) Whether it is a virtue? (4) Whether it is the will as its subject? (5) Whether it is a general virtue? (6) Whether, as a general virtue, it is essentially the same as every virtue? (7) Whether there is a particular justice? (8) Whether particular justice has a matter of its own? (9) Whether it is about passions or about operations only? (10) Whether the mean of justice is an objective mean? (11) Whether the act of justice is to render to everyone his own? (12) Whether justice is the chief of the moral virtues?

First Article

Is Justice Fittingly Defined as Being the Perpetual and Constant Will to Render to Each One His Right?

We proceed thus to the First Article:

Obj. 1. It would seem that lawyers have unfittingly defined justice as being "the perpetual and constant will to render to each one his right." For, according to the Philosopher, justice is a habit which makes a man "capable of doing what is just and of being just in action and in intention." Now, "will" denotes a power or also an act. Therefore, justice is unfittingly defined as being a will.

Obj. 2. Further, rectitude of the will is not the will; else if the will were its own rectitude, it would follow that no will is unrighteous. Yet, according to Anselm, justice is rectitude. Therefore, justice is not the will.

Obj. 3. Further, no will is perpetual save God's. If, therefore, justice is a perpetual will, in God alone will there be justice.

Obj. 4. Further, whatever is perpetual is constant, since it is unchangeable. Therefore, it is needless in defining justice to say that it is both perpetual and constant.

Obj. 5. Further, it belongs to the ruler to give each one his right. Therefore, if justice gives each one his right, it follows that it is in none but the ruler, which is absurd.

Obj. 6. Further, Augustine says that "justice is love serving God alone." Therefore, it does not render to each one what is his.

I answer that The aforesaid definition of justice is fitting if understood aright. For, since every virtue is a habit, that is, the principle of a good act, a virtue must needs be defined by means of the good act bearing on the matter proper to that virtue. Now, the proper matter of justice consists of those things that belong to our intercourse with other men, as shall be shown further on. Hence the act of justice in relation to its proper matter and object is indicated in the words, "Rendering to each one his right," since, as Isidore says, "a man is said to be just because he respects the right *[jus]* of others."

Now, in order that an act bearing upon any matter whatever be virtuous, it should be voluntary, stable, and firm, because the Philosopher says that, in order for an act to be virtuous, it needs first of all to be done knowingly; secondly, to be done by choice and for a due end; thirdly, to be done resolutely. Now the first of these is included in the second, since "what is done through ignorance is involuntary." Hence the definition of justice mentions first the will, in order to show that the act of justice must be voluntary, and mention is made afterwards of its constancy and perpetuity in order to indicate the firmness of the act.

Accordingly, this is a complete definition of justice, save that the act is mentioned instead of the habit, which takes its species from that act, because habit implies relation to act. And if anyone would reduce it to the proper form of a definition, he might say that "justice is a habit whereby a man renders to each one his due by a constant and perpetual will"; this is about the same definition as that given by the Philosopher, who says that "justice is a habit whereby a man is said to be capable of doing just actions in accordance with his choice."

Second Article

Is Justice Always towards Another?

We proceed thus to the Second Article:

Obj. 1. It would seem that justice is not always towards another. For the Apostle says that "the justice of God is by faith in Jesus Christ." Now, faith does not concern the dealings of one man with another. Neither, therefore, does justice.

Obj. 2. Further, according to Augustine, "It belongs to justice that man should direct to the service of God his authority over the things that are subject to him." Now the sensitive appetite is subject to man, according to Genesis 4:7, where it is written: "The lust thereof," viz., of sin, "shall be under you, and you shall have dominion over it." Therefore, it belongs to justice to have dominion over one's own appetite, so that justice is towards oneself.

Obj. 3. Further, the justice of God is eternal. But nothing else is co-eternal with God. Therefore, justice is not essentially towards another.

Obj. 4. Further, man's dealings with himself need to be rectified no less than his dealings with another. Now, man's dealings are rectified by justice, according to Proverbs 11:5, "The justice of the upright shall make his way prosperous." Therefore, justice is about our dealings, not only with others but also with ourselves.

On the contrary, Tully says that "the object of justice is to keep men together in society and mutual intercourse." Now this implies relationship of one man to another. Therefore, justice is concerned only about our dealings with others.

I answer that, As stated above, since justice by its name implies equality, it denotes essentially relation to another, for a thing is equal, not to itself, but to another. And, inasmuch as it belongs to justice to rectify human acts, as stated above, this otherness which justice demands must needs be between beings capable of action. Now, actions belong to ultimate objects of attribution and wholes and, properly speaking, not to parts and forms or powers, for we do not say properly that the hand strikes, but a man with his hand, nor that heat makes a thing hot, but fire by heat, although such expressions may be employed metaphorically. Hence justice, properly speaking, demands a distinction of ultimate objects of attribution and consequently is only in one man towards another. Nevertheless, in one and the same man we may speak metaphorically of his various principles of action, such as reason and the irascible and the concupiscible appetites, as though they were so many agents, so that metaphorically in one and the same man there is said to be justice insofar as the reason commands the irascible and concupiscible, and these obey reason, and in general insofar as to each part of man is ascribed what is becoming to it. Hence the Philosopher calls this "metaphorical justice."

Fifth Article

Is Justice a General Virtue?

We proceed thus to the Fifth Article:

Obj. 1. It would seem that justice is not a general virtue. For justice is specified with the other virtues, according to Wis. 8:7, "She teaches temperance and prudence and justice and fortitude." Now, the general is not specified or reckoned together with the species contained under the same general. Therefore, justice is not a general virtue.

Obj. 2. Further, as justice is accounted a cardinal virtue, so are temperance and fortitude. Now, neither temperance nor fortitude is reckoned to be a general virtue. Therefore, neither should justice in any way be reckoned a general virtue.

Obj. 3. Further, justice is always towards others, as stated above. But a sin committed against one's neighbor cannot be a general sin because it is distinguished from sin committed against oneself. Therefore, neither is justice a general virtue.

On the contrary, The Philosopher says that "justice is every virtue."

I answer that Justice, as stated above, directs man in his relations with other men. Now, this may happen in two ways: first, as regards his relation with individuals; secondly, as regards his relations with others in general, insofar as a man who serves a community serves all those who are included in that community. Accordingly, justice in its proper acceptation can be directed to another in both these senses. Now, it is evident that all who are included in a community stand in relation to that community as parts to a whole, while a part, as such, belongs to a whole, so that whatever is the good of a part can be directed to the good of the whole. It follows, therefore, that the good of any virtue, whether such virtue direct man in relation to himself or in relation to certain other individual persons, is referable to the common good, to which justice directs, so that all acts of virtue can pertain to justice insofar as it directs man to the common good. It is in this sense that justice is called a general virtue. And since it belongs to the law to direct to the common good, as stated above, it follows that the justice which is in this way styled general is called legal justice, because thereby man is in harmony with the law which directs the acts of all the virtues to the common good.

Seventh Article

Is There a Particular besides a General Justice?

We proceed thus to the Seventh Article:

Obj. 1. It would seem that there is not a particular besides a general justice. For there is nothing superfluous in the virtues, as neither is there in nature. Now, general justice directs man sufficiently in all his relations with other men. Therefore, there is no need for a particular justice.

Obj. 3. Further, between the individual and the general public stands the household community. Consequently, if in addition to general justice there is a particular justice corresponding to the individual, for the same reason there should be a domestic justice directing man to the common good of a household, and yet this is not the case. Therefore, neither should there be a particular besides a legal justice.

Reply Obj. 1. Legal justice does indeed direct man sufficiently in his relations towards others. As regards the common good, it does so immediately, but as to the good of the individual, it does so mediately. Wherefore there is need for particular justice to direct a man immediately to the good of another individual.

Reply Obj. 3. The household community, according to the Philosopher, differs in respect of a threefold fellowship namely, of husband and wife, father and son, master and slave, in each of which one person is, as it were, part of the other. Wherefore, between such persons, there is not justice simply but a species of justice, viz., domestic justice, as stated in *Ethics* V, 6.

Ninth Article

Is Justice about the Passions?

We proceed thus to the Ninth Article:

Obj. 1. It would seem that justice is about the passions. For the Philosopher says that "moral virtue is about pleasure and pain." Now, pleasure, or delight, and pain are passions, as stated above when we were treating of the passions. Therefore, justice, being a moral virtue, is about the passions.

Obj. 2. Further, justice is the means of rectifying a man's operations in relation to another man. Now, such like operations cannot be rectified unless the passions be rectified, because it is owing to disorder of the passions that there is disorder in the aforesaid operations; thus sexual lust leads to adultery, and over-much love of money leads to theft. Therefore, justice must needs be about the passions.

Obj. 3. Further, even as particular justice is towards another person, so is legal justice. Now, legal justice is about the passions, else it would not extend to all the virtues, some of which are evidently about the passions. Therefore, justice is about the passions.

On the contrary, The Philosopher says that justice is about operations.

I answer that The true answer to this question may be gathered from a twofold source. First, from the subject of justice, i.e., from the will, whose movements or acts are not passions, as stated above, for it is only the sensitive appetite whose movements are called passions. Hence justice is not about the passions, as are temperance and fortitude, which are about the irascible and concupiscible appetite. Secondly, on the part of the matter, because justice is about a man's relations with another, and we are not directed immediately to another by the internal passions. Therefore, justice is not about the passions.

Reply Obj. 1. Not every moral virtue is about pleasure and pain as its proper matter, since fortitude is about fear and daring, but every moral virtue is directed to pleasure and pain, as to ends to be acquired. For, as the Philosopher says, "pleasure and pain are the principal end in respect of which we say that this is an evil, and that a good," and in this way too they belong to justice, since "a man is not just unless he rejoice in just actions."

Reply Obj. 2. External operations are means, as it were, between external things, which are their matter, and internal passions, which are their origin. Now, it happens sometimes that there is one of these without there being a defect in the other. Thus, a man may steal another's property, not through the desire to have the thing but through the will to hurt the man; or, vice versa, a man may covet another's property without wishing to steal it. Accordingly, the directing of operations, insofar as they tend towards external things, belongs to justice, but insofar as they arise from the passions, it belongs to the other moral virtues, which are about the passions. Hence, justice hinders theft of another's property insofar as stealing is contrary to the equality that should be maintained in external things, while liberality hinders it as resulting from an immoderate desire for wealth. Since, however, external operations take their species, not from the internal passions but from external things as being their objects, it follows that external operations are essentially the matter of justice rather than of the other moral virtues.

Reply Obj. 3. The common good is the end of each individual member of a community, just as the good of the whole is the end of each part. On the other hand, the good of one individual is not the end of another individual; wherefore legal justice, which is directed to the common good, is more capable of extending to the internal passions, whereby man is disposed in some way or other in himself, than is particular justice, which is directed to the good of another individual, although legal justice extends chiefly to other virtues in the point of their external operations, insofar, to wit, as "the law commands us to perform the actions of a courageous person, . . . the actions of a temperate person . . . and the actions of a gentle person."

Tenth Article

Is the Mean of Justice an Objective Mean?

We proceed thus to the Tenth Article:

Obj. 1. It seems that the mean of justice is not the mean of some object. For the concept of a genus is preserved in all its species. But moral virtue is defined in the *Ethics* as "a willed habit which observes a mean determined in relation to us by reason." Therefore, just so, there is in justice a rational mean, not an objective one.

Obj. 2. Further, in things that are good simply, there is neither excess nor defect, and consequently neither is there a mean, as is clearly the case with the virtues according to *Ethics* II, 6. Now, justice is about things that are good simply, as stated in *Ethics* V. Therefore, in justice there is not an objective mean.

Obj. 3. Further, the reason why the other virtues are said to observe the rational and not an objective mean is because, in their case, the mean varies according to different persons, since what is too much for one is too little for another. Now, this is also the case in justice, for one who strikes a prince does not receive the same punishment as one who strikes a private individual. Therefore, just so, justice does not possess an objective mean but a rational one.

On the contrary, The Philosopher says that the mean of justice is to be taken according to arithmetical proportion, which is an objective mean.

I answer that, As stated above, the other moral virtues are chiefly concerned with the passions, the regulation of which is gauged entirely by the measure of the very man who is the subject of those passions, insofar as his anger and desire are as much as they ought to be in various circumstances. Hence, the mean in such like virtues is measured not by the proportion of one thing to another but merely by comparison with the virtuous man himself, so that with them the mean is only that which is fixed by reason in our regard.

On the other hand, the matter of justice is external operation, insofar as an operation or the thing used in that operation is duly proportonate to another person; wherefore the mean of justice consists in a certain proportion of equality between the external thing and the external person. Now, equality is the real mean between greater and less, as stated in *Metaphysics* IX; wherefore justice observes the mean objectively

Reply Obj. 1. This objective mean is also the rational mean; wherefore justice satisfies the conditions of a moral virtue.

Reply Obj. 2. We may speak in two ways of a thing being good simply. First, a thing may be good in every way; thus the virtues are good, and there is neither mean nor extremes in things that are simply good in this sense. Secondly, a thing is said to be simply good through being good absolutely, i.e., in its nature, although it may become evil through being abused. Such are riches and honors, and in the like it is possible to find excess, deficiency, and mean as regards men, who can use them well or ill, and it is in this sense that justice is about things that are good simply.

Reply Obj. 3. The injury inflicted bears a different proportion to a ruler from that which it bears to a private person; wherefore each injury needs to be equalized by punishment in a different way, and this implies an objective and not merely a rational diversity.

Eleventh Article

Is the Act of Justice to Render to Each One His Own?

We proceed thus to the Eleventh Article:

Obj. 1. It would seem that the act of justice is not to render to each one his own. For Augustine ascribes to justice the act of succoring the needy. Now, in succoring the needy, we give them what is not theirs but ours. Therefore, the act of justice does not consist in rendering to each one his own.

Obj. 2. Further, Tully says that "beneficence, which we may call kindness or liberality, belongs to justice." Now, it pertains to liberality to give to another of one's own, not of what is his. Therefore, the act of justice does not consist in rendering to each one his own.

Obj. 3. Further, it belongs to justice not only to distribute things duly but also to repress injurious actions, such as murder, adultery, and so forth. But the rendering to each one of what is his seems to belong solely to the distribution of things. Therefore, the act of justice is not sufficiently described by saying that it consists in rendering to each one his own.

On the contrary, Ambrose says, "It is justice that renders to each one what is his and claims not another's property; it disregards its own profit in order to preserve the common equity."

I answer that, As stated above, the matter of justice is an external operation, insofar as either it or the thing we use by it is made proportionate to some other person to whom we are related by justice. Now, each man's own is that which is due to him according to equality of proportion. Therefore, the proper act of justice is nothing else than to render to each one his own.

Reply Obj. 1. Since justice is a cardinal virtue, other secondary virtues, such as mercy, liberality, and the like, are connected with it, as we shall state further on. Wherefore, to succor the needy, which belongs to mercy or pity, and to be liberally beneficent, which pertains to liberality, are by a kind of reduction ascribed to justice as to their principal virtue.

This suffices for the *Reply* to the *Second Objection*.

Reply Obj. 3. As the Philosopher states, in matters of justice, the name of profit is extended to whatever is excessive, and whatever is deficient is called loss. The reason for this is that justice is first of all and more commonly exercised in voluntary interchanges of things, such as buying and selling, wherein those expressions are properly employed, and yet they are transferred to all other matters of justice. The same applies to the rendering to each one of what is his own.

Mencius, *On the Mind*
(372(?) B.C.–298(?) B.C.)

Mencius was a disciple of the great Chinese philosopher Confucius. He lived and worked about the same time as Plato and Aristotle, and he developed a sensitive and elaborate philosophy of what would much later (in the eighteenth century in Europe) be identified as a theory of the moral sentiments. He emphasized the importance of justice in a ruler and the equal importance of respect and obedience in his followers. Confucius himself lived from 551–479 B.C., and his philosophy emphasized above all the importance of "uprightness" in the prince, and obedience and "gentlemanliness" in his subjects. He introduced two key terms into Chinese thought and ordinary language: the idea of *li*—or rules of conduct, and the concept of *ren*—benevolent love or what later Western thinkers would similarly call *agape*. The ideal of justice was an upright ruler with the pious obedience and adulation of the people. It is worth noting that this ideal was, above all, personal—a matter of the character and virtue of the prince and his subjects—and not merely a matter of law and rules as such. A just state would, according to the Confucians, be the envy of surrounding tyrannical states, and people would flock from tyranny to justice. The insistence on obedience and loyalty undercut any suggestion that an unjust ruler should be overthrown, but Confucius assured us that justice would triumph, nevertheless. (In what follows, the word "mind" might better be read as "soul," suggesting deep spirituality and selfhood rather than simply our intellectual facilities.)

6.1

Mencius said, "It is a feeling common to all mankind that they cannot bear to see others suffer. The Former Kings had such feelings, and it was this that dictated their policies. One could govern the entire world with policies dictated by such feelings, as easily as though one turned it in the palm of the hand.

"I say that all men have such feelings because, on seeing a child about to fall into a well, everyone has a feeling of horror and distress. They do not have this feeling out of sympathy for the parents, or to be thought well of by friends and neighbours, or from a sense of dislike at not being thought a feeling person. Not to feel distress would be contrary to all human feeling. Just as not to feel shame and disgrace and not to defer to others and not to have a sense of right and wrong are

From *Mencius*, trans. W. A. C. H. Dobson. Copyright © University of Toronto Press. Reprinted by permission of the publisher.

contrary to all human feeling. This feeling of distress (at the suffering of others) is the first sign of Humanity. This feeling of shame and disgrace is the first sign of Justice. This feeling of deference to others is the first sign of propriety. This sense of right and wrong is the first sign of wisdom. Men have these four innate feelings just as they have four limbs. To possess these four things, and to protest that one is incapable of fulfilling them, is to deprive oneself. To protest that the ruler is incapable of doing so is to deprive him. Since all have these four capacities within themselves, they should know how to develop and to fulfil them. They are like a fire about to burst into flame, or a spring about to gush forth from the ground. If, in fact, a ruler can fully realize them, he has all that is needed to protect the entire world. But if he does not realize them fully, he lacks what is needed to serve even his own parents."

6.2

Mencius said, "All men have things they cannot tolerate, and if what makes this so can be fully developed in the things they can tolerate, the result is Humanity. All men have things they will not do, and if what makes this so can be fully developed in the things they will do, then Justice results. If a man can fully exploit the thing in his mind which makes him not wish to harm others, then Humanity will result in overwhelming measure. If a man can fully exploit the thing in his mind which makes him reluctant to break through or jump over (other people's) walls, Justice will ensue in overwhelming measure.

* * *

6.9

If the prince is a man of Humanity then nothing in his state but will be Humane.
If the prince is a man of Justice, then nothing in his state but will be Just.

6.19

The difference between a man and an animal is slight. The common man disregards it altogether, but the True Gentleman guards the distinction most carefully. Shun understood all living things, but saw clearly the relationships that exist uniquely among human beings. These relationships proceed from Humanity and Justice, it is not because of these relationships that we proceed towards Humanity and Justice.

6.20

Mencius said, "Bull Mountain was once beautifully wooded. But, because it was close to a large city, its trees all fell to the axe. What of its beauty then? However, as the days passed things grew, and with the rains and the dews it was not without greenery. Then came the cattle and goats to graze. That is why, today, it has that scoured-like appearance. On seeing it now, people imagine that nothing ever grew there. But this is surely not the true nature of a mountain? And so, too, with human beings. Can it be that any man's mind naturally lacks Humanity and Justice? If he loses his sense of the good, then he loses it as the mountain lost its trees. It has been hacked away at—day after day—what of its beauty then?

"However, as the days pass he grows, and, as with all men, in the still air of the early hours his sense of right and wrong is at work. If it is barely perceptible, it is because his actions during the day have disturbed or destroyed it. Being disturbed and turned upside down the 'night airs' can barely sustain it. If this happens he is not far removed from the animals. Seeing a man so close to an animal, people cannot imagine that once his nature was different—but this is surely not the true nature of the man? Indeed, if nurtured aright, anything will grow, but if not nurtured aright anything will wither away. Confucius said, 'Hold fast to it, and you preserve it; let it go and you destroy it; it may come and go at any time no one knows its whereabouts.' Confucius was speaking of nothing less than the mind."

6.21

Mencius said, "I am fond of fish, but, too, I am fond of bear's paws. If I cannot have both, then I prefer bear's paws. I care about life, but, too, I care about Justice. If I cannot have both, then I choose Justice. I care about life, but then there are things I care about more than life. For that reason I will not seek life improperly. I do not like death, but then there are things I dislike more than death. For that reason there are some contingencies from which I will not escape.

"If men are taught to desire life above all else, then they will seize it by all means in their power. If they are taught to hate death above all else, then they will avoid all contingencies by which they might meet it. There are times when one might save one's life, but only by means that are wrong. There are times when death can be avoided, but only by means that are improper. Having desires above life itself and having dislikes greater than death itself is a type of mind that all men possess—it is not only confined to the worthy. What distinguishes the worthy is that he ensures that he does not lose it.

"Even though it be a matter of life or death to him, a traveller will refuse a basket of rice or a dish of soup if offered in an insulting manner. But food that has been trampled upon, not even a beggar will think fit to eat. And yet a man will accept emoluments of ten thousand *chung* regardless of the claims of propriety and Justice. And what does he gain by that? Elegant palaces and houses, wives and concubines to wait on him, and the allegiance of the poor among his acquaintance! I was previously speaking of matters affecting life and death, where even there under certain conditions one will not accept relief, but this is a matter of palaces and houses, of wives and concubines, and of time-serving friends. Should we not stop such things? This is what I mean by 'losing the mind with which we originally were endowed.'"

6.42

Mencius said, "The abilities men have which are not acquired by study are part of their endowment of good. The knowledge men have which is not acquired by deep thought is part of their endowment of good. Every baby in his mother's arms knows about love for his parents. When they grow up, they know about the respect they must pay to their elder brothers. The love for parents is Humanity. The respect for elders is Justice. It is nothing more than this, and it is so all over the world."

PART II

The ancient Greeks were already fascinated by the idea of prehistory, as no doubt were all intelligent people that had any sense of history at all. So, too, the ancient Greeks were already sensitive to some sense of an implicit contract that tacitly bound citizens to their city-state, although they did not have (or miss) the concepts of individuality and legal autonomy that underlie such discussions in modern philosophy. But some sense of what is now called *the social contract* can already be found in Plato, and we also find in Plato and later in Lucretius some sensitive speculation about life "before" the formation of society. It has been argued (e.g., by Alasdair MacIntyre) that the ancient Greeks did not see such speculations as essential to the *justification* of their society or their conceptions of justice; indeed they would not have seen these as requiring justification in that sense at all. But by the time when social contract theory emerged as the central speculative doctrine of modern social philosophy in the seventeenth and eighteenth centuries, justification was the heart of the issue. What justifies the state's right to take away our property, in the form of taxes, or to induct young men (and women) into state service, for instance, in wartime? What justifies the state's right to make and enforce laws and punish those who violate them? What makes the state—what makes a government—*legitimate*? One answer, perhaps the dominant answer in Western (and Eastern) political philosophy, is that it is just. But, once again, what is justice? And how do states and governments rightfully claim to be just?

The very popular and influential modern answer (which would not have made much sense to Plato and his colleagues) is that states and governments are legitimate insofar as they are formed and supported by the mutual agreement of the citizens of such a state, illegitimate if not. But from this it seems not much of a step (especially to those who see "society" and "the state" as inseparable notions) to suggest that societies themselves are justified and legitimate only insofar as they are formed and supported by their members. Of course, whether or not a government or a form of government is in fact so "supported" is readily enough confirmed (by some sort of election of plebescite, for example), but whether it makes sense to speak of the formation of a society in such a manner— much less the formation of society *as such*—is quite another matter. What does it mean to talk about the historical "formation" of a society? We can understand, for instance, the formation of the United States of America as a nation through the drafting of a constitution, but it should be said that much of the structure of American society (if not the form of government) was already in place during colonial, prerevolutionary days. The colonialists for the most part shared a European ancestry, a language, a general if somewhat fractious sense of what social and economic life should be in this "new" world. The Constitution established a (form of) government; it did not form a society. Indeed, most societies exist for many years before drafting any such document or making any appeal to "the consent of the governed." England and France, for example, existed many centuries under hereditary monarchies justified solely by "divine right," devoid of any democratic institutions, before their modern, quasi-social contract status was established.

But the actuality of a historical as opposed to a merely hypothetical agreement has never been much of an issue for social contract theorists. What is important is the *conception* of society that such a notion gives us as a network of voluntary obligations, and it is of particular importance pertaining to questions of justice. Whatever anthropologists may tell us about the actual state of prehistorical human beings in the wild, we should conceive of society as voluntarily constructed by and composed of rational, autonomous human beings who saw and still see their way to compromising their various and often divisive self-interests in order to establish a just society in which both their own and others' self-interests would be maximized. Prior to such an agreement, there was and there is no justice.

Social contract theory as we know it had come to maturity by the eighteenth century. In his *Leviathan*, Thomas Hobbes famously contrasted the comparatively secure and happy lot of citizens in the state to their remarkably brutal and unhappy situation in "the state of nature." Before the formation of society, he tells us, life was a war of everyone against everyone and "solitary, poor, nasty, brutish and short." The "social compact," accordingly, was designed to provide mutual security and safety. John Locke developed a much more benign view of the state of nature, with industrious individuals planting and building and working to make their world a better, more comfortable place. Locke's social contract, accordingly, was designed primarily to protect the fruits of these individuals' labors and the property on which it depended. Jean-Jacques Rousseau, by way of contrast again, developed an ebullient conception of our natural state and suggested that, in nature as opposed to contemporary civil society, we were free and happy. Society has "corrupted" us and made us dependent and miserable. But despite our alleged natural happiness, human freedom and reason brought us out of nature and into the oppressive strictures of society. It is in response to this degenerate sense of society that the social contract plays a central role in Rousseau's philosophy. In opposition to the corrupted and unjust forms of society that he saw around him, Rousseau suggested a very different conception of society, based not on the power of a few and the deception and stupidity of the rest but on the social contract, in which all citizens "impose the law on themselves" and cultivate a public moral life together.

It is in conscientious contrast with Hobbes and Rousseau in particular that the German philosopher G. W. F. Hegel presents his own allegorical representation of presocial beings, as "two self-consciousnesses" meet on an abstract and unembellished stage in the absence of any social setting or any sense of mutual belonging. In the famous "master and slave" section of his book, *The Phenomenology of Spirit*, Hegel argues (or rather, shows) that two people even in such a state of abstraction could never be "indifferent" to one another, as Rousseau suggests, and if the immediate result of such a meeting is hostility, this is not to be explained, as in Hobbes, by mutual selfishness. What people want and demand from each other—indeed what makes them "self-consciousnesses" at all—is mutual recognition, some shared sense of themselves. And if the "master-slave" conception turns out to be inadequate, that is because the very idea of individual self-consciousness is inadequate as well. Hegel's point is that the very idea of autonomous individuals in a presocial state of nature makes no real sense and gives us a totally unworkable picture of the origins of society.

Today, the theory of the social contract and the state of nature form the very core of the debates about justice. John Rawls postulates an "original position," not as any possible historical circumstance but as a rational model within which we can understand and justify the central principles of justice. The idea is that, in the original position, we find ourselves behind a "veil of ignorance" and do not know any of those specific features about ourselves that would lead us to prefer one social arrangement over any other solely for the sake of our own self-interest. If we do not know whether we will be rich or poor, for example, then we will not (he argues) endorse a view of society or justice in which we would—if we turn out to be poor—be seriously handicapped or find our way to self-improvement hopelessly blocked. The original position, for Rawls, is an abstracted scenario for rational social choice rather than any hypothesis about human nature, and the social contract is a system of rational deliberations—even an attempted deductive demonstration—rather than anything like a possible historical document.

Against Rawls's view, Robert Nozick accepts the idea of a state of nature as an abstracted scenario for rational social choice but he rejects Rawls's reliance on the social contract, however hypothetical. What speculation about the state of nature teaches us, according to Nozick, is the need for some minimal state, a state that functions as a protective association to secure property, enforce contracts, and defend individual rights but not much more than this. Such a state need not arise, however, through the mutual understanding and agreement of its various members. It would rather evolve, through trial and error, possibly without the full recognition of any of its members. Borrowing a famous metaphor from Adam Smith, Nozick suggests that "an invisible hand" would lead to the creation of such a state, through a progression of unsatisfactory protective associations, without the conscientious formation of such a structure by its citizens. Finally, Michael Sandel rejects not only the idea of a social contract but the very idea of a presocial humanity as well, on grounds that are very Hegelian (but more explicitly Aristotelian). Communities are prior to individuals and contracts, not only historically but conceptually. The idea of individual rational choice that is not already circumscribed by and defined by membership in a community is, according to Sandel, utterly incomprehensible.

Plato, from *Protagoras* and *Crito*

In *Protagoras*, Plato speculates about the beginnings of humanity and society. In a particularly famous passage from *Crito*, Socrates argues that his continued, clearly voluntary residence in Athens has tacitly bound him to the laws of the city, even if those laws now turn against him and condemn him to death.

Once upon a time there were gods only, and no mortal creatures. But when the time came that these also should be created, the gods fashioned them out of earth and fire and various mixtures of both elements in the inward parts of the earth; and when they were about to bring them into the light of day, they ordered Prometheus and Epimetheus to equip them, and to distribute to them severally their proper qualities. Epimetheus said to Prometheus: "Let me distribute, and do you inspect." This was agreed, and Epimetheus made the distribution. There were some to whom he gave strength without swiftness, or again swiftness without strength; some he armed, and others he left unarmed; and devised for the latter some other means of preservation, making some large, and having their size as a protection, and others small, whose nature was to fly in the air or burrow in the ground—this was to be their way of escape. Thus did he compensate them with the view of preventing any race from becoming extinct. And when he had provided against their destruction by one another, he contrived also a means of protecting them against the seasons of heaven; clothing them against the winter cold and summer heat, and for a natural bed of their own when they wanted to rest; also he furnished them with hoofs and hair and hard and callous skins under their feet. Then he gave them varieties of food—to some herbs of the soil, to others fruits of trees, and to others roots, and to some again he gave other animals as food. And some he made to have few young ones, while those who were their prey were very prolific; and in this way the race was preserved. Thus did Epimetheus, who, not being very wise, forgot that he had distributed among the brute animals all the qualities that he had to give—and when he came to man, who was still unprovided, he was terribly perplexed. Now while he was in this perplexity, Prometheus came to inspect the distribution, and he found that the other animals were suitably furnished, but that man alone was naked and shoeless, and had neither bed nor arms of defence. The appointed hour was approaching in which man was to go forth into the light of day; and Prometheus, not knowing how he could devise his salvation, stole the mechanical arts of Hephæstus and Athene, and fire with them (they could neither have been acquired nor used without fire), and gave them to man. Thus man had the wisdom necessary to the support of life, but political wisdom he had not; for that was in the keeping of Zeus, and the power of Prometheus did not extend to entering into the castle of heaven, in

71

which Zeus dwelt, who moreover had terrible sentinels; but he did enter by stealth into the common workshop of Athene and Hephæstus, in which they used to pursue their favorite arts, and took away Hephæstus's art of working by fire, and also the art of Athene, and gave them to man. And in this way man was supplied with the means of life. But Prometheus is said to have been afterwards prosecuted for theft, owing to the blunder of Epimetheus.

Now man, having a share of the divine attributes, was at first the only one of the animals who had any gods, because he alone was of their kindred; and he would raise altars and images of them. He was not long in inventing language and names; and he also constructed houses and clothes and shoes and beds, and drew sustenance from the earth. Thus provided, mankind at first lived dispersed, and there were no cities. But the consequence was that they were destroyed by the wild beasts, for they were utterly weak in comparison of them, and their art was only sufficient to provide them with the means of life, and would not enable them to carry on war against the animals: food they had, but not as yet any art of government, of which the art of war is a part. After a while the desire of self-preservation gathered them into cities; but when they were gathered together, having no art of government, they evil entreated one another, and were again in process of dispersion and destruction. Zeus feared that the race would be exterminated, and so he sent Hermes to them, bearing reverence and justice to be the ordering principles of cities and the bonds of friendship and conciliation. Hermes asked Zeus how he should impart justice and reverence among men: should he distribute them as the arts are distributed; that is to say, to a favored few only—for one skilled individual has enough of medicine, or of any other art, for many unskilled ones? Shall this be the manner in which I distribute justice and reverence among men, or shall I give them to all? To all, said Zeus; I should like them all to have a share; for cities cannot exist, if a few only share in the virtues, as in the arts. And further, make a law by my order, that he who has no part in reverence and justice shall be put to death as a plague of the State.

And this is the reason, Socrates, why the Athenians and mankind in general, when the question relates to carpentering or any other mechanical art, allow but a few to share in their deliberations; and when anyone else interferes, then, as you say, they object, if he be not of the favored few, and that, as I say, is very natural. But when they come to deliberate about political virtue, which proceeds only by way of justice and wisdom, they are patient enough of any man who speaks of them, as is also natural, because they think that every man ought to share in this sort of virtue, and that States could not exist if this were otherwise. I have explained to you, Socrates, the reason of this phenomenon.

And that you may not suppose yourself to be deceived in thinking that all men regard every man as having a share of justice and every other political virtue, let me give you a further proof, which is this. In other cases, as you are aware, if a man says that he is a good flute-player, or skilful in any other art in which he has no skill, people either laugh at him or are angry with him, and his relations think that he is mad and go and admonish him; but when honesty is in question, or some other political virtue, even if they know that he is dishonest, yet, if the man comes publicly forward and tells the truth about his dishonesty, in this case they deem that to be madness which in the other case was held by them to be good sense. They say that men ought to profess honesty whether they are honest or not, and that a man is mad who does not make such a profession. Their notion is, that a man must have some degree of honesty; and that if he has none at all he ought not to be in the world.

* * *

S: See what follows from this: if we leave here without the city's permission, are we injuring people whom we should least injure? And are we sticking to a just agreement, or not?

C: I cannot answer your question, Socrates. I do not know.

S: Look at it this way. If, as we were planning to run away from here, or whatever one should call it, the laws and the state came and confronted us and asked: "Tell me, Socrates, what are you intending to do? Do you not by this action you are attempting intend to destroy us, the laws, and indeed the whole city, as far as you are concerned? Or do you think it possible for a city not to be destroyed if the verdicts of its courts have no force but are nullified and set at naught by private individuals?" What shall we answer to this and other such arguments? For many things could be said, especially by an orator on behalf of this law we are destroying, which orders that the judgments of the courts shall be carried out. Shall we say in answer, "The city wronged me, and its decision was not right." Shall we say that, or what?

C: Yes, by Zeus, Socrates, that is our answer.

S: Then what if the laws said: "Was that the agreement between us, Socrates, or was it to respect the judgments that the city came to?" And if we wondered at their words, they would perhaps add: "Socrates, do not wonder at what we say but answer, since you are accustomed to proceed by question and answer. Come now, what accusation do you bring against us and the city, that you should try to destroy us? Did we not, first, bring you to birth, and was it not through us that your father married your mother and begat you? Tell us, do you find anything to criticize in those of us who are concerned with marriage?" And I would say that I do not criticize them. "Or in those of us concerned with the nurture of babies and the education that you too received? Were those assigned to that subject not right to instruct your father to educate you in the arts and in physical culture?" And I would say that they were right. "Very well," they would continue, "and after you were born and nurtured and educated, could you, in the first place, deny that you are our offspring and servant, both you and your forefathers? If that is so, do you think that we are on an equal footing as regards the right, and that whatever we do to you it is right for you to do to us? You were not on an equal footing with your father as regards the right, nor with your master if you had one, so as to retaliate for anything they did to you, to revile them if they reviled you, to beat them if they beat you, and so with many other things. Do you think you have this right to retaliation against your country and its laws? That if we undertake to destroy you and think it right to do so, you can undertake to destroy us, as far as you can, in return? And will you say that you are right to do so, you who truly care for virtue? Is your wisdom such as not to realize that your country is to be honoured more than your mother, your father and all your ancestors, that it is more to be revered and more sacred, and that it counts for more among the gods and sensible men, that you must worship it, yield to it and placate its anger more than your father's? You must either persuade it or obey its orders, and endure in silence whatever it instructs you to endure, whether blows or bonds, and if it leads you into war to be wounded or killed, you must obey. To do so is right, and one must not give way or retreat or leave one's post, but both in war and in courts and everywhere else, one must obey the commands of one's city and country, or persuade it as to the nature of justice. It is impious to bring violence to bear against your mother or father, it is much more so to use it against your country." What shall we say in reply, Crito, that the laws speak the truth, or not?

C: I think they do.

S: "Reflect now, Socrates," the laws might say "that if what we say is true, you are not treating us rightly by planning to do what you are planning. We have given you birth, nurtured you, educated you, we have given you and all other citizens a share of all the good things we could. Even so, by giving every Athenian the opportunity, after he has reached manhood and observed the affairs of the city and us the laws, we proclaim that if we do not please him, he can take his possessions and go wherever he pleases. Not one of our laws raises any obstacle or forbids him, if he is not satisfied with us or the city, if one of you wants to go and live in a colony or wants to go anywhere else, and keep his property. We say, however, that whoever of you remains, when he sees how we conduct our trials and manage the city in other ways, has in fact come to an agreement with us to obey our instructions. We say that the one who disobeys does wrong in three ways, first because in us he disobeys his parents, also those who brought him up, and because, in spite of his agreement, he neither obeys us nor, if we do something wrong, does he try to persuade us to do better. Yet we only propose things, we do not issue either to persuade us or to do what we say. He does neither. We do say that you too, Socrates, are open to those charges if you do what you have in mind; you would be among, not the least, but the most guilty of the Athenians." And if I should say "Why so?" they might well be right to upbraid me and say that I am among the Athenians who most definitely came to that agreement with them. They might well say: "Socrates, we have convincing proofs that we and the city were congenial to you. You would not have dwelt here most consistently of all the Athenians if the city had not been exceedingly pleasing to you. You have never left the city, even to see a festival, nor for any other reason except military service; you have never gone to stay in any other city, as people do; you have had no desire to know another city or other laws; we and our city satisfied you.

"So decisively did you choose us and agree to be a citizen under us. Also, you have had children in this city, thus showing that it was congenial to you. Then at your trial you could have assessed your penalty at exile if you wished, and you are now attempting to do against the city's wishes what you could then have done with her consent. Then you prided yourself that you did not resent death, but you chose, as you said, death in preference to exile. Now, however, those words do not make you ashamed, and you pay no heed to us, the laws, as you plan to destroy us, and you act like the meanest type of slave by trying to run away, contrary to your undertakings and your agreement to live as a citizen under us. First then, answer us on this very point, whether we speak the truth when we say that you agreed, not only in words but by your deeds, to live in accordance with us." What are we to say to that, Crito? Must we not agree?

Lucretius, from *The Nature of Things* (99–55 B.C.)

Titus Lucretius Carus was an Epicurean poet. In *The Way Things Are* (*Rerum Naturum*), he characterizes humans in their original state, wild and free but subject to all sorts of horrors (especially being eaten by savage animals), the increasing civility and comforts that came with the use of fire and shelter and animals furs, the development of the family, and the increasing "softness" of the species. Eventually, "pacts of non-aggression" brought them into society ("Do not hurt me, please/ And I'll not hurt you") and though "they did not quite establish universal harmony, . . . some of them came reasonably close." But then, powerful men started building cities and walls, and men started dividing up nature according to their relative powers, and possessiveness, envy, ambition, and vanity undermined the happiness they had earned. Lucretius ends, accordingly, with a lament, that human foolishness has and will always eclipse our "own good sense and feelings." This tale of a formerly happier and healthier humanity, undone by the institution of property and the "civilized" vices of greed and ambition, anticipates Rousseau, who will write a similar tale seventeen centuries later.

> In the beginning, earth
> Covered the hills and all the plains with green,
> And flowering meadows shone in that rich color;
> Then into air the various kinds of trees
> Luxuriant in rivalry arose,
> And just as feathers, hair and bristles grow
> First on the bodies of all beasts and birds
> So the new earth began with grass and brush,
> And then produced the mortal animals
> Many and various. Creatures such as these
> Could not have fallen from the sky, nor come
> Out of the salt lagoons. They are earthborn,
> And truly earth deserves her title *Mother*,
> Since all things are created out of earth.

From "The Way Things Are" in *The Nature of Things* by Lucretius, translated by Rolfe Humphries, and published by Indiana University Press. Reprinted by permission of the publisher.

* * *

      I repeat, the earth
Deserves the name of Mother; by herself
She made the race of men, and in their season
The breed of beasts, those mountain stravagers,
The birds of the air in all their variousness,

* * *

The human race was tougher then: why not?
They were sons of a tough mother, and their bones
Were bigger and more solid, suitable
For stronger nerves and sinews, less distressed
By heat and cold, strange victuals, or disease.
For many centuries men led their lives
Like roving animals; no hardy soul
Steered the curved plowshare, no one understood
Planting or pruning. What the rain and sun
And earth supplied was gift enough for them.
Acorns were staple diet, or they fed
On arbute-berries, which we see today
Scarlet in wintertime—but long ago
There were more of them, much bigger, and the earth
Out of her blooming newness offered much—
No fancy fare, but adequate. The streams,
The springs, called men to quench their thirst, as now
Bright cataracts thundering over mountain-falls
Summon the thirsty and far-ranging beasts.
Then in their wanderings men came to know
The sanctuary areas which the nymphs
Considered home, where over the smooth stones
Water would ripple, brim above green moss
To even smoothness. People did not know,
In those days, how to work with fire, to use
The skins of animals for clothes; they lived
In groves and woods, and mountain-caves; they hid
Under the bushes when a sudden storm
Of wind or rain assailed their shagginess.
They had no vision of a common good,
No common law nor custom. What each man
Was given by Fortune, that he carried off,
Or, it may be, endured. He taught himself
To live and to be strong. And Venus joined
The bodies of lovers in the woods; a girl
Shared a man's appetite, or perhaps succumbed
To his insistent force, or took a bribe:
Acorns, or abute-berries, or choice pears.

Relying on their strength and speed, they'd hunt
The forest animals by throwing rocks
Or wielding clubs—there were many to bring down,
A few to hide from. When the nighttime came,
They'd lump their shaggy bodies on the ground,
Much like wild boars, under a coverlet
Of leaves or brush, and when the sun went down
They did not try to trail him across the fields
With loud lament and panic. They lay still,
Buried in slumber, patient, till the sun
Raised his red torch above the world again,
For from their earliest childhood, they had seen
Light alternate with darkness; this they took
For granted, with no wondering, no dread
Lest night hold earth in everlasting thrall
With sunlight gone. What made them much more anxious
Was that wild animals would often make
Their sleep a fatal risk; if they caught sight
Of a lion or wild boar, they'd leave their homes,
Their rocky caves, and in the dead of night
Concede their leafy beds to their savage guests.
Mortality was not much greater then
Than now in our time; individuals
Were caught and eaten by wild animals,
And moaned and groaned and filled the woods with woe
Seeing themselves entombed in living flesh;

And even those who managed to escape
With bodies gnawed and chewed on, pressed their hands
Over their pustulent sores, invoking Hell
With dreadful cries, until their agony
Ended in death—they had no way of aid,
No knowledge of the treatment of a wound.
But no one day slew thousands on the field
Of battle, nor did ocean surges sweep
Shipping and men on rocks. In vain the sea
Rose, raged and stormed, or put aside his threats
Offering false deceptive calm. No man
Was fooled, for no man was a mariner,
Nobody knew the evil art of sailing.
It was lack of food that killed men in the past;
Today it's over-indulgence and excess.
They often, in their ignorance, killed themselves
By accident with poison. We are wiser,
We dose, or treat, our patients. Or our victims.

Next, with the use of fire and huts, and hides
Or furs for clothing, when a woman stayed

Joined to one man in something like a marriage,
With offspring recognized, that was the time
When first the human race began to soften.
Fire kept their bodies from enduring cold,
Lust sapped their energies, and children broke
Their parents' haughty spirit by their wheedling,
And even neighbors started forming pacts
Of nonaggression: *Do not hurt me, please,*
And I'll not hurt you, were the terms they stammered.
Men asked protection for their little ones
As well as for their wives; with voice and gesture
They made it clear that there was nothing wrong
In pitying the weak. They did not quite
Establish universal harmony,
But some of them came reasonably close—
Enough, at least, to keep the race alive
And propagating.

* * *

So day-by-day,
They changed their former ways of living, taught
By men of lively wit and kind intent.
Kings started founding cities, building walls
Around the heights for refuge and defense.
They made division of the herds and lands
According to men's qualities, their strength,
Their wit, their beauty—virtues highly prized
In those old days; but later on, with wealth
And the discovery of gold, the strong,
The beautiful, all too easily forsook
The path of honor, more than willingly
Chasing along behind the rich man's train.
Whereas, if man would regulate his life
With proper wisdom, he would know that wealth,
The greatest wealth, is living modestly,
Serene, content with little. There's enough
Of this possession always. But men craved
Power and fame, that their fortunes might stand
On firm foundations, so they might enjoy
The rich man's blessèd life. What vanity!
To struggle toward the top, toward honor's height
They made the way a foul and deadly road,
And when they reached the summit, down they came
Like thunderbolts, for Envy strikes men down
Like thunderbolts, into most loathsome Hell,
For Envy always blasts exalted things
Above the level of the commonplace.
So it would be a better thing by far

To serve than rule; let others sweat themselves
Into exhaustion, jamming that defile
They call ambition, since their wisdom comes
Always from other mouths, and all their trust
Is put in hearsay; when do they believe
Their own good sense and feelings? Never, never.

As it was in the beginning, so it is
Now, and forever shall be.

Thomas Hobbes, from *The Leviathan* (1588–1679)

Hobbes' *Leviathan* is the classic source of modern social contract theory. In the substantial excerpt below, Hobbes describes human life in the state of nature, where justice has no place, as "solitary, poor, nasty, brutish and short." The (empirical) premise of Hobbes' argument is that people are essentially equal, at least when it comes down to their basic ability to do damage to one another. In the absence of any government or power over them, men will be naturally competitive, insecure, and mutually defensive. Moreover, perhaps most damaging of all, because all men think that they are equal, they will also be vain and insistent on respect and honor from others. (It is on this last point, in particular, that Rousseau will complain that Hobbes takes the vices that men cultivate in society and projects them back into the state of nature. It is also on this point that Hegel will insist that men in the state of nature aren't in fact presocial at all, but already bound to one another in mutual chains of recognition.) The result of this, according to Hobbes, is a war of everyone against everyone and an unhappy life in which there is no justice, because there is no injustice, either. There is no right and wrong, no right to property, no "mine or thine," no law and justice or injustice, only "force and fraud." It is in the face of this mutual insecurity and fear universally suffered in such circumstances that a mutual social compact becomes a matter of *rational* necessity, and Hobbes spells out its conditions below.

Chapter Thirteen

Of the Natural Condition of Mankind as Concerning Their Felicity and Misery

Men by nature equal. Nature has made men so equal in the faculties of the body and mind as that, though there be found one man sometimes mannifestly stronger in body or of quicker mind than another, yet, when all is reckoned together, the difference between man and man is not so considerable as that one man can thereupon claim to himself any benefit to which another may not pretend as well as he. For as to the strength of body, the weakest has strength enough to kill the strongest, either by secret machination or by confederacy with others that are in the same danger with himself.

And as to the faculties of the mind, setting aside the arts grounded upon words, and especially that skill of proceeding upon general and infallible rules called

science—which very few have and but in few things, as being not a native faculty
born with us, nor attained, as prudence, while we look after somewhat else—I find
yet a great equality among men than that of strength. For prudence is but experience, which equal time equally bestows on all men in those things they equally
apply themselves unto. That which may perhaps make such equality incredible is but
a vain conceit of one's own wisdom, which almost all men think they have in a
greater degree than the vulgar—that is, than all men but themselves and a few others
whom, by fame or for concurring with themselves, they approve. For such is the
nature of men that howsoever they may acknowledge many others to be more witty
or more eloquent or more learned, yet they will hardly believe there be many so wise
as themselves; for they see their own wit at hand and other men's at a distance. But
this proves rather that men are in that point equal than unequal. For there is not
ordinarily a greater sign of the equal distribution of anything than that every man is
contented with his share.

From equality proceeds diffidence. From this equality of ability arises equality
of hope in the attaining of our ends. And therefore if any two men desire the same
thing, which nevertheless they cannot both enjoy, they become enemies; and in the
way to their end, which is principally their own conservation, and sometimes their
delectation only, endeavor to destroy or subdue one another. And from hence it
comes to pass that where an invader has no more to fear than another man's single
power, if one plant, sow, build, or possess a convenient seat, others may probably be
expected to come prepared with forces united to dispossess and deprive him, not
only of the fruit of his labor, but also of his life or liberty. And the invader again is in
the like danger of another.

From diffidence war. And from this diffidence of one another there is no way
for any man to secure himself so reasonable as anticipation—that is, by force or
wiles to master the persons of all men he can, so long till he see no other power great
enough to endanger him; and this is no more than his own conservation requires, and
is generally allowed. Also, because there be some that take pleasure in contemplating
their own power in the acts of conquest, which they pursue farther than their security
requires, if others that otherwise would be glad to be at ease within modest bounds
should not by invasion increase their power, they would not be able, long time, by
standing only on their defense, to subsist. And by consequence, such augmentation
of dominion over men being necessary to a man's conservation, it ought to be
allowed him.

Again, men have no pleasure, but on the contrary a great deal of grief, in
keeping company where there is no power able to overawe them all. For every man
looks that his companion should value him at the same rate he sets upon himself;
and upon all signs of contempt or undervaluing naturally endeavors, as far as he
dares (which among them that have no common power to keep them in quiet is far
enough to make them destroy each other), to extort a greater value from his
contemners by damage and from others by the example.

So that in the nature of man we find three principal causes of quarrel: first,
competition; secondly, diffidence; thirdly, glory.

The first makes men invade for gain, the second for safety, and the third for
reputation. The first use violence to make themselves masters of other men's persons,
wives, children, and cattle; the second, to defend them; the third, for trifles, as a
word, a smile, a different opinion, and any other sign of undervalue, either direct in

their persons or by reflection in their kindred, their friends, their nation, their profession, or their name.

Out of civil states, there is always war of every one against every one. Hereby it is manifest that, during the time men live without a common power to keep them all in awe, they are in that condition which is called war, and such a war as is of every man against every man. For WAR consists not in battle only, or the act of fighting, but in a tract of time wherein the will to contend by battle is sufficiently known; and therefore the notion of *time* is to be considered in the nature of war as it is in the nature of weather. For as the nature of foul weather lies not in a shower or two of rain but in an inclination thereto of many days together, so the nature of war consists not in actual fighting but in the known disposition thereto during all the time there is no assurance to the contrary. All other time is peace.

The incommodities of such a war. Whatsoever, therefore, is consequent to a time of war where every man is enemy to every man, the same is consequent to the time wherein men live without other security than what their own strength and their own invention shall furnish them withal. In such condition there is no place for industry, because the fruit thereof is uncertain: and consequently no culture of the earth; no navigation nor use of the commodities that may be imported by sea; no commodious building; no instruments of moving and removing such things as require much force; no knowledge of the face of the earth; no account of time; no arts; no letters; no society; and, which is worst of all, continual fear and danger of violent death; and the life of man solitary, poor, nasty, brutish, and short.

It may seem strange to some man that has not well weighed these things that nature should thus dissociate and render men apt to invade and destroy one another; and he may therefore, not trusting to this inference made from the passions, desire perhaps to have the same confirmed by experience. Let him therefore consider with himself—when taking a journey he arms himself and seeks to go well accompanied, when going to sleep he locks his doors, when even in his house he locks his chests, and this when he knows there be laws and public officers, armed, to revenge all injuries shall be done him—what opinion he has of his fellow subjects when he rides armed, of his fellow citizens when he locks his doors, and of his children and servants when he locks his chests. Does he not there as much accuse mankind by his actions as I do by my words? But neither of us accuse man's nature in it. The desires and other passions of man are in themselves no sin. No more are the actions that proceed from those passions till they know a law that forbids them, which, till laws be made, they cannot know, nor can any law be made till they have agreed upon the person that shall make it.

It may peradventure be thought there was never such a time nor condition of war as this, and I believe it was never generally so over all the world; but there are many places where they live so now. For the savage people in many places of America, except the government of small families, the concord whereof depends on natural lust, have no government at all and live at this day in that brutish manner as I said before. Howsoever, it may be perceived what manner of life there would be where there were no common power to fear by the manner of life which men that have formerly lived under a peaceful government use to degenerate into a civil war.

But though there had never been any time wherein particular men were in a condition of war one against another, yet in all times kings and persons of sovereign authority, because of their independency, are in continual jealousies and in the state and posture of gladiators, having their weapons pointing and their eyes fixed on one another—that is, their forts, garrisons, and guns upon the frontiers of their kingdoms, and continual spies upon their neighbors—which is a posture of war. But because they uphold thereby the industry of their subjects, there does not follow from it that misery which accompanies the liberty of particular men.

In such a war nothing is unjust. To this war of every man against every man, this also is consequent: that nothing can be unjust. The notions of right and wrong, justice an injustice, have there no place. Where there is no common power, there is no law; where no law, no injustice. Force and fraud are in war the two cardinal virtues. Justice and injustice are none of the faculties neither of the body nor mind. If they were, they might be in a man that were alone in the world, as well as his senses and passions. They are qualities that relate to men in society, not in solitude. It is consequent also to the same condition that there be no propriety, no dominion, no *mine* and *thine* distinct; but only that to be every man's that he can get, and for so long as he can keep it. And thus much for the ill condition which man by mere nature is actually placed in, though with a possibility to come out of it consisting partly in the passions, partly in his reason.

The passions that incline men to peace. The passions that incline men to peace are fear of death, desire of such things as are necessary to commodious living, and a hope by their industry to obtain them. And reason suggests convenient articles of peace, upon which men may be drawn to agreement. These articles are they which otherwise are called the Laws of Nature, whereof I shall speak more particularly in the two following chapters.

Chapter Fourteen

Of the First and Second Natural Laws, and of Contracts

Right of nature what. The right of nature, which writers commonly call *jus naturale*, is the liberty each man has to use his own power, as he will himself, for the preservation of his own nature—that is to say, of his own life—and consequently of doing anything which, in his own judgment and reason, he shall conceive to be the aptest means thereunto.

Liberty what. By LIBERTY is understood, according to the proper signification of the word, the absence of external impediments; which impediments may oft take away part of a man's power to do what he would, but cannot hinder him from using the power left him according as his judgment and reason shall dictate to him.

A law of nature what. A LAW OF NATURE, *lex naturalis*, is a precept or general rule, found out by reason, by which a man is forbidden to do that which is destructive of his life or takes away the means of preserving the same and to omit that by which he thinks it may be best preserved. For though they that speak of this subject use to confound *jus* and *lex*, *right* and *law*, yet they ought to be distinguished.

Difference of right and law. RIGHT consists in liberty to do or to forbear, whereas LAW determines and binds to one of them; so that law and right differ as much as obligation and liberty, which in one and the same matter are inconsistent.

Naturally every man has right to every thing. And because the condition of man, as has been declared in the precedent chapter, is a condition of war of every one against every one—in which case everyone is governed by his own reason and there is nothing he can make use of that may not be a help unto him in preserving his life against his enemies—it follows that in such a condition every man has a right to everything, even to one another's body. And therefore, as long as this natural right of every man to everything endures, there can be no security to any man, how strong or wise soever he be, of living out the time which nature ordinarily allows men to live.

The fundamental law of nature. And consequently it is a precept or general rule of reason *that every man ought to endeavor peace, as far as he has hope of obtaining it; and when he cannot obtain it, that he may seek and use all helps and advantages of war.* The first branch of which rule contains the first and fundamental law of nature, which is *to seek peace and follow it.* The second, the sum of the right of nature, which is, *by all means we can to defend ourselves.*

The second law of nature. From this fundamental law of nature, by which men are commanded to endeavor peace, is derived this second law: *that a man be willing, when others are so too, as far forth as for peace and defense of himself he shall think it necessary, to lay down this right to all things, and be contented with so much liberty against other men as he would allow other men against himself.* For as long as every man holds this right of doing anything he likes, so long are all men in the condition of war. But if other men will not lay down their right as well as he, then there is no reason for anyone to divest himself of his, for that were to expose himself to prey, which no man is bound to, rather than to dispose himself to peace. This is that law of the gospel: *whatsoever you require that others should do to you, that do ye to them.* And that law of all men, *quod tibi fieri non vis, alteri ne feceris.*[1]

What it is to lay down a right. To *lay down* a man's *right* to anything is to *divest* himself of the *liberty* of hindering another of the benefit of his own right to the same. For he that renounces or passes away his right gives not to any other man a right which he had not before—because there is nothing to which every man had not right by nature—but only stands out of his way, that he may enjoy his own original right without hindrance from him, not without hindrance from another. So that the effect which redounds to one man by another man's defect of right is but so much diminution of impediments to the use of his own right original. Right is laid aside either by simply renouncing it or by transferring it to another.

Renouncing a right, what it is. By *simply* RENOUNCING, when he cares not to whom the benefit thereof redounds.

Transferring right what. Obligation. By TRANSFERRING, when he intends the benefit thereof to some certain person or persons.

Duty. Injustice. And when a man has in either manner abandoned or granted away his right, then he is said to be OBLIGED or BOUND not to hinder those to whom such right is granted or abandoned from the benefit of it; and that he *ought*, and it is his DUTY, not to make void that voluntary act of his own; and that such hindrance is INJUSTICE and INJURY as being *sine jure,*[2] the right being before renounced or transferred. So that *injury* or *injustice* in the controversies of the world is somewhat like to that which in the disputations of scholars is called *absurdity.* For as it is there called an absurdity to contradict what one maintained in the beginning, so in the

---

[1][Matt. 7:12; Luke 6:31. The Latin expresses the same rule negatively: "What you would not have done to you, do not do to others."]

[2][Without legal basis.]

world it is called injustice and injury voluntarily to undo that which from the beginning he had voluntarily done. The way by which a man either simply renounces or transfers his right is a declaration or signification by some voluntary and sufficient sign or signs that he does so renounce or transfer, or has so renounced or transferred, the same to him that accepts it. And these signs are either words only or actions only; or as it happens most often, both words and actions. And the same are the BONDS by which men are bound and obliged—bonds that have their strength, not from their own nature, for nothing is more easily broken than a man's word, but from fear of some evil consequence upon the rupture.

Not all rights are alienable. Whensoever a man transfers his right or renounces it, it is either in consideration of some right reciprocally transferred to himself or for some other good he hopes for thereby. For it is a voluntary act; and of the voluntary acts of every man, the object is some *good to himself.* And therefore there be some rights which no man can be understood by any words or other signs to have abandoned or transferred. As, first, a man cannot lay down the right of resisting them that assault him by force to take away his life, because he cannot be understood to aim thereby at any good to himself. The same may be said of wounds and chains and imprisonment, both because there is no benefit consequent to such patience as there is to the patience of suffering another to be wounded or imprisoned, as also because a man cannot tell, when he sees men proceed against him by violence, whether they intend his death or not. And, lastly, the motive and end for which this renouncing and transferring of right is introduced is nothing else but the security of a man's person in his life and in the means of so preserving life as not to be weary of it. And therefore if a man by words or other signs seem to despoil himself of the end for which those signs were intended, he is not to be understood as if he meant it or that it was his will, but that he was ignorant of how such words and actions were to be interpreted.

Contract what. The mutual transferring of right is that which men call CONTRACT.

There is difference between transferring of right to the thing and transferring, or tradition—that is, delivery—of the thing itself. For the thing may be delivered together with the translation of the right, as in buying and selling with ready money or exchange of goods or lands, and it may be delivered some time after.

Covenant what. Again, one of the contractors may deliver the thing contracted for on his part and leave the other to perform his part at some determinate time after and in the meantime be trusted, and then the contract on his part is called PACT or COVENANT; or both parts may contract now to perform hereafter, in which cases he that is to perform in time to come, being trusted, his performance is called *keeping of promise* or faith, and the failing of performance, if it be voluntary, *violation of faith.*

Free gift. When the transferring of right is not mutual, but one of the parties transfers in hope to gain thereby friendship or service from another or from his friends, or in hope to gain the reputation of charity or magnanimity, or to deliver his mind from the pain of compassion, or in hope of reward in heaven—this is not contract but GIFT, FREE GIFT, GRACE, which words signify one and the same thing.

* * *

Covenants of mutual trust, when invalid. If a covenant be made wherein neither of the parties perform presently but trust one another, in the condition of mere nature, which is a condition of war of every man against every man, upon any reasonable suspicion, it is void; but if there be a common power set over them both, with right and force sufficient to compel performance, it is not void. For he that performs first has no assurance the other will perform after, because the bonds of words are too weak to bridle men's ambition, avarice, anger, and other passions without the fear of some coercive power which in the condition of mere nature, where all men are equal and judges of the justness of their own fears, cannot possibly be supposed. And therefore he which performs first does but betray himself to his enemy, contrary to the right he can never abandon of defending his life and means of living.

But in a civil estate, where there is a power set up to constrain those that would otherwise violate their faith, that fear is no more reasonable; and for that cause, he which by the covenant is to perform first is obliged so to do.

The cause of fear which makes such a covenant invalid must be always something arising after the covenant made, as some new fact or other sign of the will not to perform; else it cannot make the covenant void. For that which could not hinder a man from promising ought not to be admitted as a hindrance of performing.

* * *

Covenants how made void. Men are freed of their covenants two ways: by performing or by being forgiven. For performance is the natural end of obligation, and forgiveness the restitution of liberty, as being a retransferring of that right in which the obligation consisted.

Covenants extorted by fear are valid. Covenants entered into by fear, in the condition of mere nature, are obligatory. For example, if I covenant to pay a ransom or service for my life to an enemy, I am bound by it; for it is a contract, wherein one receives the benefit of life, the other is to receive money or service for it; and consequently, where no other law, as in the condition of mere nature, forbids the performance, the covenant is valid. Therefore prisoners of war, if trusted with the payment of their ransom, are obliged to pay it; and if a weaker prince make a disadvantageous peace with a stronger, for fear, he is bound to keep it; unless, as has been said before, there arises some new and just cause of fear to renew the war. And even in commonwealths, if I be forced to redeem myself from a thief by promising him money, I am bound to pay it till the civil law discharge me. For whatsoever I may lawfully do without obligation, the same I may lawfully covenant to do through fear; and what I lawfully covenant, I cannot lawfully break.

The former covenant to one makes void the later to another. A former covenant makes void a later. For a man that has passed away his right to one man today has it not to pass tomorrow to another; and therefore the later promise passes no right, but is null.

A man's covenant not to defend himself is void. A covenant not to defend myself from force by force is always void. For, as I have showed before, no man can transfer or lay down his right to save himself from death, wounds, and imprisonment, the avoiding whereof is the only end of laying down any right; and therefore the promise of not resisting force in no covenant transfers any right, nor is obliging. For though a man may covenant thus: *unless I do so or so, kill me,* he cannot covenant thus: *unless I do so or so, I will not resist you when you come to kill me.* For man by nature chooses the lesser evil, which is danger of death in resisting, rather than the greater, which is certain and present death in not resisting. And this is granted to be true by all men, in that they lead criminals to execution and prison with armed men, notwithstanding that such criminals have consented to the law by which they are condemned.

No man obliged to accuse himself. A covenant to accuse oneself, without assurance of pardon, is likewise invalid. For in the condition of nature, where every man is judge, there is no place for accusation; and in the civil state, the accusation is followed with punishment, which, being force, a man is not obliged not to resist. The same is also true of the accusation of those by whose condemnation a man falls into misery, as of a father, wife, or benefactor. For the testimony of such an accuser, if it be not willingly given, is presumed to be corrupted by nature, and therefore not to be received; and where a man's testimony is not to be credited, he is not bound to give it. Also accusations upon torture are not to be reputed as testimonies. For torture is to be used but as means of conjecture and light in the further examination and search of truth; and what is in that case confessed tends to the ease of him that is tortured, not to the informing of the torturers, and therefore ought not to have the credit of a sufficient testimony; for whether he deliver himself by true or false accusation, he does it by the right of preserving his own life.

The end of an oath. The force of words being, as I have formerly noted, too weak to hold men to the performance of their covenants, there are in man's nature but two imaginable helps to strengthen it. And those are either a fear of the consequence of breaking their word, or a glory or pride in appearing not to need to break it. This latter is a generosity too rarely found to be presumed on, especially in the pursuers of wealth, command, or sensual pleasure—which are the greatest part of mankind. The passion to be reckoned upon is fear, whereof there be two very general objects: one, the power of spirits invisible; the other, the power of those men they shall therein offend. Of these two, though the former be the greater power, yet the fear of the latter is commonly the greater fear. The fear of the former is in every man his own religion, which has place in the nature of man before civil society. The latter has not so, at least not place enough to keep men to their promises, because in the condition of mere nature the inequality of power is not discerned but by the event of battle. So that before the time of civil society, or in the interruption thereof by war, there is nothing can strengthen a covenant of peace agreed on against the temptations of avarice, ambition, lust, or other strong desire but the fear of that invisible power, which they everyone worship as God and fear as a revenger of their perfidy.

The form of an oath. All therefore that can be done between two men not subject to civil power is to put one another to swear by the God he fears, which *swearing* or OATH is a *form of speech, added to a promise, by which he that promises signifies that, unless he perform, he renounces the mercy of his God, or calls to him for vengeance on himself.* Such was the heathen form, *Let Jupiter kill me else, as I kill this beast.* So is our form, *I shall do thus and thus, so help me God.* And this, with the rites and ceremonies which everyone uses in his own religion, that the fear of breaking faith might be the greater.

No oath but by God. By this it appears that an oath taken according to any other form or rite than his that swears is in vain and no oath, and that there is no swearing by anything which the swearer thinks not God. For though men have sometimes used to swear by their kings, for fear or flattery, yet they would have it thereby understood they attributed to them divine honor. And that swearing unnecessarily by God is but profaning of his name; and swearing by other things, as men do in common discourse, is not swearing but an impious custom gotten by too much vehemence of talking.

An oath adds nothing to the obligation. It appears also that the oath adds nothing to the obligation. For a covenant, if lawful, binds in the sight of God without the oath as much as with it; if unlawful, binds not at all, though it be confirmed with an oath.

Chapter Fifteen

Of Other Laws of Nature

The third law of nature, justice. From that law of nature by which we are obliged to transfer to another such rights as, being retained, hinder the peace of mankind, there follows a third, which is this: *that men perform their covenants made;* without which covenants are in vain and but empty words, and, the right of all men to all things remaining, we are still in the condition of war.

Justice and injustice what. And in this law of nature consists the fountain and original of JUSTICE. For where no covenant has preceded there has no right been transferred, and every man has right to every thing; and consequently no action can be unjust. But when a covenant is made, then to break it is *unjust*; and the definition of INJUSTICE is no other than *the not performance of covenant.* And whatsoever is not unjust is *just.*

Justice and propriety begin with the constitution of commonwealth. But because covenants of mutual trust, where there is a fear of not performance on either part, as has been said in the former chapter, are invalid, though the original of justice be the making of covenants, yet injustice actually there can be none till the cause of such fear be taken away, which, while men are in the natural condition of war, cannot be done. Therefore, before the names of just and unjust can have place, there must be some coercive power to compel men equally to the performance of their covenants by the terror of some punishment greater than the benefit they expect by the breach of their covenant, and to make good that propriety which by mutual contract men acquire in recompense of the universal right they abandon; and such power there is none before the erection of a commonwealth. And this is also to be gathered out of the ordinary definition of justice in the Schools, for they say that *justice is the constant will of giving to every man his own.* And therefore where there is no *own*—that is, no propriety—there is no injustice; and where there is no coercive power erected—that is, where there is no commonwealth—there is no propriety, all men having right to all things; therefore, where there is no commonwealth, there nothing is unjust. So that the nature of justice consists in keeping of valid covenants; but the validity of covenants begins not but with the constitution of a civil power sufficient to compel men to keep them; and then it is also that propriety begins.

Justice not contrary to reason. The fool hath said in his heart, there is no such thing as justice; and sometimes also with his tongue, seriously alleging that, every man's conservation and contentment being committed to his own care, there could be no reason why every man might not do what he thought conduced thereunto; and therefore also to make or not make, keep or not keep covenants was not against reason when it conduced to one's benefit. He does not therein deny that there be covenants and that they are sometimes broken, sometimes kept, and that such breach of them may be called injustice and the observance of them justice; but he questions whether injustice, taking away the fear of God—for the same fool hath said in his heart which dictates to every man his own good, and particularly then when it conduces to such a benefit as shall put a man in a condition to neglect not only the dispraise and revilings, but also the power of other men. The kingdom of God is gotten by violence; but what if it could be gotten by unjust violence? Were it against reason so to get it, when it is impossible to receive hurt by it? And if it be not against reason, it is not against justice, or else justice is not to be approved for good. From such reasoning as this, successful wickedness has obtained the name of virtue.

* * *

This specious reasoning is nevertheless false.

For the question is not of promises mutual where there is no security of performance on either side—as when there is no civil power erected over the parties promising—for such promises are no covenants; but either where one of the parties has performed already or where there is a power to make him perform, there is the question whether it be against reason—that is, against the benefit of the other—to perform or not. And I say it is not against reason. For the manifestation whereof we are to consider, first, that when a man does a thing which, notwithstanding anything can be foreseen and reckoned on, tends to his own destruction, howsoever some accident which he could not expect, arriving, may turn it to his benefit, yet such events do not make it reasonably or wisely done. Secondly, that in a condition of war, wherein every man to every man, for want of a common power to keep them all in awe, is an enemy, there is no man who can hope by his own strength or wit to defend himself from destruction without the help of confederates, where everyone expects the same defense by the confederation that anyone else does; and therefore he which declares he thinks it reason to deceive those that help him can in reason expect no other means of safety than what can be had from his own single power. He, therefore, that breaks his covenant, and consequently declares that he thinks he may with reason do so, cannot be received into any society that unite themselves for peace and defense, but by the error of them that receive him; nor, when he is received, be retained in it without seeing the danger of their error, which errors a man cannot reasonably reckon upon as the means of his security; and therefore if he be left or cast out of society he perishes, and if he live in society, it is by the errors of other men, which he could not foresee nor reckon upon, and consequently against the reason of his preservation; and so, as all men that contribute not to his destruction, forbear him only out of ignorance of what is good for themselves.

As for the instance of gaining the secure and perpetual felicity of heaven by any way, it is frivolous, there being but one way imaginable, and that is not breaking but keeping of covenant.

And for the other instance of attaining sovereignty by rebellion, it is manifest that, though the event follow, yet because it cannot reasonably be expected, but rather the contrary, and because by gaining it so others are taught to gain the same in like manner the attempt thereof is against reason. Justice, therefore—that is to say, keeping of covenant—is a rule of reason by which we are forbidden to do anything destructive to our life, and consequently a law of nature.

* * *

Justice of men and justice of actions what. The names of just and unjust, when they are attributed to men, signify one thing, and when they are attributed to actions, another. When they are attributed to men, they signify conformity or inconformity of manners to reason. But when they are attributed to actions, they signify the conformity or inconformity to reason, not of manners or manner of life, but of particular actions. A just man, therefore, is he that takes all the care he can that his actions may be all just; and an unjust man is he that neglects it. And such men are more often in our language styled by the names of righteous and unrighteous than just and unjust, though the meaning be the same. Therefore a righteous man does not lose that title by one or a few unjust actions that proceed from sudden passion or mistake of things or persons; nor does an unrighteous man lose his character for such actions as he does or forbears to do for fear, because his will is not framed by the justice but by the apparent benefit of what he is to do. That which gives to human actions the relish of justice is a certain nobleness or gallantness of courage, rarely found, by which a man scorns to be beholden for the contentment of his life to fraud or breach of promise. This justice of the manners is that which is meant where justice is called a virtue and injustice a vice.

But the justice of actions denominates men, not just, but *guiltless;* and the injustice of the same, which is also called injury, gives them but the name of *guilty.*

Justice of manners, and justice of actions. Again, the injustice of manners is the disposition or aptitude to do injury, and is injustice before it proceed to act and without supposing any individual person injured. But the injustice of an action—that is to say, injury—supposes an individual person injured—namely, him to whom the covenant was made—and therefore many times the injury is received by one man when the damage redounds to another. As when the master commands his servant to give money to a stranger: if it be not done, the injury is done to the master, whom he had before covenanted to obey; but the damage redounds to the stranger, to whom he had no obligation and therefore could not injure him. And so also in commonwealths private men may remit to one another their debts but not robberies or other violences whereby they are endamaged; because the detaining of debt is an injury to themselves, but robbery and violence are injuries to the person of the commonwealth.

Nothing done to a man by his own consent can be injury. Whatsoever is done to a man, conformable to his own will signified to the doer, is no injury to him. For if he that does it has not passed away his original right to do what he please by some antecedent covenant, there is no breach of covenant and therefore no injury done him. And if he have, then his will to have it done, being signified, is a release of that covenant, and so again there is no injury done him.

Justice commutative and distributive. Justice of actions is by writers divided into *commutative* and *distributive;* and the former they say consists in proportion arithmetical, the latter in proportion geometrical. Commutative, therefore, they place in the equality of value of the things contracted for, and distributive in the distribution of equal benefit to men of equal merit. As if it were injustice to sell dearer than we buy, or to give more to a man than he merits. The value of all things contracted for is measured by the appetite of the contractors, and therefore the just value is that which they be contented to give. And merit (besides that which is by covenant, where the performance on one part merits the performance of the other part, and falls under justice commutative, not distributive) is not due by justice, but is rewarded of grace only. And therefore this distinction, in the sense wherein it uses to be expounded, is not right. To speak properly, commutative justice is the justice of a contractor—that is, a performance of covenant in buying and selling, hiring and letting to hire, lending and borrowing, exchanging, bartering, and other acts of contract.

And distributive justice, the justice of an arbitrator—that is to say, the act of defining what is just. Wherein, being trusted by them that make him arbitrator, if he perform his trust, he is said to distribute to every man his own; and this is indeed just distribution, and may be called, though improperly, distributive justice, but more properly equity, which also is a law of nature, as shall be shown in due place. [P. 128.]

* * *

The sixth, facility to pardon. A sixth law of nature is this: *that upon caution of the future time, a man ought to pardon the offenses past of them that, repenting, desire it.* For PARDON is nothing but granting of peace, which, though granted to them that persevere in their hostility, be not peace but fear, yet, not granted to them that give caution of the future time, is sign of an aversion to peace, and therefore contrary to the law of nature.

The seventh, that in revenges men respect only the future good. A seventh is *that in revenges*—that is, retribution of evil for evil—*men look not at the greatness of the evil past, but the greatness of the good to follow.* Whereby we are forbidden to inflict punishment with any other design than for correction of the offender or direction of others. For this law is consequent to the next before it that commands pardon upon security of the future time. Besides, revenge without respect to the example and profit to come is a triumph or glorying in the hurt of another, tending to no end; for the end is always somewhat to come, and glorying to no end is vainglory and contrary to reason; and to hurt without reason tends to the introduction of war, which is against the law of nature and is commonly styled by the name of *cruelty.*

The eighth, against contumely. And because all signs of hatred or contempt provoke to fight, insomuch as most men choose rather to hazard their life than not to be revenged, we may in the eighth place for a law of nature set down this precept: *that no man by deed, word, countenance, or gesture declare hatred or contempt of another.* The breach of which law is commonly called *contumely.*

The ninth, against pride. The question who is the better man has no place in the condition of mere nature, where, as has been shown before, all men are equal. The inequality that now is has been introduced by the laws civil. I know that Aristotle in the first book of his *Politics*, for a foundation of his doctrine, makes men by nature some more worthy to command, meaning the wiser sort such as he thought himself to be for his philosophy, others to serve, meaning those that had strong bodies but were not philosophers as he; as if master and servant were not introduced by consent of men but by difference of wit, which is not only against reason but also against experience. For there are very few so foolish that had not rather govern themselves than be governed by others; nor when the wise in their own conceit contend by force with them who distrust their own wisdom, do they always, or often, or almost at any time, get the victory. If nature therefore have made men equal, that equality is to be acknowledged; or if nature have made men unequal, yet because men that think themselves equal will not enter into conditions of peace but upon equal terms, such equality must be admitted. And therefore for the ninth law of nature, I put this: *that every man acknowledge another for his equal by nature.* The breach of this precept is *pride.*

* * *

The eleventh, equity. Also if *a man be trusted to judge between man and man,* it is a precept of the law of nature *that he deal equally between them.* For without that, the controversies of men cannot be determined but by war. He, therefore, that is partial in judgment does what in him lies to deter men from the use of judges and arbitrators, and consequently, against the fundamental law of nature, is the cause of war.

The observance of this law, from the equal distribution to each man of that which in reason belongs to him, is called EQUITY and, as I have said before, distributive justice.

John Locke, from *The Second Treatise on Government* (1632–1704)

John Locke wrote his two treatises on government during the years of the English civil war. One of his main aims was to defend the notion of property as a natural right rather than a mere legal convention of modern society. Thus he endows the state of nature with a good deal more by way of "natural law" and rights than Hobbes did earlier in that century, and his state of nature shows little of the utter selfishness and mutual hostility of Hobbes's horrifying portrait. Indeed, he carefully distinguishes the state of nature from a state of war and accuses Hobbes of confusing them. The social contract, accordingly, does not so much offer new rights and impose new strictures as it implements the natural law more fully and formally. The primary purpose of the contract is to unify individuals into a community, but it is a much looser and more individually free arrangement than Hobbes could or would have allowed.

Of the State of Nature

4. TO UNDERSTAND political power right and derive it from its original, we must consider what state all men are naturally in, and that is a state of perfect freedom to order their actions and dispose of their possessions and persons as they think fit, within the bounds of the law of nature, without asking leave or depending upon the will of any other man.

A state also of equality, wherein all the power and jurisdiction is reciprocal, no one having more than another; there being nothing more evident than that creatures of the same species and rank, promiscuously born to all the same advantages of nature and the use of the same faculties, should also be equal one amongst another without subordination or subjection; unless the lord and master of them all should, by any manifest declaration of his will, set one above another, and confer on him by an evident and clear appointment an undoubted right to dominion and sovereignty.

5. This equality of men by nature the judicious Hooker looks upon as so evident in itself and beyond all question that he makes it the foundation of that obligation to mutual love amongst men on which he builds the duties we owe one another, and from whence he derives the great maxims of justice and charity. His words are:

> The like natural inducement hath brought men to know that it is no less
> their duty to love others than themselves; for seeing those things which are equal

must needs all have one measure; if I cannot but wish to receive good, even as
much at every man's hands as any man can wish unto his own soul, how should I
look to have any part of my desire herein satisfied unless myself be careful to
satisfy the like desire, which is undoubtedly in other men, being of one and the
same nature? To have anything offered them repugnant to this desire must needs
in all respects grieve them as much as me; so that, if I do harm, I must look to
suffer, there being no reason that others should show greater measure of love to
me than they have by me showed unto them; my desire therefore to be loved of
my equals in nature, as much as possibly may be, imposeth upon me a natural
duty of bearing to them-ward fully the like affection; from which relation of
equality between ourselves and them that are as ourselves, what several rules and
canons natural reason hath drawn, for direction of life, no man is ignorant. (*Eccl.
Pol.* lib. i.)

6. But though this be a state of liberty, yet it is not a state of license; though
man in that state have an uncontrollable liberty to dispose of his person or possessions, yet he has not liberty to destroy himself, or so much as any creature in his
possession, but where some nobler use than its bare preservation calls for it. The
state of nature has a law of nature to govern it, which obliges every one; and reason,
which is that law, teaches all mankind who will but consult it that, being all equal
and independent, no one ought to harm another in his life, health, liberty, or
possessions; for men being all the workmanship of one omnipotent and infinitely
wise Maker—all the servants of one sovereign master, sent into the world by his
order, and about his business—they are his property whose workmanship they are,
made to last during his, not one another's, pleasure; and being furnished with like
faculties, sharing all in one community of nature, there cannot be supposed any such
subordination among us that may authorize us to destroy another, as if we were
made for one another's uses as the inferior ranks of creatures are for ours. Every one,
as he is bound to preserve himself and not to quit his station wilfully, so by the like
reason, when his own preservation comes not in competition, ought he, as much as
he can, to preserve the rest of mankind, and may not, unless it be to do justice to an
offender, take away or impair the life, or what tends to the preservation of the life,
the liberty, health, limb, or goods of another.

7. And that all men may be restrained from invading others' rights and from
doing hurt to one another, and the law of nature be observed, which wills the peace
and preservation of all mankind, the execution of the law of nature is, in that state,
put into every man's hands, whereby everyone has a right to punish the transgressors
of that law to such a degree as may hinder its violation; for the law of nature would,
as all other laws that concern men in this world, be in vain if there were nobody that
in that state of nature had a power to execute that law and thereby preserve the
innocent and restrain offenders. And if anyone in the state of nature may punish
another for any evil he has done, everyone may do so; for in that state of perfect
equality, where naturally there is no superiority or jurisdiction of one over another,
what any may do in prosecution of that law, everyone must needs have a right to do.

8. And thus in the state of nature one man comes by a power over another; but
yet no absolute or arbitrary power to use a criminal, when he has got him in his
hands, according to the passionate heats or boundless extravagance of his own will;
but only to retribute to him, so far as calm reason and conscience dictate, what is
proportionate to his transgression, which is so much as may serve for reparation and

restraint; for these two are the only reasons why one man may lawfully do harm to another, which is that we call punishment. In transgressing the law of nature, the offender declares himself to live by another rule than that of reason and common equity, which is that measure God has set to the actions of men for their mutual security; and so he becomes dangerous to mankind, the tie which is to secure them from injury and violence being slighted and broken by him. Which being a trespass against the whole species and the peace and safety of it provided for by the law of nature, every man upon this score, by the right he has to preserve mankind in general, may restrain, or, where it is necessary, destroy things noxious to them, and so may bring such evil on any one who has transgressed that law, as may make him repent the doing of it and thereby deter him, and by his example others, from doing the like mischief. And in this case, and upon this ground, *every man has a right to punish the offender and be executioner of the law of nature.*

* * *

10. Besides the crime which consists in violating the law and varying from the right rule of reason, whereby a man so far becomes degenerate and declares himself to quit the principles of human nature and to be a noxious creature, there is commonly injury done to some person or other, and some other man receives damage by his transgression; in which case he who has received any damage has, besides the right of punishment common to him with other men, a particular right to seek reparation from him that has done it; and any other person, who finds it just, may also join with him that is injured and assist him in recovering from the offender so much as may make satisfaction for the harm he has suffered.

11. From these two distinct rights—the one of punishing the crime for restraint and preventing the like offense, which right of punishing is in everybody; the other of taking reparation, which belongs only to the injured party—comes it to pass that the magistrate, who by being magistrate has the common right of punishing put into his hands, can often, where the public good demands not the execution of the law, remit the punishment of criminal offenses by his own authority, but yet cannot remit the satisfaction due to any private man for the damage he has received. That he who has suffered the damage has a right to demand in his own name, and he alone can remit; the damnified person has this power of appropriating to himself the goods or service of the offender by right of self-preservation, as every man has a power to punish the crime to prevent its being committed again, by the right he has of preserving all mankind and doing all reasonable things he can in order to that end; and thus it is that every man, in the state of nature, has a power to kill a murderer, both to deter others from doing the like injury, which no reparation can compensate, by the example of the punishment that attends it from everybody, and also to secure men from the attempts of a criminal who, having renounced reason—the common rule and measure God has given to mankind—has, by the unjust violence and slaughter he has committed upon one, declared war against all mankind, and therefore may be destroyed as a lion or a tiger, one of those wild savage beasts with whom men can have no society nor security. And upon this is grounded that great law of nature, "Whoso sheddeth man's blood, by man shall his blood be shed." And Cain was so fully convinced that every one had a right to destroy such a criminal that, after the murder of his brother, he cries out, "Every one that findeth me, shall slay me"; so plain was it written in the hearts of mankind.

12. By the same reason may a man in the state of nature punish the lesser breaches of that law. It will perhaps be demanded: with death? I answer: Each transgression may be punished to that degree and with so much severity as will suffice to make it an ill bargain to the offender, give him cause to repent, and terrify others from doing the like. Every offense that can be committed in the state of nature may in the state of nature be also punished equally, and as far forth as it may in a commonwealth; for though it would be beside my present purpose to enter here into the particulars of the law of nature, or its measures of punishment, yet it is certain there is such a law, and that, too, as intelligible and plain to a rational creature and a studier of that law as the positive laws of commonwealths, nay, possibly plainer, as much as reason is easier to be understood than the fancies and intricate contrivances of men, following contrary and hidden interests put into words; for so truly are a great part of the municipal laws of countries, which are only so far right as they are founded on the law of nature, by which they are to be regulated and interpreted.

* * *

14. It is often asked as a mighty objection, "Where are or ever were there any men in such a state of nature?" To which it may suffice as an answer at present that since all princes and rulers of independent governments all through the world are in a state of nature, it is plain the world never was, nor ever will be, without numbers of men in that state. I have named all governors of independent communities, whether they are, or are not, in league with others; for it is not every compact that puts an end to the state of nature between men, but only this one of agreeing together mutually to enter into one community and make one body politic; other promises and compacts men may make one with another and yet still be in the state of nature. The promises and bargains for truck, etc., between the two men in the desert island, mentoned by Garcilasso de la Vega, in his history of Peru, or between a Swiss and an Indian in the woods of America, are binding to them, though they are perfectly in a state of nature in reference to one another; for truth and keeping of faith belongs to men as men, and not as members of society.

* * *

Of the State of War

16. THE STATE of war is a state of enmity and destruction; and, therefore, declaring by word or action, not a passionate and hasty but a sedate, settled design upon another man's life, puts him in a state of war with him against whom he has declared such an intention, and so has exposed his life to the other's power to be taken away by him or anyone that joins with him in his defense and espouses his quarrel; it being reasonable and just I should have a right to destroy that which threatens me with destruction; for, by the fundamental law of nature, man being to be preserved as much as possible when all cannot be preserved, the safety of the innocent is to be preferred; and one may destroy a man who makes war upon him, or has discovered an enmity to his being, for the same reason that he may kill a wolf or a lion, because such men are not under the ties of the common law of reason, have no other rule but

that of force and violence, and so may be treated as beasts of prey, those dangerous and noxious creatures that will be sure to destroy him whenever he falls into their power.

17. And hence it is that he who attempts to get another man into his absolute power does thereby put himself into a state of war with him, it being to be understood as a declaration of a design upon his life; for I have reason to conclude that he who would get me into his power without my consent would use me as he pleased when he got me there, and destroy me, too, when he had a fancy to it; for nobody can desire to have me in his absolute power unless it be to compel me by force to that which is against the right of my freedom, i.e., make me a slave. To be free from such force is the only security of my preservation; and reason bids me look on him as an enemy to my preservation who would take away that freedom which is the fence to it; so that he who makes an attempt to enslave me thereby puts himself into a state of war with me. He that, in the state of nature, would take away the freedom that belongs to any one in that state must necessarily be supposed to have a design to take away everything else, that freedom being the foundation of all the rest; as he that, in the state of society, would take away the freedom belonging to those of that society or commonwealth must be supposed to design to take away from them everything else, and so be looked on as in a state of war.

18. This makes it lawful for a man to kill a thief who has not in the least hurt him, nor declared any design upon his life any farther than, by the use of force, so to get him in his power as to take away his money, or what he pleases, from him; because using force where he has no right to get me into his power, let his pretense be what it will, I have no reason to suppose that he who would take away my liberty would not, when he had me in his power, take away everything else. And therefore it is lawful for me to treat him as one who has put himself into a state of war with me, i.e., kill him if I can; for to that hazard does he justly expose himself whoever introduces a state of war and is aggressor in it.

19. And here we have the plain difference between the state of nature and the state of war which, however some men have confounded, are as far distant as a state of peace, good-will, mutual assistance, and preservation, and a state of enmity, malice, violence, and mutual destruction are one from another. Men living together according to reason, without a common superior on earth with authority to judge between them, is properly the state of nature. But force, or a declared design of force, upon the person of another, where there is no common superior on earth to appeal to for relief, is the state of war; and it is the want of such an appeal [that] gives a man the right of war even against an aggressor, though he be in society and a fellow subject. Thus a thief, whom I cannot harm but by appeal to the law for having stolen all that I am worth, I may kill when he sets on men to rob me but of my horse or coat; because the law, which was made for my preservation, where it cannot interpose to secure my life from present force, which, if lost, is capable of no reparation, permits me my own defense and the right of war, a liberty to kill the aggressor, because the aggressor allows not time to appeal to our common judge, nor the decision of the law, for remedy in a case where the mischief may be irreparable. Want of a common judge with authority puts all men in a state of nature; force without right upon a man's person makes a state of war both where there is and is not a common judge.

20. But when the actual force is over, the state of war ceases between those that are in society and are equally on both sides subjected to the fair determination of the law, because then there lies open the remedy of appeal for the past injury and to prevent future harm. But where no such appeal is, as in the state of nature, for want of positive laws and judges with authority to appeal to, the state of war once begun continues with a right to the innocent party to destroy the other whenever he can, until the aggressor offers peace and desires reconciliation on such terms as may repair any wrongs he has already done and secure the innocent for the future; nay, where an appeal to the law and constituted judges lies open, but the remedy is denied by a manifest perverting of justice and a barefaced wresting of the laws to protect or indemnify the violence or injuries of some men, or party of men, there it is hard to imagine anything but a state of war; for wherever violence is used and injury done, though by hands appointed to administer justice, it is still violence and injury, however colored with the name, pretenses, or forms of law, the end whereof being to protect and redress the innocent by an unbiased application of it to all who are under it; wherever that is not bona fide done, war is made upon the sufferers, who having no appeal on earth to right them, they are left to the only remedy in such cases—an appeal to heaven.

21. To avoid this state of war—wherein there is no appeal but to heaven, and wherein every the least difference is apt to end, where there is no authority to decide between the contenders—is one great reason of men's putting themselves into society and quitting the state of nature; for where there is an authority, a power on earth from which relief can be had by appeal, there the continuance of the state of war is excluded, and the controversy is decided by that power. Had there been any such court, any superior jurisdiction on earth, to determine the right between Jephthah and the Ammonites, they had never come to a state of war; but we see he was forced to appeal to heaven: "The Lord the Judge," says he, "be judge this day between the children of Israel and the children of Ammon" (Judges xi. 27.), and then prosecuting and relying on his appeal, he leads out his army to battle. And, therefore, in such controversies where the question is put, "Who shall be judge?" it cannot be meant, "who shall decide the controversy"; every one knows what Jephthah here tells us, that "the Lord the Judge" shall judge. Where there is no judge on earth, the appeal lies to God in heaven. That question then cannot mean: who shall judge whether another has put himself in a state of war with me, and whether I may, as Jephthah did, appeal to heaven in it? Of that I myself can only be judge in my own conscience, as I will answer it, at the great day, to the supreme Judge of all men.

* * *

Of the Beginning of Political Societies

95. MEN BEING, as has been said, by nature all free, equal, and independent, no one can be put out of this estate and subjected to the political power of another without his own consent. The only way whereby any one divests himself of his natural liberty and puts on the bonds of civil society is by agreeing with other men to join and unite into a community for their comfortable, safe, and peaceable living one amongst another, in a secure enjoyment of their properties and a greater security against any

that are not of it. This any number of men may do, because it injures not the freedom of the rest; they are left as they were in the liberty of the state of nature. When any number of men have so consented to make one community or government, they are thereby presently incorporated and make one body politic wherein the majority have a right to act and conclude the rest.

96. For when any number of men have, by the consent of every individual, made a community, they have thereby made that community one body, with a power to act as one body, which is only by the will and determination of the majority; for that which acts any community being only the consent of the individuals of it, and it being necessary to that which is one body to move one way, it is necessary the body should move that way whither the greater force carries it, which is the consent of the majority; or else it is impossible it should act or continue one body, one community, which the consent of every individual that united into it agreed that it should; and so every one is bound by that consent to be concluded by the majority. And therefore we see that in assemblies impowered to act by positive laws, where no number is set by that positive law which impowers them, the act of the majority passes for the act of the whole and, of course, determines, as having by the law of nature and reason the power of the whole.

97. And thus every man, by consenting with others to make one body politic under one government, puts himself under an obligation to every one of that society to submit to the determination of the majority and to be concluded by it; or else this original compact, whereby he with others incorporates into one society, would signify nothing, and be no compact, if he be left free and under no other ties than he was in before in the state of nature. For what appearance would there be of any compact? What new engagement if he were no further tied by any decrees of the society than he himself thought fit and did actually consent to? This would be still as great a liberty as he himself had before his compact, or any one else in the state of nature has who may submit himself and consent to any acts of it if he thinks fit.

98. For if the consent of the majority shall not in reason be received as the act of the whole and conclude every individual, nothing but the consent of every individual can make anything to be the act of the whole; but such a consent is next to impossible ever to be had if we consider the infirmities of health and avocations of business which in a number, though much less than that of a commonwealth, will necessarily keep many away from the public assembly. To which, if we add the variety of opinions and contrariety of interests which unavoidably happen in all collections of men, the coming into society upon such terms would be only like Cato's coming into the theatre only to go out again. Such a constitution as this would make the mighty leviathan of a shorter duration than the feeblest creatures, and not let it outlast the day it was born in; which cannot be supposed till we can think that rational creatures should desire and constitute societies only to be dissolved; for where the majority cannot conclude the rest, there they cannot act as one body, and consequently will be immediately dissolved again.

99. Whosoever, therefore, out of a state of nature unite into a community must be understood to give up all the power necessary to the ends for which they unite into society to the majority of the community, unless they expressly agreed in any number greater than the majority. And this is done by barely agreeing to unite into one political society, which is all the compact that is, or needs be, between the individuals

that enter into or make up a commonwealth. And thus that which begins and actually constitutes any political society is nothing but the consent of any number of freemen capable of a majority to unite and incorporate into such a society. And this is that, and that only, which did or could give beginning to any lawful government in the world.

Jean-Jacques Rousseau, from
The Discourse on the Origins of Inequality and *The Social Contract* (1712–1778)

Rouseau wrote two *Discourses* for the Academy in Dijon in the early 1750s. The first, on the arts and sciences, argued the eccentric thesis that the development of the arts and sciences actually contributed to the degeneracy of human civilization, and yet the Academy gave him a much-coveted prize for it. The second *Discourse, On The Origins of Inequality*, did not win a prize but established its young Swiss author as one of the foremost social critics in France. The second *Discourse* offers us an exuberant portrait of life in the state of nature, where happy and healthy individuals wander the face of the earth, picking up plentiful food as they wish, sleeping comfortably where they will, and taking what they need from life's natural abundance, hardly aware of one another. It is quite a contrast with Hobbes' unhappy creatures, who feel compelled to enter into society to protect themselves and their meager provisions. But Rousseau's theory bears at least one essential similarity to Locke's theory, for he, too, agrees that it is the protection of private property that forces these happy natural individuals into the conventions of society. But Rousseau thought that this was not entirely an improvement in the human situation. The invention of private property, and the grotesque inequalities that followed from it, resulted in a catastrophe for humanity, the end of our primeval happiness and independence. It is from the institution of private property that all of our unhappiness arises, the artificiality and competitiveness of contemporary society, the grotesque differences between the rich and the poor, between those in power and those without it. But we do live in society, and there is no "going back" to nature. So what kind of a society should it be, and what are its foundations? It is here that Rousseau offers his own version of "the social contract," not as a vehicle for controlling each other or protecting ourselves or our property, but as a way of (each of us) "giving the law to ourselves," and thereby elevating us from mere men to the morally illustrious state of *citizens*.

As long as we are ignorant of natural man, it is futile for us to attempt to determine the law he has received or which is best suited to his constitution. All that we can see very clearly regarding this law is that, for it to be law, not only must the will of him who is obliged by it be capable of knowing submission to it, but also, for it to be natural, it must speak directly by the voice of nature.

Leaving aside therefore all the scientific books which teach us only to see men as they have made themselves, and meditating on the first and most simple operations of the human soul, I believe I perceive in it two principles that are prior to reason, of which one makes us ardently interested in our well-being and our self-preservation, and the other inspires in us a natural repugnance to seeing any sentient being, especially our fellow man, perish or suffer. It is from the conjunction and combination that our mind is in a position to make regarding these two principles, without the need for introducing that of sociability, that all the rules of natural right appear to me to flow; rules which reason is later forced to reestablish on other foundations, when, by its successive developments, it has succeeded in smothering nature.

In this way one is not obliged to make a man a philosopher before making him a man. His duties toward others are not uniquely dictated to him by the belated lessons of wisdom; and as long as he does not resist the inner impulse of compassion, he will never harm another man or even another sentient being, except in the legitimate instance where, if his preservation were involved, he is obliged to give preference to himself. By this means, an end can also be made to the ancient disputes regarding the participation of animals in the natural law. For it is clear that, lacking intelligence and liberty, they cannot recognize this law; but since they share to some extent in our nature by virtue of the sentient quality with which they are endowed, one will judge that they should also participate in natural right, and that man is subject to some sort of duties toward them. It seems, in effect, that if I am obliged not to do any harm to my fellow man, it is less because he is a rational being than because he is a sentient being: a quality that, since it is common to both animals and men, should at least give the former the right not to be needlessly mistreated by the latter.

* * *

In considering human society from a tranquil and disinterested point of view it seems at first to manifest merely the violence of powerful men and the oppression of the weak. The mind revolts against the harshness of the former; one is inclined to deplore the blindness of the latter. And since nothing is less stable among men than those external relationships which chance brings about more often than wisdom, and which are called weakness or power, wealth or poverty, human establishments appear at first glance to be based on piles of shifting sand. It is only in examining them closely, only after having cleared away the dust and sand that surround the edifice, that one learns to respect its foundations.

* * *

I conceive of two kinds of inequality in the human species: one which I call natural or physical, because it is established by nature and consists in the difference of age, health, bodily strength, and qualities of mind or soul. The other may be called moral or political inequality, because it depends on a kind of convention and is established, or at least authorized, by the consent of men. This latter type of inequality consists in the different privileges enjoyed by some at the expense of others, such as being richer, more honored, more powerful than they, or even causing themselves to be obeyed by them.

There is no point in asking what the source of natural inequality is, because the answer would be found enunciated in the simple definition of the word. There is still less of a point in asking whether there would not be some essential connection between the two inequalities, for that would amount to asking whether those who command are necessarily better than those who obey, and whether strength of body or mind, wisdom or virtue are always found in the same individuals in proportion to power or wealth. Perhaps this is a good question for slaves to discuss within earshot of their masters, but it is not suitable for reasonable and free men who seek the truth.

Precisely what, then, is the subject of this discourse? To mark, in the progress of things, the moment when, right taking the place of violence, nature was subjected to the law. To explain the sequence of wonders by which the strong could resolve to serve the weak, and the people to buy imaginary repose at the price of real felicity.

The philosophers who have examined the foundations of society have all felt the necessity of returning to the state of nature, but none of them has reached it. Some have not hesitated to ascribe to man in that state the notion of just and unjust, without bothering to show that he had to have that notion, or even that it was useful to him. Others have spoken of the natural right that everyone has to preserve what belongs to him, without explaining what they mean by "belonging." Others started out by giving authority to the stronger over the weaker, and immediately brought about government, without giving any thought to the time that had to pass before the meaning of the words "authority" and "government" could exist among men. Finally, all of them, speaking continually of need, avarice, oppression, desires, and pride, have transferred to the state of nature the ideas they acquired in society. They spoke about savage man, and it was civil man they depicted. It did not even occur to most of our philosophers to doubt that the state of nature had existed, even though it is evident from reading the Holy Scriptures that the first man, having received enlightenment and precepts immediately from God, was not himself in that state; and if we give the writings of Moses the credence that every Christian owes them, we must deny that, even before the flood, men were ever in the pure state of nature, unless they had fallen back into it because of some extraordinary event: a paradox that is quite awkward to defend and utterly impossible to prove.

Let us therefore begin by putting aside all the facts, for they have no bearing on the question. The investigations that may be undertaken concerning this subject should not be taken for historical truths, but only for hypothetical and conditional reasonings, better suited to shedding light on the nature of things than on pointing out their true origin, like those our physicists make everyday with regard to the formation of the world.

* * *

When I strip that being, thus constituted, of all the supernatural gifts he could have received and of all the artificial faculties he could have acquired only through long progress; when I consider him, in a word, as he must have left the hands of nature, I see an animal less strong than some, less agile than others, but all in all, the most advantageously organized of all. I see him satisfying his hunger under an oak tree, quenching his thirst at the first stream, finding his bed at the foot of the same tree that supplied his meal; and thus all his needs are satisfied.

Accustomed from childhood to inclement weather and the rigors of the seasons, acclimated to fatigue, and forced, naked and without arms, to defend their lives and their prey against other ferocious beasts, or to escape them by taking flight, men develop a robust and nearly unalterable temperament.

Since the savage man's body is the only instrument he knows, he employs it for a variety of purposes that, for lack of practice, ours are incapable of serving. And our industry deprives us of the force and agility that necessity obliges him to acquire. If he had had an axe, would his wrists break such strong branches? If he had had a sling, would he throw a stone with so much force? If he had had a ladder, would he climb a tree so nimbly? If he had had a horse, would he run so fast? Give a civilized man time to gather all his machines around him, and undoubtedly he will easily overcome a savage man. But if you want to see an even more unequal fight, pit them against each other naked and disarmed, and you will soon realize the advantage of constantly having all of one's forces at one's disposal, of always being ready for any event, and of always carrying one's entire self, as it were, with one.

Hobbes maintains that man is naturally intrepid and seeks only to attack and to fight. On the other hand, an illustrious philosopher thinks . . . that nothing is as timid as man in the state of nature, and that he is always trembling and ready to take flight at the slightest sound he hears or at the slightest movement he perceives. That may be the case with regard to objects with which he is not acquainted. And I do not doubt that he is frightened by all the new sights that present themselves to him every time he can neither discern the physical good and evil he may expect from them nor compare his forces with the dangers he must run: rare circumstances in the state of nature, where everything takes place in such a uniform manner and where the face of the earth is not subject to those sudden and continual changes caused by the passions and inconstancy of peoples living together. But since a savage man lives dispersed among the animals and, finding himself early on in a position to measure himself against them, he soon makes the comparison; and, aware that he surpasses them in skillfulness more than they surpass him in strength, he learns not to fear them any more. Pit a bear or a wolf against a savage who is robust, agile, and courageous, as they all are, armed with stones and a hefty cudgel, and you will see that the danger will be at least equal on both sides, and that after several such experiences, ferocious beasts, which do not like to attack one another, will be quite reluctant to attack a man, having found him to be as ferocious as themselves.

* * *

Therefore we must take care not to confuse savage man with the men we have before our eyes. Nature treats all animals left to their own devices with a predilection that seems to show how jealous she is of that right. The horse, the cat, the bull, even the ass, are usually taller, and all of them have a more robust constitution, more vigor, more strength, and more courage in the forests than in our homes. They lose half of these advantages in becoming domesticated; it might be said that all our efforts at feeding them and treating them well only end in their degeneration. It is the same for man himself. In becoming habituated to the ways of society and a slave, he becomes weak, fearful, and servile; his soft and effeminate lifestyle completes the enervation of both his strength and his courage. Let us add that the difference between the savage man and the domesticated man should be still greater than that between the savage animal and the domesticated animal; for while animal and man have been treated equally by nature, man gives more comforts to himself than to the animals he tames, and all of these comforts are so many specific causes that make him degenerate more noticeably.

* * *

So far I have considered only physical man. Let us now try to look at him from a metaphysical and moral point of view.

* * *

Every animal has ideas, since it has senses; up to a certain point it even combines its ideas, and in this regard man differs from an animal only in degree. Some philosophers have even suggested that there is a greater difference between two given men than between a given man and an animal. Therefore it is not so much understanding which causes the specific distinction of man from all other animals as it is his being a free agent. Nature commands every animal, and beasts obey. Man feels the same impetus, but he knows he is free to go along or to resist; and it is above all in the awareness of this freedom that the spirituality of his soul is made manifest.

* * *

. . . But if the difficulties surrounding all these questions should leave some room for dispute on this difference between man and animal, there is another very specific quality which distinguishes them and about which there can be no argument: the faculty of self-perfection, a faculty which, with the aid of circumstances, successively develops all the others, and resides among us as much in the species as in the individual. On the other hand, an animal, at the end of a few months, is what it will be all its life; and its species, at the end of a thousand years, is what it was in the first of those thousand years. Why is man alone subject to becoming an imbecile? Is it not that he thereby returns to his primitive state, and that, while the animal which has acquired nothing and which also has nothing to lose, always retains its instinct, man, in losing through old age or other accidents all that his *perfectibility* has enabled him to acquire, thus falls even lower than the animal itself? It would be sad for us to be forced to agree that this distinctive and almost unlimited faculty is the source of all man's misfortunes; that this is what, by dint of time, draws him out of that original condition in which he would pass tranquil and innocent days; that this is what, through centuries of giving rise to his enlightenment and his errors, his vices and his virtues, eventually makes him a tyrant over himself and nature.

* * *

Whatever these origins may be, it is clear, from the little care taken by nature to bring men together through mutual needs and to facilitate their use of speech, how little she prepared them for becoming habituated to the ways of society, and how little she contributed to all that men have done to establish the bonds of society. In fact, it is impossible to imagine why, in that primitive state, one man would have a greater need for another man than a monkey or a wolf has for another of its respective species; or, assuming this need, what motive could induce the other man to satisfy it; or even, in this latter instance, how could they be in mutual agreement regarding the conditions. I know that we are repeatedly told that nothing would have been so miserable than man in that state; and if it is true, as I believe I have proved, that it is only after many centuries that men could have had the desire and the opportunity to leave that state, that would be a charge to bring against nature, not

against him whom nature has thus constituted. But if we understand the word
miserable properly, it is a word which is without meaning or which signifies merely a
painful privation and suffering of the body or the soul. Now I would very much like
someone to explain to me what kind of misery can there be for a free being whose
heart is at peace and whose body is in good health? I ask which of the two, civil or
natural life, is more likely to become insufferable to those who live it? We see about
us practically no people who do not complain about their existence; many even
deprive themselves of it to the extent they are able, and the combination of divine
and human laws is hardly enough to stop this disorder. I ask if anyone has ever heard
tell of a savage who was living in liberty ever dreaming of complaining about his life
and of killing himself. Let the judgment therefore be made with less pride on which
side real misery lies. On the other hand, nothing would have been so miserable as
savage man, dazzled by enlightenment, tormented by passions, and reasoning about
a state different from his own. It was by a very wise providence that the latent
faculties he possessed should develop only as the occasion to exercise them presents
itself, so that they would be neither superfluous nor troublesome to him beforehand,
nor underdeveloped and useless in time of need. In instinct alone, man had everything he needed in order to live in the state of nature; in a cultivated reason, he has
only what he needs to live in society.

* * *

Above all, let us not conclude with Hobbes that because man has no idea of
goodness he is naturally evil; that he is vicious because he does not know virtue; that
he always refuses to perform services for his fellow men he does not believe he owes
them; or that, by virtue of the right, which he reasonably attributes to himself, to
those things he needs, he foolishly imagines himself to be the sole proprietor of the
entire universe. Hobbes has very clearly seen the defect of all modern definitions of
natural right, but the consequences he draws from his own definition show that he
takes it in a sense that is no less false. Were he to have reasoned on the basis of the
principles he establishes, this author should have said that since the state of nature is
the state in which the concern for our self-preservation is the least prejudicial to that
of others, that state was consequently the most appropriate for peace and the best
suited for the human race. He says precisely the opposite, because he had wrongly
injected into the savage man's concern for self-preservation the need to satisfy a
multitude of passions which are the product of society and which have made laws
necessary.

* * *

Hobbes did not see that the same cause preventing savages from using their
reason, as our jurists claim, is what prevents them at the same time from abusing
their faculties, as he himself maintains. Hence we could say that savages are not evil
precisely because they do not know what it is to be good; for it is neither the
development of enlightenment nor the restraint imposed by the law, but the calm of
the passions and the ignorance of vice which prevents them from doing evil. *So much
more profitable to these is the ignorance of vice than the knowledge of virtue is to
those.* Moreover, there is another principle that Hobbes failed to notice, and which,
having been given to man in order to mitigate, in certain circumstances, the ferocity
of his egocentrism or the desire for self-preservation before this egocentrism of his

came into being, tempers the ardor he has for his own well-being by an innate repugnance to seeing his fellow men suffer. I do not believe I have any contradiction to fear in granting the only natural virtue that the most excessive detractor of human virtues was forced to recognize. I am referring to pity, a disposition that is fitting for beings that are as weak and as subject to ills as we are; a virtue all the more universal and all the more useful to man in that it precedes in him any kind of reflection, and so natural that even animals sometimes show noticeable signs of it.

* * *

Reason is what engenders egocentrism, and reflection strengthens it. Reason is what turns man in upon himself. Reason is what separates him from all that troubles him and afflicts him. Philosophy is what isolates him and what moves him to say in secret, at the sight of a suffering man, "Perish if you will; I am safe and sound." No longer can anything but danger to the entire society trouble the tranquil slumber of the philosopher and yank him from his bed. His fellow man can be killed with impunity underneath his window. He has merely to place his hands over his ears and argue with himself a little in order to prevent nature, which rebels within him, from identifying him with the man being assassinated. Savage man does not have this admirable talent, and for lack of wisdom and reason he is always seen thoughtlessly giving in to the first sentiment of humanity. When there is a riot or a street brawl, the populace gathers together; the prudent man withdraws from the scene. It is the rabble, the women of the marketplace, who separate the combatants and prevent decent people from killing one another.

It is therefore quite certain that pity is a natural sentiment, which, by moderating in each individual the activity of the love of oneself, contributes to the mutual preservation of the entire species. Pity is what carries us without reflection to the aid of those we see suffering. Pity is what, in the state of nature, takes the place of laws, mores, and virtue, with the advantage that no one is tempted to disobey its sweet voice. Pity is what will prevent every robust savage from robbing a weak child or an infirm old man of his hard-earned subsistence, if he himself expects to be able to find his own someplace else. Instead of the sublime maxim of reasoned justice, *Do unto others as you would have them do unto you*, pity inspires all men with another maxim of natural goodness, much less perfect but perhaps more useful than the preceding one: *Do what is good for you with as little harm as possible to others*. In a word, it is in this natural sentiment, rather than in subtle arguments that one must search for the cause of the repugnance at doing evil that every man would experience, even independently of the maxims of education. Although it might be appropriate for Socrates and minds of his stature to acquire virtue through reason, the human race would long ago have ceased to exist, if its preservation had depended solely on the reasonings of its members.

* * *

The first person who, having enclosed a plot of land, took it into his head to say *this is mine* and found people simple enough to believe him, was the true founder of civil society. What crimes, wars, murders, what miseries and horrors would the human race have been spared, had someone pulled up the stakes or filled in the ditch and cried out to his fellow men: "Do not listen to this impostor. You are lost if you forget that the fruits of the earth belong to all and the earth to no one!" But it is quite likely that by then things had already reached the point where they could no longer continue as they were. For this idea of property, depending on many prior ideas which could only have arisen successively, was not formed all at once in the human mind. It was necessary to make great progress, to acquire much industry and enlightenment, and to transmit and augment them from one age to another, before arriving at this final stage in the state of nature. Let us therefore take things farther back and try to piece together under a single viewpoint that slow succession of events and advances in knowledge in their most natural order.

* * *

Having previously wandered about the forests and having assumed a more fixed situation, men slowly came together and united into different bands, eventually forming in each country a particular nation, united by mores and characteristic features, not by regulations and laws, but by the same kind of life and foods and by the common influence of the climate. Eventually a permanent proximity cannot fail to engender some intercourse among different families. Young people of different sexes live in neighboring huts; the passing intercourse demanded by nature soon leads to another, through frequent contact with one another, no less sweet and more permanent. People become accustomed to consider different objects and to make comparisons. Imperceptibly they acquire the ideas of merit and beauty which produce feelings of preference. By dint of seeing one another, they can no longer get along without seeing one another again. A sweet and tender feeling insinuates itself into the soul and at the least opposition becomes an impetuous fury. Jealousy awakens with love; discord triumphs, and the sweetest passion receives sacrifices of human blood.

In proportion as ideas and sentiments succeed one another and as the mind and heart are trained, the human race continues to be tamed, relationships spread and bonds are tightened. People grew accustomed to gather in front of their huts or around a large tree; song and dance, true children of love and leisure, became the amusement or rather the occupation of idle men and women who had flocked together. Each one began to look at the others and to want to be looked at himself, and public esteem had a value. The one who sang or danced the best, the handsomest, the strongest, the most adroit or the most eloquent became the most highly regarded. And this was the first step toward inequality and, at the same time, toward vice. From these first preferences were born vanity and contempt on the one hand, and shame and envy on the other. And the fermentation caused by these new leavens eventually produced compounds fatal to happiness and innocence.

As soon as men had begun mutually to value one another, and the idea of esteem was formed in their minds, each one claimed to have a right to it, and it was no longer possible for anyone to be lacking it with impunity. From this came the first duties of civility, even among savages; and from this very voluntary wrong became an outrage, because along with the harm that resulted from the injury, the offended party saw in it contempt for his person, which often was more insufferable than the harm itself. Hence each man punished the contempt shown him in a manner proportionate to the esteem in which he held himself; acts of revenge became terrible, and men became bloodthirsty and cruel. This is precisely the stage reached by most of the savage people known to us; and it is for want of having made adequate distinctions among their ideas or of having noticed how far these peoples already were from the original state of nature that many have hastened to conclude that man is naturally cruel, and that he needs civilization in order to soften him. On the contrary, nothing is so gentle as man in his primitive state, when, placed by nature at an equal distance from the stupidity of brutes and the fatal enlightenment of civil man, and limited equally by instinct and reason to protecting himself from the harm that threatens him, he is restrained by natural pity from needlessly harming anyone himself, even if he has been harmed. For according to the axiom of the wise Locke, *where there is no property, there is no injury.*

* * *

From the cultivation of land, there necessarily followed the division of land; and from property once recognized, the first rules of justice. For in order to render everyone what is his, it is necessary that everyone can have something. Moreover, as men began to look toward the future and as they saw that they all saw they had goods to lose, there was not one of them who did not have to fear reprisals against himself for wrongs he might do to another. This origin is all the more natural as it is impossible to conceive of the idea of property arising from anything but manual labor, for it is not clear what man can add, beyond his own labor, in order to appropriate things he has not made. It is labor alone that, in giving the cultivator a right to the product of the soil he has tilled, consequently gives him a right, at least until the harvest, and thus from year to year. With this possession continuing uninterrupted, it is easily transformed into property.

* * *

Things in this state could have remained equal, if talents had been equal, and if the use of iron and the consumption of foodstuffs had always been in precise balance. But this proportion, which was not maintained by anything, was soon broken. The strongest did the most work; the most adroit turned theirs to better advantage: the most ingenious found ways to shorten their labor. The farmer had a greater need for iron, or the blacksmith had a greater need for wheat; and in laboring equally, the one earned a great deal while the other barely had enough to live. Thus it is that natural inequality imperceptibly manifests itself together with inequality occasioned by the socialization process. Thus it is that the differences among men, developed by those of circumstances, make themselves more noticeable, more permanent in their effects, and begin to influence the fate of private individuals in the same proportion.

Thus we find here all our faculties developed, memory and imagination in play, egocentrism looking out for its interests, reason rendered active, and the mind having nearly reached the limit of the perfection of which it is capable. We find here all the natural qualities put into action, the rank and fate of each man established not only on the basis of the quantity of goods and the power to serve or harm, but also on the basis of mind, beauty, strength or skill, on the basis of merit or talents. And since these qualities were the only ones that could attract consideration, he was soon forced to have them or affect them. It was necessary, for his advantage, to show himself to be something other than what he in fact was. Being something and appearing to be something became two completely different things; and from this distinction there arose grand ostentation, deceptive cunning, and all the vices that follow in their wake. On the other hand, although man had previously been free and independent, we find him, so to speak, subject, by virtue of a multitude of fresh needs, to all of nature and particularly to his fellowmen, whose slave in a sense he becomes even in becoming their master; rich, he needs their services; poor, he needs their help; and being midway between wealth and poverty does not put him in a position to get along without them. It is therefore necessary for him to seek incessantly to interest them in his fate and to make them find their own profit, in fact or in appearance, in working for his. This makes him two-faced and crooked with some, imperious and harsh with others, and puts him in the position of having to abuse everyone he needs when he cannot make them fear them and does not find it in his interests to be of useful service to them. Finally, consuming ambition, the zeal for raising the relative level of his fortune, less out of real need than in order to put himself above others, inspires in all men a wicked tendency to harm one another, a secret jealousy all the more dangerous because, in order to strike its blow in greater safety, it often wears the mask of benevolence; in short, competition and rivalry on the one hand, opposition of interest[s] on the other, and always the hidden desire to profit at the expense of someone else. All these ills are the first effect of property and the inseparable offshoot of incipient inequality.

* * *

I want to inquire whether there can be some legitimate and sure rule of administration in the civil order, taking men as they are and laws as they might be. I will always try in this inquiry to bring together what right permits with what interest prescribes, so that justice and utility do not find themselves at odds with one another.

* * *

Man is born free, and everywhere he is in chains. He who believes himself the master of others does not escape being more of a slave than they. How did this change take place? I have no idea. What can render it legitimate? I believe I can answer this question.

Were I to consider only force and the effect that flows from it, I would say that so long as a people is constrained to obey and does obey, it does well. As soon as it can shake off the yoke and does shake it off, it does even better. For by recovering its liberty by means of the same right that stole it, either the populace is justified in getting it back or else those who took it away were not justified in their actions. But the social order is a sacred right which serves as a foundation for all other rights. Nevertheless, this right does not come from nature. It is therefore founded upon convention. Before coming to that, I ought to substantiate what I just claimed.

The most ancient of all societies and the only natural one, is that of the family. Even so children remain bound to their father only so long as they need him to take care of them. As soon as the need ceases, the natural bond is dissolved. Once the children are freed from the obedience they owed the father and their father is freed from the care he owed his children, all return equally to independence. If they continue to remain united, this no longer takes place naturally but voluntarily, and the family maintains itself only by means of convention.

This common liberty is one consequence of the nature of man. Its first law is to see to his maintenance; its first concerns are those he owes himself; and, as soon as he reaches the age of reason, since he alone is the judge of the proper means of taking care of himself, he thereby becomes his own master.

The family therefore is, so to speak, the prototype of political societies; the leader is the image of the father, the populace is the image of the children, and, since all are born equal and free, none give up their liberty except for their utility. The entire difference consists in the fact that in the family the love of the father for his children repays him for the care he takes for them, while in the state, where the leader does not have love for his peoples, the pleasure of commanding takes the place of this feeling.

* * *

The strongest is never strong enough to be master all the time, unless he transforms force into right and obedience into duty. Hence the right of the strongest, a right that seems like something intended ironically and is actually established as a basic principle. But will no one explain this word to me? Force is a physical power; I fail to see what morality can result from its effects. To give in to force is an act of necessity, not of will. At most, it is an act of prudence. In what sense could it be a duty?

Let us suppose for a moment that there is such a thing as this alleged right. I maintain that all that results from it is an inexplicable mish-mash. For once force produces the right, the effect changes places with the cause. Every force that is superior to the first succeeds to its right. As soon as one can disobey with impunity, one can do so legitimately; and since the strongest is always right, the only thing to do is to make oneself the strongest. For what kind of right is it that perishes when the force on which it is based ceases? If one must obey because of force, one need not do so out of duty; and if one is no longer forced to obey one is no longer obliged. Clearly then, this word "right" adds nothing to force. It is utterly meaningless here.

Let us then agree that force does not bring about right, and that one is obliged to obey only legitimate powers. Thus my original question keeps returning.

Since no man has a natural authority over his fellow man, and since force does not give rise to any right, conventions therefore remain the basis of all legitimate authority among men.

If, says Grotius, a private individual can alienate his liberty and turn himself into the slave of a master, why could not an entire people alienate its liberty and turn itself into the subject of a king? There are many equivocal words here which need explanation, but let us confine ourselves to the word *alienate*. To alienate is to give or to sell. A man who makes himself the slave of someone else does not give himself; he sells himself, at least for his subsistence. But why does a people sell itself? Far from furnishing his subjects with their subsistence, a king derives his own from them alone, and, according to Rabelais, a king does not live cheaply. Do subjects then give their persons on the condition that their estate will also be taken? I fail to see what remains for them to preserve.

It will be said that the despot assures his subjects of civil tranquility. Very well. But what do they gain, if the wars his ambition drags them into, if his insatiable greed, if the oppressive demands caused by his ministers occasion more grief for his subjects than their own dissensions would have done? What do they gain, if this very tranquility is one of their miseries? A tranquil life is also had in dungeons; is that enough to make them desirable? The Greeks who were locked up in the Cyclops' cave lived a tranquil existence as they awaited their turn to be devoured.

To say that a man gives himself gratuitously is to say something absurd and inconceivable. Such an act is illegitimate and null, if only for the fact that he who commits it does not have his wits about him. To say the same thing of an entire populace is to suppose a populace composed of madmen. Madness does not bring about right.

Even if each person can alienate himself, he cannot alienate his children. They are born men and free. Their liberty belongs to them; they alone have the right to dispose of it. Before they have reached the age of reason, their father can, in their name, stipulate conditions for their maintenance and for their well-being. But he cannot give them irrevocably and unconditionally, for such a gift is contrary to the ends of nature and goes beyond the rights of paternity. For an arbitrary government to be legitimate, it would therefore be necessary in each generation for the people to be master of its acceptance or rejection. But in that event this government would no longer be arbitrary.

Renouncing one's liberty is renouncing one's dignity as a man, the rights of humanity and even its duties. There is no possible compensation for anyone who renounces everything. Such a renunciation is incompatible with the nature of man. Removing all morality from his actions is tantamount to taking away all liberty from his will. Finally, it is a vain and contradictory convention to stipulate absolute authority on one side and a limitless obedience on the other. Is it not clear that no commitments are made to a person from whom one has the right to demand everything? And does this condition alone not bring with it, without equivalent or exchange, the nullity of the act? For what right would my slave against me, given that all he has belongs to me, and that, since his right is my right, my having a right against myself makes no sense?

* * *

Even if I were to grant all that I have thus far refuted, the supporters of despotism would not be any better off. There will always be a great difference between subduing a multitude and ruling a society. If scattered men, however many they may be, were successively enslaved by a single individual, I see nothing there but a master and slaves; I do not see a people and its leader. It is, if you will, an aggregation, but not an association. There is neither a public good nor a body politic there. Even if that man had enslaved half the world, he is always just a private individual. His interest, separated from that of others, is never anything but a private interest. If this same man is about to die, after his passing his empire remains scattered and disunited, just as an oak tree dissolves and falls into a pile of ashes after fire has consumed it.

* * *

In fact, if there were no prior convention, then, unless the vote were unanimous, what would become of the minority's obligation to submit to the majority's choice, and where do one hundred who want a master get the right to vote for ten who do not? The law of majority rule is itself an established convention, and presupposes unanimity on at least one occasion.

I suppose that men have reached the point where obstacles that are harmful to their maintenance in the state of nature gain the upper hand by their resistance to the forces that each individual can bring to bear to maintain himself in that state. Such being the case, that original state cannot subsist any longer, and the human race would perish if it did not alter its mode of existence.

For since men cannot engender new forces, but merely unite and direct existing ones, they have no other means of maintaining themselves but to form by aggregation a sum of forces that could gain the upper hand over the resistance, so that their forces are directed by means of a single moving power and made to act in concert.

This sum of forces cannot come into being without the cooperation of many. But since each man's force and liberty are the primary instruments of his maintenance, how is he going to engage them without hurting himself and without neglecting the care that he owes himself? This difficulty, seen in terms of my subject, can be stated in the following terms:

"Find a form of association which defends and protects with all common forces the person and goods of each associate, and by means of which each one, while uniting with all, nevertheless obeys only himself and remains as free as before?" This is the fundamental problem for which the social contract provides the solution.

The clauses of this contract are so determined by the nature of the act that the least modification renders them vain and ineffectual, that, although perhaps they have never been formally promulgated, they are everywhere the same, everywhere tacitly accepted and acknowledged. Once the social compact is violated, each person then regains his first rights and resumes his natural liberty, while losing the conventional liberty for which he renounced it.

These clauses, properly understood, are all reducible to a single one, namely the total alienation of each associate, together with all of his rights, to the entire community. For first of all, since each person gives himself whole and entire, the condition is equal for everyone; and since the condition is equal for everyone, no one has an interest in making it burdensome for the others.

Moreover, since the alienation is made without reservation, the union is as perfect as possible, and no associate has anything further to demand. For if some rights remained with private individuals, in the absence of any common superior who could decide between them and the public, each person would eventually claim to be his own judge in all things, since he is on some point his own judge. The state of nature would subsist and the association would necessarily become tyrannical or hollow.

Finally, in giving himself to all, each person gives himself to no one. And since there is no associate over whom he does not acquire the same right that he would grant others over himself, he gains the equivalent of everything he loses, along with a greater amount of force to preserve what he has.

If, therefore, one eliminates from the social compact whatever is not essential to it, one will find that it is reducible to the following terms. *Each of us places his person and all his power in common under the supreme direction of the general will; and as one we receive each member as an indivisible part of the whole.*

At once, in place of the individual person of each contracting party, this act of association produces a moral and collective body composed of as many members as there are voices in the assembly, which receives from this same act its unity, its common *self*, its life and its will. This public person, formed thus by union of all the others formerly took the name *city*, and at present takes the name *republic* or *body politic*, which is called *state* by its members when it is passive, *sovereign* when it is active, *power* when compared to others like itself. As to the associates, they collectively take the name *people*; individually they are called *citizens*, insofar as participants in the sovereign authority, and *subjects*, insofar as they are subjected to the laws of the state. But these terms are often confused and mistaken for one another. It is enough to know how to distinguish them when they are used with absolute precision.

This formula shows that the act of association includes a reciprocal commitment between the public and private individuals, and that each individual, contracting, as it were, with himself, finds himself under a twofold commitment: namely as a member of the sovereign to private individuals, and as a member of the state toward the sovereign. But the maxim of civil law that no one is held to commitments made to himself cannot be applied here, for there is a considerable difference between being obligated to oneself, or to a whole of which one is a part.

It must be further noted that the public deliberation that can obligate all the subjects to the sovereign, owing to the two different relationships in which each of them is viewed, cannot, for the opposite reason, obligate the sovereign to itself, and that consequently it is contrary to the nature of the body politic that the sovereign impose upon itself a law it could not break. Since the sovereign can be considered under but one single relationship, it is then in the position of a private individual contracting with himself. Whence it is apparent that there neither is nor can be any type of fundamental law that is obligatory for the people as a body, not even the social contract. This does not mean that the whole body cannot perfectly well commit itself to another body with respect to things that do not infringe on this contract. For in regard to the foreigner, it becomes a simple being, an individual.

However, since the body politic or the sovereign derives its being exclusively from the sanctity of the contract, it can never obligate itself, not even to another power, to do anything that derogates from the original act, such as alienating some portion of itself or submitting another sovereign. Violation of the act whereby it exists would be self-annihilation, and whatever is nothing produces nothing.

As soon as this multitude is thus united in a body, one cannot harm one of the members without attacking the whole body. It is even less likely that the body can be harmed without the members feeling it. Thus duty and interest equally obligate the two parties to come to one another's aid, and the same men should seek to combine in this two-fold relationship all the advantages that result from it.

For since the sovereign is formed entirely from the private individuals who make it up, it neither has nor could have an interest contrary to theirs. Hence, the sovereign power has no need to offer a guarantee to its subjects, since it is impossible for a body to want to harm all of its members, and, as we will see later, it cannot harm any one of them in particular. The sovereign, by the mere fact that it exists, is always all that it should be.

But the same thing cannot be said of the subjects in relation to the sovereign, for which, despite their common interest, their commitments would be without substance if it did not find ways of being assured of their fidelity.

In fact, each individual can, as a man, have a private will contrary to or different from the general will that he has as a citizen. His private interest can speak to him in an entirely different manner than the common interest. His absolute and naturally independent existence can cause him to envisage what he owes the common cause as a gratuitous contribution, the loss of which will be less harmful to others than its payment is burdensome to him. And in viewing the moral person which constitutes the state as a being of reason because it is not a man, he would enjoy the rights of a citizen without wanting to fulfill the duties of a subject, an injustice whose growth would bring about the ruin of the body politic.

Thus, in order for the social compact to avoid being an empty formula, it tacitly entails the commitment—which alone can give force to the others—that whoever refuses to obey the general will will be forced to do so by the entire body. This means merely that he will be forced to be free. For this is the sort of condition that, by giving each citizen to the homeland, guarantees him against all personal dependence—a condition that produces the skill and the performance of the political machine, and which alone bestows legitimacy upon civil commitments. Without it such commitments would be absurd, tyrannical and subject to the worst abuses.

This passage from the state of nature to the civil state produces quite a remarkable change in man, for it substitutes justice for instinct in his behavior and gives his actions a moral quality they previously lacked. Only then, when the voice of duty replaces physical impulse and right replaces appetite, does man, who had hitherto taken only himself into account, find himself forced to act upon other principles and to consult his reason before listening to his inclinations. Although in this state he deprives himself of several of the advantages belonging to him in the state of nature, he regains such great ones. His faculties are exercised and developed, his ideas are broadened, his feelings are ennobled, his entire soul is elevated to such a height that, if the abuse of this new condition did not often lower his status to beneath the level he left, he ought constantly to bless the happy moment that pulled him away from it forever and which transformed him from a stupid, limited animal into an intelligent being and a man.

Let us summarize this entire balance sheet so that the credits and debits are easily compared. What man loses through the social contract is his natural liberty and an unlimited right to everything that tempts him and that he can acquire. What he gains is civil liberty and the proprietary ownership of all he possesses. So as not to be in error in these compensations, it is necessary to draw a careful distinction between natural liberty (which is limited solely by the force of the individual involved) and civil liberty (which is limited by the general will), and between possession (which is merely the effect of the force or the right of the first occupant) and proprietary ownership (which is based solely on a positive title).

To the preceding acquisitions could be added the acquisition in the civil state of moral liberty, which alone makes man truly the master of himself. For to be driven by appetite alone is slavery, and obedience to the law one has prescribed for oneself is liberty. But I have already said too much on this subject, and the philosophical meaning of the word *liberty* is not my subject here.

The first and most important consequence of the principles established above is that only the general will can direct the forces of the state according to the purpose for which it was instituted, which is the common good. For if the opposition of private interests made necessary the establishment of societies, it is the accord of these same interests that made it possible. It is what these different interests have in common that forms the social bond, and, were there no point of agreement among all these interests, no society could exist. For it is utterly on the basis of this common interest that society ought to be governed.

I therefore maintain that sovereignty is merely the exercise of the general will, it can never be alienated, and that the sovereign, which is only a collective being, cannot be represented by anything but itself.

G. W. F. Hegel, "Master and Slave," from
The Phenomenology of Spirit (1770–1831)

G. W. F. Hegel finished writing his monumental *Phenomenology of Spirit* in 1806, as he watched "the world spirit" in action as Napoleon completed his destruction of the Holy Roman Empire just a few miles away from the University of Jena, where Hegel was teaching. That sense of a new world beginning inspired Hegel, and so many other young philosophers and poets, and the *Phenomenology* was above all an expression of the new sense of internationalism and the hopes for justice that accompanied it. The "master and slave" parable occurs early on in the book, as a thought-experiment concerning the most primitive possible human relationship, abstracted from all social concerns, nothing but the unembellished meeting of two primal people, "two self-consciousnesses." What would it be like, a mutually defensive Hobbesian attack, born of selfishness and insecurity? A mere matter of Rousseauian indifference, as each of them went about its merry way? Hegel suggests that the primary concern of any two self-conscious beings would be their mutual *recognition*, a need to establish what they are—in their own and in each other's eyes. But then the very idea of two autonomous, independent beings is already undermined, for they gain their existence, their sense of selfhood, not from within themselves, but from and through each other. That is the point of the Hegelian parable, and throughout *The Phenomenology* we are treated to repeated refutations of the various claims of "individualism," the very idea that it makes sense to talk about individuals prior to or apart from the contexts of society. As the parable develops, the two beings struggle, each to be recognized as the superior of the other. Fighting for mutual recognition, one loses and, rather than lose his life (and, for the other, rather than lose the one witness to his victory) he becomes a slave, wholly dependent on the other, the master. But the parable takes on a perverse twist, which would have considerable influence on other philosophers to follow in their own speculations about power and dependency. The master, who is supposedly independent, becomes dependent on the slave for the very means of his existence, and the slave, who is supposedly dependent on the master, finds that he is, in fact, very much in control of his life in a way that the master is no longer. In the language of justice, it is now the slave who deserves recognition, but this will be possible only if the Spartan framework of the master-slave relationship is relinquished in favor of a much more sociable *and natural* conception of human community. (We will see more of Hegel's view in Part III.)

From Hegel's *Phenomenology of Spirit*, translated by A. V. Miller and published by Oxford University Press, 1977. Reprinted by permission of the publisher.

178. Self-consciousness exists in and for itself when, and by the fact that, it so exists for another; that is, it exists only in being acknowledged. The Notion of this its unity in its duplication embraces many and varied meanings. Its moments, then, must on the one hand be held strictly apart, and on the other hand must in this differentiation at the same time also be taken and known as not distinct, or in their opposite significance. The twofold significance of the distinct moments has in the nature of self-consciousness to be infinite, or directly the opposite of the determinateness in which it is posited. The detailed exposition of the Notion of this spiritual unity in its duplication will present us with the process of Recognition.

* * *

186. Self-consciousness is, to begin with, simple being-for-self, self-equal through the exclusion from itself of everything else. For it, its essence and absolute object is 'I'; and in this immediacy, or in this [mere] being, of its being-for-self, it is an *individual.* What is 'other' for it is an unessential, negatively characterized object. But the 'other' is also a self-consciousness; one individual is confronted by another individual. Appearing thus immediately on the scene, they are for one another like ordinary objects, *independent* shapes, individuals submerged in the being [or immediacy] of *Life*—for the object in its immediacy is here determined as Life. They are, *for each other,* shapes of consciousness which have not yet accomplished the movement of absolute abstraction, of rooting-out all immediate being, and of being merely the purely negative being of self-identical consciousness; in other words, they have not as yet exposed themselves to each other in the form of pure being-for-self, or as self-consciousnesses. Each is indeed certain of its own self, but not of the other, and therefore its own self-certainty still has no truth. For it would have truth only if its own being-for-self had confronted it as an independent object, or, what is the same thing, if the object had presented itself as this pure self-certainty. But according to the Notion of recognition this is possible only when each is for the other what the other is for it, only when each in its own self through its own action, and again through the action of the other, achieves this pure abstraction of being-for-self.

187. The presentation of itself, however, as the pure abstraction of self-consciousness consists in showing itself as the pure negation of its objective mode, or in showing that it is not attached to any specific *existence,* not to the individuality common to existence as such, that it is not attached to life. This presentation is a twofold action: action on the part of the other, and action on its own part. In so far as it is the action of the *other,* each seeks the death of the other. But in doing so, the second kind of action, action on its own part, is also involved; for the former involves the staking of its own life. Thus the relation of the two self-conscious individuals is such that they prove themselves and each other through a life-and-death struggle. They must engage in this struggle, for they must raise their certainty of being *for themselves* to truth, both in the case of the other and in their own case. And it is only through staking one's life that freedom is won; only thus is it proved that for self-consciousness, its essential being is not [just] being, not the *immediate* form in which it appears, not its submergence in the expanse of life, but rather that there is nothing present in it which could not be regarded as a vanishing moment, that it is only pure *being-for-self.* The individual who has not risked his life may well be recognized as a *person,* but he has not attained to the truth of this recognition as an independent self-consciousness. Similarly, just as each stakes his own life, so each

must seek the other's death, for it values the other no more than itself; its essential being is present to it in the form of an 'other,' it is outside of itself and must rid itself of its self-externality. The other is an *immediate* consciousness entangled in a variety of relationships, and it must regard its otherness as a pure being-for-self or as an absolute negation.

188. This trial by death, however, does away with the truth which was supposed to issue from it, and so, too, with the certainty of self generally. For just as life is the *natural* setting of consciousness, independence without absolute negativity, so death is the *natural* negation of consciousness, negation without independence, which thus remains without the required significance of recognition. Death certainly shows that each staked his life and held it of no account, both in himself and in the other; but that is not for those who survived this struggle. They put an end to their consciousness in its alien setting of natural existence, that is to say, they put an end to themselves, and are done away with as *extremes* wanting to be *for themselves*, or to have an existence of their own. But with this there vanishes from their interplay the essential moment of splitting into extremes with opposite characteristics; and the middle term collapses into a lifeless unity which is split into lifeless, merely immediate, unopposed extremes; and the two do not reciprocally give and receive one another back from each other consciously, but leave each other free only indifferently, like things. Their act is an abstract negation, not the negation coming from consciousness, which supersedes in such a way as to preserve and maintain what is superseded, and consequently survives its own supersession.

189. In this experience, self-consciousness learns that life is as essential to it as pure self-consciousness. In immediate self-consciousness the simple 'I' is absolute mediation, and has as its essential moment lasting independence. The dissolution of that simple unity is the result of the first experience; through this there is posited a pure self-consciousness, and a consciousness which is not purely for itself but for another, i.e. is a merely *immediate* consciousness, or consciousness in the form of *thinghood*. Both moments are essential. Since to begin with they are unequal and opposed, and their reflection into a unity has not yet been achieved, they exist as two opposed shapes of consciousness; one is the independent consciousness whose essential nature is to be for itself, the other is the dependent consciousness whose essential nature is simply to live or to be for another. The former is lord, the other is bondsman.

190. The lord is the consciousness that exists *for itself*, but no longer merely the Notion of such a consciousness. Rather, it is a consciousness existing *for itself* which is mediated with itself through another consciousness, i.e. through a consciousness whose nature it is to be bound up with an existence that is independent, or thinghood in general. The lord puts himself into relation with both of these moments, to a *thing* as such, the object of desire, and to the consciousness for which thinghood is the essential characteristic. And since he is (a) *qua* the Notion of self-consciousness an immediate relation of *being-for-self*, but (b) is now at the same time mediation, or a being-for-self which is for itself only through another, he is related (a) immediately to both, and (b) mediately to each through the other. The lord relates himself mediately to the bondsman through a being [a thing] that is independent, for it is just this which holds the bondsman in bondage; it is his chain from which he could not break free in the struggle, thus proving himself to be dependent, to possess his independence in thinghood. But the lord is the power over this thing, for he proved

in the struggle that it is something merely negative; since he is the power over this thing and this again is the power over the other [the bondsman], it follows that he holds the other in subjection. Equally, the lord relates himself mediately to the thing through the bondsman; the bondsman, *qua* self-consciousness in general, also relates himself negatively to the thing, and takes away its independence; but at the same time the thing is independent *vis-à-vis* the bondsman, whose negating of it, therefore, cannot go the length of being altogether done with it to the point of annihilation; in other words, he only *works* on it. For the lord, on the other hand, the *immediate* relation becomes through this mediation the sheer negation of the thing, or the enjoyment of it. What desire failed to achieve, he succeeds in doing, viz. to have done with the thing altogether, and to achieve satisfaction in the enjoyment of it. Desire failed to do this because of the thing's independence; but the lord, who has interposed the bondsman between it and himself, takes to himself only the dependent aspect of the thing and has the pure enjoyment of it. The aspect of its independence he leaves to the bondsman, who works on it.

191. In both of these moments the lord achieves his recognition through another consciousness; for in them, that other consciousness is expressly something unessential, both by its working on the thing, and by its dependence on a specific existence. In neither case can it be lord over the being of the thing and achieve absolute negation of it. Here, therefore, is present this moment of recognition, viz. that the other consciousness sets aside its own being-for-self, and in so doing itself does what the first does to it. Similarly, the other moment too is present, that this action of the second is the first's own action; for what the bondsman does is really the action of the lord. The latter's essential nature is to exist only for himself; he is the sheer negative power for whom the thing is nothing. Thus he is the pure, essential action in this relationship, while the action of the bondsman is impure and unessential. But for recognition proper the moment is lacking, that what the lord does to the other he also does to himself, and what the bondsman does to himself he should also do to the other. The outcome is a recognition that is one-sided and unequal.

192. In this recognition the unessential consciousness is for the lord the object, which constitutes the *truth* of his certainty of himself. But it is clear that this object does not correspond to its Notion, but rather that the object in which the lord has achieved his lordship has in reality turned out to be something quite different from an independent consciousness. What now really confronts him is not an independent consciousness, but a dependent one. He is, therefore, not certain of *being-for-self* as the truth of himself. On the contrary, his truth is in reality the unessential consciousness and its unessential action.

193. The *truth* of the independent consciousness is accordingly the servile consciousness of the bondsman. This, it is true, appears at first *outside* of itself and not as the truth of self-consciousness. But just as lordship showed that its essential nature is the reverse of what it wants to be, so too servitude in its consummation will really turn into the opposite of what it immediately is; as a consciousness forced back into itself, it will withdraw into itself and be transformed into a truly independent consciousness.

194. We have seen what servitude is only in relation to lordship. But it is a self-consciousness, and we have now to consider what as such it is in and for itself. To begin with, servitude has the lord for its essential reality; hence the *truth* for it is the independent consciousness that is *for itself*. However, servitude is not yet aware that

this truth is implicit in it. But it does in fact contain within itself this truth of pure negativity and being-for-self, for it has experienced this its own essential nature. For this consciousness has been fearful, not of this or that particular thing or just at odd moments, but its whole being has been seized with dread. In that experience it has been quite unmanned, has trembled in every fibre of its being, and everything solid and stable has been shaken to its foundations. But this pure universal movement, the absolute melting-away of everything stable, is the simple, essential nature of self-consciousness, absolute negativity, *pure being-for-self*, which consequently is *implicit* in this consciousness. This moment of pure being-for-self is also *explicit* for the bondsman, for in the lord it exists for him as his *object*. Furthermore, his consciousness is not this dissolution of everything stable merely in principle; in his service he *actually* brings this about. Through his service he rids himself of his attachment to natural existence in every single detail; and gets rid of it by working on it.

195. However, the feeling of absolute power both in general, and in the particular form of service, is only implicitly this dissolution, and although the fear of the lord is indeed the beginning of wisdom, consciousness is not therein aware that it is a being-for-self. Through work, however, the bondsman becomes conscious of what he truly is. In the moment which corresponds to desire in the lord's consciousness, it did seem that the aspect of unessential relation to the thing fell to the lot of the bondsman, since in that relation the thing retained its independence. Desire has reserved to itself the pure negating of the object and thereby its unalloyed feeling of self. But that is the reason why this satisfaction is itself only a fleeting one, for it lacks the side of objectivity and permanence. Work, on the other hand, is desire held in check, fleetingness staved off; in other words, work forms and shapes the thing. The negative relation to the object becomes its *form* and something *permanent*, because it is precisely for the worker that the object has independence. This *negative* middle term or the formative *activity* is at the same time the individuality or pure being-for-self of consciousness which now, in the work outside of it, acquires an element of permanence. It is in this way, therefore, that consciousness, *qua* worker, comes to see in the independent being [of the object] its *own* independence.

196. But the formative activity has not only this positive significance that in it the pure being-for-self of the servile consciousness acquires an existence; it also has, in contrast with its first moment, the negative significance of *fear*. For, in fashioning the thing, the bondsman's own negativity, his being-for-self, becomes an object for him only through his setting at nought the existing *shape* confronting him. But this objective *negative* moment is none other than the alien being before which it has trembled. Now, however, he destroys this alien negative moment, posits *himself* as a negative in the permanent order of things, and thereby becomes *for himself*, someone existing on his own account. In the lord, the being-for-self is an 'other' for the bondsman, or is only *for* him [i.e. is not his own]; in fear, the being-for-self is present in the bondsman himself; in fashioning the thing, he becomes aware that being-for-self belongs to *him*, that he himself exists essentially and actually in his own right. The shape does not become something other than himself through being made external to him; for it is precisely this shape that is his pure being-for-self, which in this externality is seen by him to be the truth. Through this rediscovery of himself by himself, the bondsman realizes that it is precisely in his work wherein he seemed to have only an alienated existence that he acquires a mind of his own. For this reflection, the two moments of fear and service as such, as also that of formative

activity, are necessary, both being at the same time in a universal mode. Without the discipline of service and obedience, fear remains at the formal stage, and does not extend to the known real world of existence. Without the formative activity, fear remains inward and mute, and consciousness does not become explicitly *for itself*. If consciousness fashions the thing without that initial absolute fear, it is only an empty self-centred attitude; for its form or negativity is not negativity *per se*, and therefore its formative activity cannot give it a consciousness of itself as essential being. If it has not experienced absolute fear but only some lesser dread, the negative being has remained for it something external, its substance has not been infected by it through and through. Since the entire contents of its natural consciousness have not been jeopardized, determinate being still *in principle* attaches to it; having a 'mind of one's own' is self-will, a freedom which is still enmeshed in servitude. Just as little as the pure form can become essential being for it, just as little is that form, regarded as extended to the particular, a universal formative activity, an absolute Notion; rather it is a skill which is master over some things, but not over the universal power and the whole of objective being.

John Rawls, from *A Theory of Justice* (1921-)

John Rawls rejuvenated the metaphor of the state of nature as well as the theory of the social contract in his *Theory of Justice*, published to great acclaim (and with much anticipation) in 1971. Rawls's version of the state of nature, however, has nothing to do with men and women tripping around in primeval forests, whether indifferent or hostile to one another. Rawls's "original position" is rather a problem in rational decision-making, and the question is, simply put, what sort of a world would we choose to live in if we had no idea—before we actually found ourselves in it—what positions in that world we would occupy? One might imagine, as Bruce Ackerman does in a book following Rawls's pursuit of a liberal theory, a group of interplanetary travelers on a spaceship, trying to design a society which they—or their offspring—would then have to live in. Or, one might imagine, in the simplest sort of analogy, a group of children making up a game, in which different positions would no doubt have very different advantages and disadvantages but, before the game actually begins, no one knows in exactly what position they might find themselves. So, in Rawls's ingenious theory, we imagine ourselves behind a "veil of ignorance," where we do not know whether we are born rich or poor, healthy or chronically ill, white, brown, red, or black. We do not know who our parents are. We do not know what our sex is. We do not know our own children, and so on. What we do know is a good deal about moral psychology and what the consequences would be given any particular social arrangement. For instance, we know that if some people are given great wealth and power and uncontested, continuing opportunities for more, while others are born in poverty and are powerless, the rich will get richer and the poor will get poorer, even allowing for a certain optimism about charity and compassion. What we have to do is to choose, on the basis of the various possibilities, the ideally rational arrangement. What if I should be one of the poor as opposed to one of the rich? Thus the original position is the imaginative stage on which we design the terms of the social contract.

———

The circumstances of justice may be described as the normal conditions under which human cooperation is both possible and necessary. Thus, as I noted at the outset, although a society is a cooperative venture for mutual advantage, it is typically marked by a conflict as well as an identity of interests. There is an identity of interests since social cooperation makes possible a better life for all than any would have if each were to try to live solely by his own efforts. There is a conflict of interests since men are not indifferent as to how the greater benefits produced by their collaboration are distributed, for in order to pursue their ends they each prefer a larger to a lesser share. Thus principles are needed for choosing among the various social arrangements which determine this division of advantages and for underwriting an agreement on the proper distributive shares. These requirements define the role of justice. The background conditions that give rise to these necessities are the circumstances of justice.

These conditions may be divided into two kinds. First, there are the objective circumstances which make human cooperation both possible and necessary. Thus, many individuals coexist together at the same time on a definite geographical territory. These individuals are roughly similar in physical and mental powers; or at any rate, their capacities are comparable in that no one among them can dominate the rest. They are vulnerable to attack, and all are subject to having their plans blocked by the united force of others. Finally, there is the condition of moderate scarcity understood to cover a wide range of situations. Natural and other resources are not so abundant that schemes of cooperation become superfluous, nor are conditions so harsh that fruitful ventures must inevitably break down. While mutually advantageous arrangements are feasible, the benefits they yield fall short of the demands men put forward.

The subjective circumstances are the relevant aspects of the subjects of cooperation, that is, of the persons working together. Thus while the parties have roughly similar needs and interests, or needs and interests in various ways complementary, so that mutually advantageous cooperation among them is possible, they nevertheless have their own plans of life. These plans, or conceptions of the good, lead them to have different ends and purposes, and to make conflicting claims on the natural and social resources available. Moreover, although the interests advanced by these plans are not assumed to be interests in the self, they are the interests of a self that regards its conception of the good as worthy of recognition and that advances claims in its behalf as deserving satisfaction. I shall emphasize this aspect of the circumstances of justice by assuming that the parties take no interest in one another's interests. I also suppose that men suffer from various shortcomings of knowledge, thought, and judgment. Their knowledge is necessarily incomplete, their powers of reasoning, memory, and attention are always limited, and their judgment is likely to be distorted by anxiety, bias, and a preoccupation with their own affairs. Some of these defects spring from moral faults, from selfishness and negligence; but to a large degree, they are simply part of men's natural situation. As a consequence individuals not only have different plans of life but there exists a diversity of philosophical and religious belief, and of political and social doctrines.

Now this constellation of conditions I shall refer to as the circumstances of justice. Hume's account of them is especially perspicuous and the preceding summary adds nothing essential to his much fuller discussion. For simplicity I often stress the condition of moderate scarcity (among the objective circumstances), and that of mutual disinterest, or individuals taking no interest in one another's interests (among the subjective circumstances). Thus, one can say, in brief, that the circumstances of justice obtain whenever mutually disinterested persons put forward conflicting claims to the division of social advantages under conditions of moderate scarcity. Unless these circumstances existed there would be no occasion for the virtue of justice, just as in the absence of threats of injury to life and limb there would be no occasion for physical courage.

Several clarifications should be noted. First of all, I shall, of course, assume that the persons in the original position know that these circumstances of justice obtain. This much they take for granted about the conditions of their society. A further assumption is that the parties try to advance their conception of the good as best they can, and that in attempting to do this they are not bound by prior moral ties to each other.

* * *

The situation of the persons in the original position reflects certain constraints. The alternatives open to them and their knowledge of their circumstances are limited in various ways. These restrictions I refer to as the constraints of the concept of right since they hold for the choice of all ethical principles and not only for those of justice. If the parties were to acknowledge principles for the other virtues as well, these constraints would also apply.

I shall consider first the constraints on the alternatives. There are certain formal conditions that it seems reasonable to impose on the conceptions of justice that are to be allowed on the list presented to the parties. I do not claim that these conditions follow from the concept of right, much less from the meaning of morality. I avoid an appeal to the analysis of concepts at crucial points of this kind. There are many constraints that can reasonably be associated with the concept of right, and different selections can be made from these and counted as definitive within a particular theory. The merit of any definition depends upon the soundness of the theory that results; by itself, a definition cannot settle any fundamental question.

The propriety of these formal conditions is derived from the task of principles of right in adjusting the claims that persons make on their institutions and one another. If the principles of justice are to play their role, that of assigning basic rights and duties and determining the division of advantages, these requirements are natural enough. Each of them is suitably weak and I assume that they are satisfied by the traditional conceptions of justice. These conditions do, however, exclude the various forms of egoism, as I note below, which shows that they are not without moral force. This makes it all the more necessary that the conditions not be justified by definition or the analysis of concepts, but only by the reasonableness of the theory of which they are a part. I arrange them under five familiar headings.

First of all, principles should be general. That is, it must be possible to formulate them without the use of what would be intuitively recognized as proper names, or rigged definite descriptions. Thus the predicates used in their statement should express general properties and relations. Unfortunately deep philosophical difficulties seem to bar the way to a satisfactory account of these matters. I shall not try to deal with them here. In presenting a theory of justice one is entitled to avoid the problem of defining general properties and relations and to be guided by what seems reasonable. Further, since the parties have no specific information about themselves or their situation, they cannot identify themselves anyway. Even if a person could get others to agree, he does not know how to tailor principles to his advantage. The parties are effectively forced to stick to general principles, understanding the notion here in an intuitive fashion.

The naturalness of this condition lies in part in the fact that first principles must be capable of serving as a public charter of a well-ordered society in perpetuity. Being unconditional, they always hold (under the circumstances of justice), and the knowledge of them must be open to individuals in any generation. Thus, to understand these principles should not require a knowledge of contingent particulars, and surely not a reference to individuals or associations. Traditionally the most obvious test of this condition is the idea that what is right is that which accords with God's will. But in fact this doctrine is normally supported by an argument from general principles. For example, Locke held that the fundamental principle of morals is the following: if one person is created by another (in the theological sense), then that person has a duty to comply with the precepts set to him by his creator. This principle is perfectly general and given the nature of the world on Locke's view, it singles out God as the legitimate moral authority. The generality condition is not violated, although it may appear so at first sight.

Next, principles are to be universal in application. They must hold for everyone in virtue of their being moral persons. Thus I assume that each can understand these principles and use them in his deliberations. This imposes an upper bound of sorts on how complex they can be, and on the kinds and number of distinctions they draw. Moreover, a principle is ruled out if it would be self-contradictory, or self-defeating, for everyone to act upon it. Similarly, should a principle be reasonable to follow only when others conform to a different one, it is also inadmissible. Principles are to be chosen in view of the consequences of everyone's complying with them.

* * *

A third condition is that of publicity, which arises naturally from a contractarian standpoint. The parties assume that they are choosing principles for a public conception of justice. They suppose that everyone will know about these principles all that he would know if their acceptance were the result of an agreement. Thus the general awareness of their universal acceptance should have desirable effects and support the stability of social cooperation. The difference between this condition and that of universality is that the latter leads one to assess principles on the basis of their being intelligently and regularly followed by everyone. But it is possible that all should understand and follow a principle and yet this fact not be widely known or explicitly recognized. The point of the publicity condition is to have the parties evaluate conceptions of justice as publicly acknowledged and fully effective moral constitutions of social life. The publicity condition is clearly implicit in Kant's doctrine of the categorical imperative insofar as it requires us to act in accordance with principles that one would be willing as a rational being to enact as law for a kingdom of ends. He thought of this kingdom as an ethical commonwealth, as it were, which has such moral principles for its public charter.

A further condition is that a conception of right must impose an ordering on conflicting claims. This requirement springs directly from the role of its principles in adjusting competing demands. There is a difficulty, however, in deciding what counts as an ordering. It is clearly desirable that a conception of justice be complete, that is, able to order all the claims that can arise (or that are likely to in practice). And the ordering should in general be transitive; if, say, a first arrangement of the basic structure is ranked more just than a second, and the second more just than a third, then the first should be more just than the third. These formal conditions are natural enough, though not always easy to satisfy. But is trial by combat a form of adjudication? After all, physical conflict and resort to arms result in an ordering; certain claims do win out over others. The main objection to this ordering is not that it may be intransitive. Rather, it is to avoid the appeal to force and cunning that the principles of right and justice are accepted. Thus I assume that to each according to his threat advantage is not a conception of justice. It fails to establish an ordering in the required sense, an ordering based on certain relevant aspects of persons and their situation which are independent from their social position, or their capacity to intimidate and coerce.

The fifth and last condition is that of finality. The parties are to assess the system of principles as the final court of appeal in practical reasoning. There are no higher standards to which arguments in support of claims can be addressed; reasoning successfully from these principles is conclusive. If we think in terms of the fully general theory which has principles for all the virtues, then such a theory specifies the totality of relevant considerations and their appropriate weights, and its requirements are decisive. They override the demands of law and custom, and of social rules generally. We are to arrange and respect social institutions as the principles of right and justice direct. Conclusions from these principles also override considerations of prudence and self-interest. This does not mean that these principles insist upon self-sacrifice; for in drawing up the conception of right the parties take their interests into account as best they can. The claims of personal prudence are already given an appropriate weight within the full system of principles. The complete scheme is final in that when the course of practical reasoning it defines has reached its conclusion, the question is settled. The claims of existing social arrangements and of self-interest have been duly allowed for. We cannot at the end count them a second time because we do not like the result.

Taken together, then, these conditions on conceptions of right come to this: a conception of right is a set of principles, general in form and universal in application, that is to be publicly recognized as a final court of appeal for ordering the conflicting claims of moral persons.

* * *

The idea of the original position is to set up a fair procedure so that any principles agreed to will be just. The aim is to use the notion of pure procedural justice as a basis of theory. Somehow we must nullify the effects of specific contingencies which put men at odds and tempt them to exploit social and natural circumstances to their own advantage. Now in order to do this I assume that the parties are situated behind a veil of ignorance. They do not know how the various alternatives will affect their own particular case and they are obliged to evaluate principles solely on the basis of general considerations.

It is assumed, then, that the parties do not know certain kinds of particular facts. First of all, no one knows his place in society, his class position or social status; nor does he know his fortune in the distribution of natural assets and abilities, his intelligence and strength, and the like. Nor, again, does anyone know his conception of the good, the particulars of his rational plan of life, or even the special features of his psychology such as his aversion to risk or liability to optimism or pessimism. More than this, I assume that the parties do not know the particular circumstances of their own society. That is, they do not know its economic or political situation, or the level of civilization and culture it has been able to achieve. The persons in the original position have no information as to which generation they belong. These broader restrictions on knowledge are appropriate in part because questions of social justice arise between generations as well as within them, for example, the question of the appropriate rate of capital saving and of the conservation of natural resources and the environment of nature. There is also, theoretically anyway, the question of a reasonable genetic policy. In these cases too, in order to carry through the idea of the original position, the parties must not know the contingencies that set them in opposition. They must choose principles the consequences of which they are prepared to live with whatever generation they turn out to belong to.

As far as possible, then, the only particular facts which the parties know is that their society is subject to the circumstances of justice and whatever this implies. It is taken for granted, however, that they know the general facts about human society. They understand political affairs and the principles of economic theory; they know the basis of social organization and the laws of human psychology. Indeed, the parties are presumed to know whatever general facts affect the choice of the principles of justice. There are no limitations on general information, that is, on general laws and theories, since conceptions of justice must be adjusted to the characteristics of the systems of social cooperation which they are to regulate, and there is no reason to rule out these facts. It is, for example, a consideration against a conception of justice that, in view of the laws of moral psychology, men would not acquire a desire to act upon it even when the institutions of their society satisfied it. For in this case there would be difficulty in securing the stability of social cooperation. It is an important feature of a conception of justice that it should generate its own support. That is, its principles should be such that when they are embodied in the basic structure of society men tend to acquire the corresponding sense of justice. Given the principles of moral learning, men develop a desire to act in accordance with its principles. In this case a conception of justice is stable. This kind of general information is admissible in the original position.

The notion of the veil of ignorance raises several difficulties. Some may object that the exclusion of nearly all particular information makes it difficult to grasp what is meant by the original position. Thus it may be helpful to observe that one or more persons can at any time enter this position, or perhaps, better, simulate the deliberations of this hypothetical situation, simply by reasoning in accordance with the appropriate restrictions.

* * *

These remarks show that the original position is not to be thought of as a general assembly which includes at one moment everyone who will live at some time; or, much less, as an assembly of everyone who could live at some time. It is not a gathering of all actual or possible persons. To conceive of the original position in either of these ways is to stretch fantasy too far; the conception would cease to be a natural guide to intuition. In any case, it is important that the original position be interpreted so that one can at any time adopt its perspective. It must make no difference when one takes up this viewpoint, or who does so: the restrictions must be such that the same principles are always chosen. The veil of ignorance is a key condition in meeting this requirement. It insures not only that the information available is relevant, but that it is at all times the same.

* * *

I have assumed throughout that the persons in the original position are rational. In choosing between principles each tries as best he can to advance his interests. But I have also assumed that the parties do not know their conception of the good. This means that while they know that they have some rational plan of life, they do not know the details of this plan, the particular ends and interests which it is calculated to promote. How, then, can they decide which conceptions of justice are most to their advantage? Or must we suppose that they are reduced to mere guessing? To meet this difficulty, I postulate that they accept the account of the good touched upon in the preceding chapter: they assume that they would prefer more primary social goods rather than less. Of course, it may turn out, once the veil of ignorance is removed, that some of them for religious or other reasons may not, in fact, want more of these goods.

But from the standpoint of the original position, it is rational for the parties to suppose that they do want a larger share, since in any case they are not compelled to accept more if they do not wish to, nor does a person suffer from a greater liberty. Thus even though the parties are deprived of information about their particular ends, they have enough knowledge to rank the alternatives. They know that in general they must try to protect their liberties, widen their opportunities, and enlarge their means for promoting their aims whatever these are. Guided by the theory of the good and the general facts of moral psychology, their deliberations are no longer guesswork. They can make a rational decision in the ordinary sense.

The concept of rationality invoked here, with the exception of one essential feature, is the standard one familiar in social theory. Thus in the usual way, a rational person is thought to have a coherent set of preferences between the options open to him. He ranks these options according to how well they further his purposes; he follows the plan which will satisfy more of his desires rather than less, and which has the greater chance of being successfully executed. The special assumption I make is that a rational individual does not suffer from envy. He is not ready to accept a loss for himself if only others have less as well. He is not downcast by the knowledge or perception that others have a larger index of primary social goods. Or at least this is true as long as the differences between himself and others do not exceed certain limits, and he does not believe that the existing inequalities are founded on injustice or are the result of letting chance work itself out for no compensating social purpose.

Robert Nozick, from *Anarchy, State and Utopia* (1938–)

Nozick's *Anarchy, State and Utopia* was published only 3 years after Rawls's *Theory of Justice*, and in addition to the notoriety it achieved in its own right, it formed an obvious pole of opposition and a staging area for attacks on Rawls and the welfare liberal tradition. Nozick also goes back to the state of nature imagery, but for him, too, this is more than anything else a thought-experiment in which to establish the various rational possibilities for the construction or and justification of society. Or, more properly, it is a question of the justification for the *state*, not society as such, for Nozick is wisely abstemious on the question of the natural state of society and human communities. What justifies the existence of state power, and how much state power is justified? To answer these questions, Nozick raises some neo-Lockean issues about the protection of property and whether we could assure ourselves of mutual security without the imposition of a state. He considers the possibility of various "protective associations" and whether they might do the job, and argues that, of necessity but not by design, a state—if only a *minimal* state—would be the inevitable product of these various efforts.

Individuals in Locke's state of nature are in "a state of perfect freedom to order their actions and dispose of their possessions and persons as they think fit, within the bounds of the law of nature, without asking leave or dependency upon the will of any other man" (sect. 4). The bounds of the law of nature require that "no one ought to harm another in his life, health, liberty, or possessions" (sect. 6). Some persons transgress these bounds, "invading others' rights and . . . doing hurt to one another," and in response people may defend themselves or others against such invaders of rights (chap. 3). The injured party and his agents may recover from the offender "so much as may make satisfaction for the harm he has suffered" (sect. 10); "everyone has a right to punish the transgressors of that law to such a degree as may hinder its violation" (sect. 7); each person may, and may only "retribute to [a criminal] so far as calm reason and conscience dictate, what is proportionate to his transgression, which is so much as may serve for reparation and restraint" (sect. 8).

There are "inconveniences of the state of nature" for which, says Locke, "I easily grant that civil government is the proper remedy" (sect. 13). To understand precisely what civil government remedies, we must do more than repeat Locke's list of the inconveniences of the state of nature. We also must consider what arrangements might be made within a state of nature to deal with these inconveniences—to avoid them or to make them less likely to arise or to make them less serious on the occasions when they do arise. Only after the full resources of the state of nature are brought into play, namely all those voluntary arrangements and agreements persons might reach acting within their rights, and only after the effects of these are estimated, will we be in a position to see how serious are the inconveniences that yet remain to be remedied by the state, and to estimate whether the remedy is worse than the disease.

In a state of nature, the understood natural law may not provide for every contingency in a proper fashion, and men who judge in their own case will always give themselves the benefit of the doubt and assume that they are in the right. They will overestimate the amount of harm or damage they have suffered, and passions will lead them to attempt to punish others more than proportionately and to exact excessive compensation. Thus private and personal enforcement of one's rights (including those rights that are violated when one is excessively punished) leads to feuds, to an endless series of acts of retaliation and exactions of compensation. And there is no firm way to *settle* such a dispute, to *end* it and to have both parties know it is ended. Even if one party *says* he'll stop his acts of retaliation, the other can rest secure only if he knows the first still does not feel entitled to gain recompense or to exact retribution, and therefore entitled to try when a promising occasion presents itself. Any method a single individual might use in an attempt irrevocably to bind himself into ending his part in a feud would offer insufficient assurance to the other party; tacit agreements to stop also would be unstable. Such feelings of being mutually wronged can occur even with the clearest right and with joint agreement on the facts of each person's conduct; all the more is there opportunity for such retaliatory battle when the facts or the rights are to some extent unclear. Also, in a state of nature a person may lack the power to enforce his rights; he may be unable to punish or exact compensation from a stronger adversary who has violated them (sects. 123, 126).

Protective Associations

How might one deal with these troubles within a state of nature? Let us begin with the last. In a state of nature an individual may himself enforce his rights, defend himself, exact compensation, and punish (or at least try his best to do so). Others may join with him in his defense, at his call. They may join with him to repulse an attacker or to go after an aggressor because they are public spirited, or because they are his friends, or because he has helped them in the past, or because they wish him to help them in the future, or in exchange for something. Groups of individuals may form mutual-protection associations: all will answer the call of any member for defense or for the enforcement of his rights. In union there is strength.

* * *

A mutual-protection association might attempt to deal with conflict among its own members by a policy of nonintervention. But this policy would bring discord within the association and might lead to the formation of subgroups who might fight among themselves and thus cause the breakup of the association. This policy would also encourage potential aggressors to join as many mutual-protection associations as possible in order to gain immunity from retaliatory or defensive action, thus placing a great burden on the adequacy of the initial screening procedure of the association. Thus protective associations (almost all of those that will survive which people will join) will not follow a policy of nonintervention; they will use some procedure to determine how to act when some members claim that other members have violated their rights. Many arbitrary procedures can be imagined (for example, act on the side of that member who complains first), but most persons will want to join associations that follow some procedure to find out which claimant is correct. When a member of the association is in conflict with nonmembers, the association also will want to determine in some fashion who is in the right, if only to avoid constant and costly involvement in each member's quarrels, whether just or unjust. The inconvenience of everyone's being on call, whatever their activity at the moment or inclinations or comparative advantage, can be handled in the usual manner by division of labor and exchange. Some people will be *hired* to perform protective functions, and some entrepreneurs will go into the business of selling protective services. Different sorts of protective policies would be offered, at different prices, for those who may desire more extensive or elaborate protection.

An individual might make more particular arrangements or commitments short of turning over to a private protective agency all functions of detection, apprehension, judicial determination of guilt, punishment, and exaction of compensation. Mindful of the dangers of being the judge in his own case, he might turn the decision as to whether he has indeed been wronged, and to what extent, to some other neutral or less involved party. In order for the occurrence of the social effect of justice's being seen to be done, such a party would have to be generally respected and thought to be neutral and upright. Both parties to a dispute may so attempt to safeguard themselves against the appearance of partiality, and both might even agree upon the *same* person as the judge between them, and agree to abide by his decision. (Or there might be a specified process through which one of the parties dissatisfied with the decision could appeal it.) But, for obvious reasons, there will be strong tendencies for the above-mentioned functions to converge in the same agent or agency.

Presumably what drives people to use the state's system of justice is the issue of ultimate enforcement. Only the state can enforce a judgment against the will of one of the parties. For the state does not *allow* anyone else to enforce another system's judgment. So in any dispute in which both parties cannot agree upon a method of settlement, or in any dispute in which one party does not trust another to abide by the decision (if the other contracts to forfeit something of enormous value if he doesn't abide by the decision, by what agency is *that* contract to be enforced?), the parties who wish their claims put into effect will have no recourse permitted by the state's legal system other than to use that very legal system. This may present persons greatly opposed to a given state system with particularly poignant and painful choices. (If the state's legal system enforces the results of certain arbitration procedures, people may come to agree—supposing they abide by this agreement—without any actual direct contact with what they perceive to be officers or institutions of the state. But this holds as well if they sign a contract that is enforced only by the state.)

The Dominant Protective Association

Initially, several different protective associations or companies will offer their services in the same geographical area. What will occur when there is a conflict between clients of different agencies? Things are relatively simple if the agencies reach the same decision about the disposition of the case. (Though each might want to exact the penalty.) But what happens if they reach different decisions as to the merits of the case, and one agency attempts to protect its client while the other is attempting to punish him or make him pay compensation? Only three possibilities are worth considering:

> 1. In such situations the forces of the two agencies do battle. One of the agencies always wins such battles. Since the clients of the losing agency are ill protected in conflicts with clients of the winning agency, they leave their agency to do business with the winner.
> 2. One agency has its power centered in one geographical area, the other in another. Each wins the battles fought close to its center of power, with some gradient being established. People who deal with one agency but live under the power of the other either move closer to their own agency's home headquarters or shift their patronage to the other protective agency. (The border is about as conflictful as one between states.)

In neither of these two cases does there remain very much geographical interspersal. Only one protective agency operates over a given geographical area.

> 3. The two agencies fight evenly and often. They win and lose about equally, and their interspersed members have frequent dealings and disputes with each other. Or perhaps without fighting or after only a few skirmishes the agencies realize that such battling will occur continually in the absence of preventive measures. In any case, to avoid frequent, costly, and wasteful battles the two agencies, perhaps through their executives, agree to resolve peacefully those cases about which they reach differing judgments. They agree to set up, and abide by the decisions of, some third judge or court to which they can turn when their respective judgments differ. (Or they might establish rules determining which agency has jurisdiction under which circumstances.) Thus emerges a system of appeals courts and agreed upon rules about jurisdiction and the conflict of laws. Though different agencies operate, there is one unified federal judicial system of which they all are components.

In each of these cases, almost all the persons in a geographical area are under some common system that judges between their competing claims and *enforces* their rights. Out of anarchy, pressed by spontaneous groupings, mutual-protection associations, division of labor, market pressures, economies of scale, and rational self-interest there arises something very much resembling a minimal state or a group of geographically distinct minimal states. Why is this market different from all other markets? Why would a virtual monopoly arise in this market without the government intervention that elsewhere creates and maintains it? The worth of the product purchased, protection against others, is *relative:* it depends upon how strong the others are. Yet unlike other goods that are comparatively evaluated, maximal competing protective services cannot coexist; the nature of the service brings different agencies not only into competition for customers' patronage, but also into violent conflict with each other. Also, since the worth of the less than maximal product declines disproportionately with the number who purchase the maximal product, customers will not stably settle for the lesser good, and competing companies are caught in a declining spiral. Hence the three possibilities we have listed.

Our story above assumes that each of the agencies attempts in good faith to act within the limits of Locke's law of nature. But one "protective association" might aggress against other persons. Relative to Locke's law of nature, it would be an outlaw agency. What actual counterweights would there be to its power? (What actual counterweights are there to the power of a state?) Other agencies might unite to act against it. People might refuse to deal with the outlaw agency's clients, boycotting them to reduce the probability of the agency's intervening in their own affairs. This might make it more difficult for the outlaw agency to get clients; but this boycott will seem an effective tool only on very optimistic assumptions about what cannot be kept secret, and about the costs to an individual of partial boycott as compared to the benefits of receiving the more extensive coverage offered by an "outlaw" agency. If the "outlaw" agency simply is an *open* aggressor, pillaging, plundering, and extorting under no plausible claim of justice, it will have a harder time than states. For the state's claim to legitimacy induces its citizens to believe they have some duty to obey its edicts, pay its taxes, fight its battles, and so on; and so some persons cooperate with it voluntarily. An openly aggressive agency could not depend upon, and would not receive, any such voluntary cooperation, since persons would view themselves simply as its victims rather than as its citizens.

Invisible-Hand Explanations

How, if at all, does a dominant protective association differ from the state? Was Locke wrong in imagining a compact necessary to establish civil society? As he was wrong in thinking (sects. 46, 47, 50) that an "agreement," or "mutual consent," was needed to establish the "invention of money." Within a barter system, there is great inconvenience and cost to searching for someone who has what you want and wants what you have, even at a marketplace, which, we should note, needn't become a marketplace by everyone's expressly agreeing to deal there. People will exchange their goods for something they know to be more generally wanted than what they have. For it will be more likely that they can exchange this for what they want. For the same reasons others will be more willing to take in exchange this more generally desired thing. Thus persons will converge in exchanges on the more marketable goods, being willing to exchange their goods for them; the more willing, the more they know others who are also willing to do so, in a mutually reinforcing process. (This process will be reinforced and hastened by middlemen seeking to profit in facilitating exchanges, who themselves will often find it most expedient to offer more marketable goods in exchange.) For obvious reasons, the goods they converge on, via their individual decisions, will have certain properties: initial independent value (else they wouldn't begin as more marketable), physically enduring, non-perishable, divisible, portable, and so forth. No express agreement and no social contract fixing a medium of exchange is necessary.

Is the Dominant Protective Association a State?

Have we provided an invisible-hand explanation of the state? There are at least two ways in which the scheme of private protective associations might be thought to differ from a minimal state, might fail to satisfy a minimal conception of a state: (1) it appears to allow some people to enforce their own rights, and (2) it appears not to protect all individuals within its domain.

* * *

A state claims a monopoly on deciding who may use force when; it says that only it may decide who may use force and under what conditions; it reserves to itself the sole right to pass on the legitimacy and permissibility of any use of force within its boundaries; furthermore it claims the right to punish all those who violate its claimed monopoly. The monopoly may be violated in two ways: (1) a person may use force though unauthorized by the state to do so, or (2) though not themselves using force a group or person may set themselves up as an alternative authority (and perhaps even claim to be the sole legitimate one) to decide when and by whom the use of force is proper and legitimate.

* * *

We may proceed, for our purposes, by saying that a necessary condition for the existence of a state is that it (some person or organization) announce that, to the best of its ability (taking into account costs of doing so, the feasibility, the more important alternative things it should be doing, and so forth), it will punish everyone whom it discovers to have used force without its express permission. (This permission may be a particular permission or may be granted via some general regulation or authorization.) This still won't quite do: the state may reserve the right to forgive someone, *ex post facto;* in order to punish they may have not only to discover the "unauthorized" use of force but also prove via a certain specified procedure of proof that it occurred, and so forth. But it enables us to proceed. The protective agencies, it seems, do not make such an announcement, either individually or collectively. *Nor does it seem morally legitimate for them to do so.* So the system of private protective associations, if they perform no morally illegitimate action, appears to lack any monopoly element and so appears not to constitute or contain a state. To examine the question of the monopoly element, we shall have to consider the situation of some group of persons (or some one person) living within a system of private protective agencies who refuse to join any protective society; who insist on judging for themselves whether their rights have been violated, and (if they so judge) on personally enforcing their rights by punishing and/or exacting compensation from those who infringed them.

The second reason for thinking the system described is not a state is that, under it (apart from spillover effects) only those paying for protection get protected; furthermore, differing degrees of protection may be purchased. External economies again to the side, no one pays for the protection of others except as they choose to; no one is required to purchase or contribute to the purchasing of protection for others. Protection and enforcement of people's rights is treated as an economic good to be provided by the market, as are other important goods such as food and clothing. However, under the usual conception of a state, each person living within (or even sometimes traveling outside) its geographical boundaries gets (or at least, is entitled to get) its protection. Unless some private party donated sufficient funds to cover the costs of such protection (to pay for detectives, police to bring criminals into custody, courts, and prisons), or unless the state found some service it could charge for that would cover these costs, one would expect that a state which offered protection so broadly would be redistributive. It would be a state in which some persons paid more so that others could be protected. And indeed the most minimal state seriously discussed by the mainstream of political theorists, the night-watchman state of classical liberal theory, appears to be redistributive in this fashion. Yet how can a protection agency, a business, charge some to provide its product to others? (We ignore things like some partially paying for others because it is too costly for the agency to refine its classification of, and charges to, customers to mirror the costs of the services to them.)

Thus it appears that the dominant protective agency in a territory not only lacks the requisite monopoly over the use of force, but also fails to provide protection for all in its territory; and so the dominant agency appears to fall short of being a state. But these appearances are deceptive.

Behavior in the Process

We have argued that even someone who foresees that a protective association will become dominant may not forbid others to join up. But though no one may be forbidden to join up, might not everyone *choose* to stay out, in order to avoid the state at the end of the process? Might not a population of anarchists realize how individual efforts at hiring protection will lead, by an invisible-hand process, to a state, and because they have historical evidence and theoretical grounds for the worry that the state is a Frankenstein monster that will run amuck and will not stay limited to minimal functions, might not they each prudentially choose not to begin along that path? If told to anarchists, is the invisible-hand account of how the state arises a self-defeating prophecy?

* * *

We have described a process whereby individuals in an area separately sign up for personal protection with different business enterprises which provide protective services, all but one of the agencies being extinguished or all coming to some *modus vivendi*, and so on. To what degree, if any, does this process fit what Locke envisioned as individuals "agreeing with other men to join and unite into a community," consenting "to make one community or government" (sect. 95), compacting to make up a commonwealth (sect. 99)? The process looks nothing like unanimous joint agreement to create a government or state. No one, as they buy protective services from their local protective agency, has in mind anything so grand. But perhaps joint agreement where each has in mind that the others will agree and each intends to bring about the end result of this is not necessary for a Lockean compact. I myself see little point to stretching the notion of "compact" so that each pattern or state of affairs that arises from the disparate voluntary actions of separately acting individuals is viewed as arising from a *social compact*, even though no one had the pattern in mind or was acting to achieve it. Or, if the notion is so stretched, this should be made clear that the notion is such that each of the following arises from a social compact: the total state of affairs constituted by who is married to, or living with, whom; the distribution on a given evening in a given city of who is in what movie theater, sitting where; the particular traffic pattern on a state's highways on a given day; the set of customers of a given grocery store on a given day and the particular pattern of purchase they make, and so on. Far be it from me to claim that this wider notion is of no interest; that a state can arise by a process that fits this wider notion (without fitting the narrower one) is of very great interest indeed!

The view we present here should not be confused with other views. It differs from social compact views in its invisible-hand structure. It differs from views that "*de facto* might makes state (legal) right" in holding that enforcement rights and rights to oversee this enforcement exist independently and are held by all rather than confined to one or a small group, and that the process of accumulating sole effective enforcement and overseeing power may take place without anyone's rights being violated; that a state may arise by a process in which no one's rights are violated. Shall we say that a state which has arisen from a state of nature by the process described has replaced the state of nature which therefore no longer exists, or shall we say that it exists within a state of nature and hence is compatible with one? No doubt, the first would better fit the Lockean tradition; but the state arises so gradually and imperceptibly out of Locke's state of nature, without any great or fundamental breach of continuity, that one is *tempted* to take the second option, disregarding Locke's incredulousness: ". . . unless any one will say the state of nature and civil society are one and the same thing, which I have never yet found any one so great a patron of anarchy as to affirm" (sect. 94).

Michael Sandel, from *Liberalism and the Limits of Justice* (1953–)

What Rawls and Nozick share in common with virtually every one of the modern authors included before them (notably excepting Hegel) is a conception of the individual human being in some sense prior to society and the state. Thus the question becomes, how and why do people get together to form society and a government? Hegel argued that this was nonsense. People belong to a community and grow up in states before they become individuals, and this is not only an obvious factual claim but also an important conceptual truth. It is unintelligible, Sandel suggests, to talk any other way. But if this is so, then the very idea of a state of nature or an original position and the concept of a social contract makes no sense. Communities are prior to individuals, and it is the fallacy of what Sandel refers to as the "deonotological position"—in particular the moral philosophy of Immanuel Kant and the social philosophy of John Rawls—to presuppose the very opposite.

What Really Goes on Behind the Veil of Ignorance

What goes on in the original position is first of all a choice, or more precisely, a choosing together, an agreement among parties. What the parties agree to are the principles of justice. Unlike most actual contracts, which cannot justify, the hypothetical contract the parties agree to does justify; the principles they choose are just in virtue of their choosing them. As the voluntarist account of justification would suggest, the principles of justice are the products of choice.

> The guiding idea is that the principles of justice for the basic structure of society are the object of the original *agreement* [emphasis added] (all references to Rawls, *A Theory of Justice*)

> Thus we are to imagine that those who engage in social cooperation *choose together, in one joint act*, the principles which are to assign basic rights and duties and to determine the division of social benefits. Men are to *decide* in advance how they are to regulate their claims against one another [emphasis added] (11).

> Just as each person must *decide* by rational reflection what constitutes his good, that is, the system of ends which it is rational for him to pursue, so *a group of persons must decide* once and for all what is to count among them as just and unjust. The *choice* which rational men would make in this hypothetical situation of equal liberty, assuming for the present that this *choice* problem has a solution, determines the principles of justice [emphasis added] (11–12).

Since all are similarly situated and no one is able to design principles to favor his particular condition, the principles of justice are the result of a fair *agreement or bargain* [emphasis added] (12).

Justice as fairness begins, as I have said, with one of the most general of all *choices* which persons might make together, namely with the *choice* of the first principles of a conception of justice [emphasis added] (13).

The principles of justice are those which would be *chosen* in the original position. They are the outcome of a certain *choice* situation [emphasis added] (41–2).

Justice as fairness the principles of justice are not thought of as self-evident, but *have their justification in the fact that they would be chosen* [emphasis added] (42).

On a contract doctrine the moral facts are determined by the principles which would be *chosen* in the original position. . . . *[I]t is up to the persons* in the original position *to choose* these principles [emphasis added] (45).

Justice as fairness differs from traditional contract theories in that 'the relevant agreement is not to enter a given society or to adopt a given form of government, but to accept certain moral principles' (16). The result of the agreement is not a set of obligations applying to individuals, at least not directly, but principles of justice applying to the basic structure of society. Still, the voluntarist aspect of justification corresponds in some sense to the notion of society as a voluntary agreement. Rawls writes that living in a society governed by principles of justice derived from a voluntary account of justification is, in effect, the next best thing to living in a society we have actually chosen.

> No society can, of course, be a scheme of co-operation which men enter voluntarily in a literal sense; each person finds himself placed at birth in some particular position in some particular society, and the nature of this position materially affects his life prospects. Yet a society satisfying the principles of justice as fairness comes as close as a society can to being a *voluntary scheme*, for it meets the principles which free and equal persons *would assent to* under circumstances which are fair. In this sense its members are autonomous and the obligations they recognize *self-imposed* [emphasis added] (13).

As our reconstruction suggests, the voluntarist nature of Rawls' contract view is bound up with the essential plurality of human subjects and the need to resolve conflicting claims. Without plurality, contracts, and for that matter principles of justice, would be neither possible nor necessary. 'Principles of justice deal with conflicting claims upon the advantages won by social co-operation; they apply to the relations among several persons or groups. The word "contract" suggests this plurality as well as the condition that the appropriate division of advantages must be in accordance with principles acceptable to all parties' (16).

As previously seen, justice as fairness differs from utilitarianism in its emphasis on the plurality and distinctiveness of individuals, and this difference is embodied in the role contract plays in justification.

> Whereas the utilitarian extends to society the principles of choice for one man, justice as fairness, *being a contract view*, assumes that the principles of social choice, and so the principles of justice, are themselves the object of an original *agreement* [emphasis added] (28).

> From the standpoint of *contract theory* one cannot arrive at a principle of social choice merely by extending the principle of rational prudence to the system of desires constructed by the impartial spectator. To do this is not to take seriously the *plurality and distinctness of individuals,* nor to recognize as the basis of justice that to which men would *consent* [emphasis added] (29).

In basing the principles of justice on an agreement among parties, Rawls emphasizes two characteristics that the hypothetical contract shares with actual ones, namely choice and plurality. But we have already seen that the ingredients of choice and plurality are not sufficient to make justice; actual contracts, which include both, cannot justify. This is due to the problems we have described as contingency and conventionalism. Actual agreements often turn out unfairly because of the various (coercive and non-coercive) contingencies associated with the inevitable differences of power and knowledge among persons differently situated. But in the original position, such contingencies are cured. Due to the veil of ignorance and other conditions of equality, all are similarly situated, and so none can take advantage, even inadvertently, of a more favorable bargaining position.

The original position is designed to overcome the problem of conventionalism as well. Where actual contracts are inescapably embedded in the practices and conventions of some particular society, the agreement in the original position is not implicated in the same way. It is not an actual contract, only a hypothetical one. Since it is imagined to occur before the principles of justice arrive on the scene, it may be thought of as 'pre-situated' in the relevant sense, a status quo antecedent to the arrival of justice such that no prior moral principles are available by which its results might be impugned. In this way it is able to realize the ideal of pure procedural justice. (Ironically, where the hypothetical nature of the original agreement at first appeared to weaken its justificatory force, it now appears as a positive, perhaps indispensable, advantage. Where Rawls emphasizes that 'nothing resembling [the original agreement] need ever have taken place' (120), it might be the case that no such agreement ever *could* take place and still overcome the problem of conventionalism.)

> Since *all are similarly situated* and no one is able to design principles to favor his particular condition, the principles of justice are the result of a *fair* agreement or bargain [emphasis added] (12).

> The original position is, one might say, the *appropriate initial status quo,* and thus the fundamental agreements reached in it are *fair.* This explains the propriety of the name 'justice as fairness': it conveys the idea that the principles of justice are agreed to in *an initial situation that is fair* [emphasis added] (12).

> It is a state of affairs in which the parties are equally represented as moral persons and *the outcome is not conditioned by arbitrary contingencies* or the relative balance of social forces. Thus justice as fairness is able to use the idea of pure procedural justice from the beginning [emphasis added] (120).

By imposing the veil of ignorance it is possible to 'nullify the effects of specific contingencies which put men at odds and tempt them to exploit social and natural circumstances to their own advantage' (136).

> If a knowledge of particulars is allowed, then the outcome is biased by *arbitrary contingencies.* As already observed, to each according to his threat advantage is

not a principle of justice. If the original position is to yield agreements that are just, the parties must be *fairly situated* and treated equally as moral persons. The arbitrariness of the world must be corrected for by adjusting the circumstances of the initial contractual situation [emphasis added] (141).

Once the parties to an agreement are assumed to be similarly situated in all relevant respects, differences of power and knowledge disappear, and the possible sources of unfairness are thus eradicated. Since no one is able to choose on the basis of contingently-given attributes, the ideal of autonomy, implicit but imperfect in actual contracts, is fulfilled, the ideal of reciprocity is realized as a matter of course, and the vulnerability of contract to the 'further question' ('But is it fair?') is eliminated. 'The veil of ignorance deprives the persons in the original position of the knowledge that would enable them to choose heteronomous principles. The parties arrive at their choice together as free and equal rational persons knowing only that those circumstances obtain which give rise to the need for principles of justice' (252).

Once the 'further question' of fairness loses its independent moral force, owing to the fact that the parties are situated in such a way that no unfairness conceivably could result, any agreement reached becomes a case of pure procedural justice; its outcome is fair, 'whatever it is,' in virtue of its agreement alone. Under such conditions, a contract ceases to be a constitutive convention and becomes instead an instrument of justification.

> The aim is to characterize this situation so that the principles that would be chosen, *whatever they turn out to be*, are acceptable from a moral point of view. The original position is defined in such a way that it is a status quo in which *any agreements reached are fair* [emphasis added] (120).

> The idea of the original position is to set up a fair procedure so that *any principles agreed to will be just*. The aim is to use the notion of pure procedural justice as a basis of theory [emphasis added] (136).

But at this point a crucial ambiguity arises, for it is not clear what exactly it means 'to use the notion of pure procedural justice as a basis of theory.' Rawls claims that *once* the situation is appropriately characterized, *then* the principles chosen, *whatever they turn out to be*, are acceptable from a moral point of view; *once* the original position is properly defined, *then any agreements reached* in it are fair; *once* a fair procedure is established, *then any principles agreed to* will be just.

What is unclear is how generous these provisions are to the choosers. On one reading, the terms seem generous indeed, the very embodiment of the voluntarist provisions suggested above. Once the parties find themselves in a fair situation, anything goes; the scope for their choice is unlimited. The results of their deliberations will be morally acceptable 'whatever they turn out to be.' No matter what principles they choose, those principles will count as just.

But there is another, less expansive reading of their situation, which gives considerably less scope to their enterprise. On this interpretation, what it means to say that the principles chosen will be just 'whatever they turn out to be' is simply that, given their situation, the parties are guaranteed to choose the *right* principles. While it may be true that, strictly speaking, they can choose any principles they wish, their situation is designed in such a way that they are guaranteed to 'wish' to choose only certain principles. On this view, 'any agreements reached' in the original position are fair, not because the procedure sanctifies just any outcome, but because the situation guarantees a particular outcome. But if the principles agreed to are just because only (the) just principles can be agreed to, the voluntarist aspect of the enterprise is not as spacious as would first appear. The distinction between pure and perfect procedural justice fades, and it becomes unclear whether the procedure 'translates its fairness to the outcome,' or whether the fairness of the procedure is given by the fact that it necessarily leads to the right result.

Rawls confirms the less voluntarist reading when he writes, 'The acceptance of these principles is not conjectured as a psychological law or probability. Ideally anyway, I should like to show that their acknowledgement is the only choice [sic] consistent with the full description of the original position. The argument aims eventually to be strictly deductive' (121). The notion that the full description of the original position determines a single 'choice' which the parties cannot but acknowledge seems to introduce a cognitive element to justification after all and to call into question the priority of procedure over principle which the contract view—and the deontological project generally—seemed to require. But a more immediate consequence of this reading is that it complicates our account of what goes on in the original position.

* * *

For Rawls, the consequences of taking seriously the distinction between persons are not directly moral but more decisively epistemological. What the bounds between persons confine is less the reach of our sentiments—this they do not prejudge—than the reach of our understanding, of our cognitive access to others. And it is this *epistemic* deficit (which derives from the nature of the subject) more than any shortage of benevolence (which is in any case variable and contingent) that requires justice for its remedy and so accounts for its pre-eminence. Where for Hume, we need justice because we do not *love* each other well enough, for Rawls we need justice because we cannot *know* each other well enough for even love to serve alone.

But as our discussion of agency and reflection suggests, we are neither as transparent to ourselves nor as opaque to others as Rawls' moral epistemology requires. If our agency is to consist in something more than the exercise in 'efficient administration' which Rawls' account implies, we must be capable of deeper introspection than a 'direct self-knowledge' of our immediate wants and desires allows. But to be capable of a more thoroughgoing reflection, we cannot be wholly unencumbered subjects of possession, individuated in advance and given prior to our ends, but must be subjects constituted in part by our central aspirations and attachments, always open, indeed vulnerable, to growth and transformation in the light of revised self-understandings. And in so far as our constitutive self-understandings comprehend a wider subject than the individual alone, whether a family or tribe or city or class or nation or people, to this extent they define a community in the constitutive sense. And what marks such a community is not merely a spirit of benevolence, or the prevalence of communitarian vocabulary of discourse and a background of implicit practices and understandings within which the opacity of the participants is reduced if never finally dissolved. In so far as justice depends for its pre-eminence on the separateness or boundedness of persons in the cognitive sense, its priority would diminish as that opacity faded and this community deepened.

Justice and Community

Of any society it can always be asked to what extent it is just, or 'well-ordered' in Rawls' sense, and to what extent it is a community, and the answer can in neither case fully be given by reference to the sentiments and desires of the participants alone. As Rawls observes, to ask whether a particular society is just is not simply to ask whether a large number of its members happen to have among their various desires the desire to act justly—although this may be one feature of a just society— but whether the society is itself a society of a certain kind, ordered in a certain way, such that justice describes its 'basic structure' and not merely the dispositions of persons within the structure. Thus Rawls writes that although we call the attitudes and dispositions of persons just and unjust, for justice as fairness the 'primary subject of justice is the basic structure of society' (7). For a society to be just in this strong sense, justice must be constitutive of its framework and not simply an attribute of certain of the participants' plans of life.

Similarly, to ask whether a particular society is a community is not simply to ask whether a large number of its members happen to have among their various desires the desire to associate with others or to promote communitarian aims— although this may be one feature of a community—but whether the society is itself a society of a certain kind, ordered in a certain way, such that community describes its basic structure and not merely the dispositions of persons within the structure. For a society to be a community in this strong sense, community must be constitutive of the shared self-understandings of the participants and embodied in their institutional arrangements, not simply an attribute of certain of the participants' plans of life.

Rawls might object that a constitutive conception of community such as this should be rejected 'for reasons of clarity among others,' or on the grounds that it supposes society to be 'an organic whole with a life of its own distinct from and superior to that of all its members in their relations with one another' (264). But a constitutive conception of community is no more metaphysically problematic than a constitutive conception of justice such as Rawls defends. For if this notion of community describes a framework of self-understandings that is distinguishable from and in some sense prior to the sentiments and dispositions of individuals within the framework, it is only in the same sense that justice as fairness describes a 'basic structure' or framework that is likewise distinguishable from and prior to the sentiments and dispositions of individuals within it.

If utilitarianism fails to take seriously our distinctness, justice as fairness fails to take seriously our commonality. In regarding the bounds of the self as prior, fixed once and for all, it relegates our commonality to an aspect of the good, and relegates the good to a mere contingency, a product of indiscriminate wants and desires 'not relevant from a moral standpoint.' Given a conception of the good that is diminished in this way, the priority of right would seem an unexceptionable claim indeed. But utilitarianism gave the good a bad name, and in adopting it uncritically, justice as fairness wins for deontology a false victory.

PART III

What is the role of justice and its justification in modern society? Is it, as Hobbes insisted, an attempt to guarantee security and safety? Is it, as John Locke insisted, primarily to safeguard our hard-earned property? Or is it rather to maximize and assure the public good? How is that good to be determined, and how is it to be distributed—but isn't that just the question that justice raises? Does it presuppose, as so much of our modern rhetoric and ideology insist, that everyone should be treated as equals? Does this mean that they are *in fact* equals? This would seem to fly in the face of the most obvious common sense. People are different, and some of those differences at least are relevant to justice. True, they are all human beings. (Do we have to limit "justice" to human beings?) And isn't the question of justice precisely to what extent and for what reasons should we distinguish between people—in terms of their needs, their abilities, their contributions, and their individual rights? Does equality mean that everyone should be treated similarly? But this would seem to be the height of injustice, to give two students the same grade despite the fact that one studied and one did not; to give two employees the same raise although one produced great results and the other nearly sent the company into bankruptcy; to give two criminals the same sentence although one robbed a bank with a sawed-off shotgun and the other took two newspapers for the price of one. Perhaps equality means no more than that we should not arbitrarily discriminate, and that we should treat like cases alike and be consistent in our judgments. But isn't this a demand simply of logic rather than one of justice, and, moreover, which are the relevant similarities that make two cases alike?

In this part of the book, our aim is to present a spectrum of views on such issues as the role of private property in society; the nature and the role of "the public good" in consideration of justice and the well-known philosophy of "utilitarianism";—the special concerns of our central economic (and social?) institution the free market; and the always problematic concept of equality, which, for many thinkers and theorists today, remains the crucial if not the sole criterion for the most basic considerations of justice. The central tension in these various topics and concerns is the very modern antagonism between equality and rights. The right to own private property, for example, has often been blamed for the grotesque inequalities of wealth in modern society (for example, by Rousseau, but also, obviously, by Karl Marx and his followers). So, too, the free market, which begins by presupposing the right to private property, has as one of its most obvious consequences the various outcomes that depend on one's commercial insight and intelligence, one's ingenuity and inventiveness, one's being in the right place at the right time, and in other ways having just plain good luck. Not surprisingly, people who have been raised and educated in the wiles of the market or in the professions that support and participate in the market (e.g.,

law) do quite well in such a society; those who are deprived of any such
education and experience or never get the opportunity to enter into the supposedly "free" market in the first place are doomed to a life of poverty and
deprivation. But it is this uncomfortable observation that leads defenders of the
market to insist all the more on the importance of equality, for without equal
opportunity to enter into it and compete, the market can hardly be "free" and is
therefore unfair as well. In other words, the idea of a free market both presupposes the necessity of equality as opportunity and accepts the inevitable inequality that results from the various exchanges and peoples' very different abilities
and fortunes. And yet, it can be and often has been argued that the free market,
even with its inevitable inequalities, assures prosperity for all, even those who are
"least advantaged."

This claim, that the free market maximizes mutual good, is quite different
from the claim that the market rests upon an inalienable right, the right to
private property. First of all, it is an empirical claim, and one can (at least in
theory) actually test and check to see whether a market society is or is not as
prosperous as, say, a socialist society. But how does one measure such prosperity? In the days before Adam Smith—in the economic philosophy known as
merchantilism—the wealth of a nation was measured by the amount of money in
the royal treasury. After Adam Smith and his radical revision of the idea of "the
wealth of nations," the prosperity of a people was to be measured by the comfort
of the individual citizens. But how do we do this? It is one thing to talk in general
about the "utility" of this or that social program—including, perhaps, the idea of
justice itself—but something else to talk about the precise distribution of that
utility. Do we, as John Stuart Mill suggests, count "each person as one and as no
more than one" and then aim at "the greatest good of the greatest number?" But
does this mean that extremely unequal distributions of wealth are to be tolerated
if (and only if) they altogether add up to the greatest amount of good? What,
then, happens to the idea of the "greatest number"? Or are we to distribute the
goods of society as evenly and as equally as possible (presumably without
sacrificing overall prosperity)? But, then, what happens to the freedom of the
market, and how do we redistribute the goods of society while at the same time
respecting the right to private property? Indeed, what happens to rights in
general, in the utilitarian concern for the aggregate public good? What if the
prosperity of the majority of citizens could be assured at the expense of an
indisputable injustice to be suffered by a small minority? Such is one of the
classic objections to utilitarianism and John Stuart Mill takes it up in his essay,
insisting that, properly conceived, utility and justice are not incompatible but
mutually necessary.

The readings that follow represent both the classic defenses and the classic
objections to private property and the free market, as well as Mill's classic
defense (following that of David Hume) of justice as a mode of social utility. The
question of equality weaves its way through virtually all of the readings, and so,
too, the matter of rights—in particular property rights—underlies them as well.
But we have ended this part of the book with a selection that does not try to
reconcile these central debates but rather throws them into cross-cultural perspective. Both the ideas of equality and rights are peculiar to certain sorts of
societies, but there are others as well.

Thomas Jefferson et al., The Declaration of Independence and Amendments to the Constitution of the United States of America (1776, 1887)

The Declaration of Independence is perhaps the single most famous example of the social contract theory at work in practical politics. It states outright, "that governments are instituted among men, deriving their just powers from the consent of the governed." It further states that when a government fails to perform its duties, "it is the right of the people to alter or abolish" and even "their duty to throw off such a government, and to provide new guards for their future security," and that, of course, is just what this bold and elegant document does. But the Declaration is also the source of our best-known statement about natural rights, indicating a considerable eclecticism on the part of Jefferson and the other authors of the document (see, for example, Garry Wills, *Inventing America*, 1978). But the statement itself is a remarkable *tour de force*; "*We hold these truths to be self-evident*"—blocking out any room for philosophical disagreement and indicating what would today be called a "non-negotiable set of demands"—"*that all men are created equal*"—as if this had been a matter of routine recognition all through the ages—"*that they are endowed by their creator with certain inalienable rights*"—"inalienable" is a central enlightenment concept which means, again, absolutely unconditional and non-negotiable— "*that among these are life, liberty and the pursuit of happiness.*" The Declaration has a Lockean ring, particularly with reference to natural laws and natural rights, but it omits mention of the primary Lockean right, the right to private property. According to historical accounts (e.g. Becker in *The Declaraton of Independence*), that phrase was an original part of the opening paragraph, but it was deleted after much debate and replaced by the more innocuous and egalitarian "pursuit of happiness." What is unmistakable and virtually definitive of our outlook as a nation, however, is the uncompromising stand on the primacy of individual rights. This is no mere declaration of national sovereignty or cultural autonomy. It does not indicate just another rupture in government or disastrous disagreement in policy. It is stated, through and through, that it is first of all the break of a naturally free people from a government that they have come to see as despotic, and it is primarily the multiple violation of individual rights of representation and fair treatment that justifies the breach.

The Constitution was "framed" and written, after much argument and debate both public and private, and completed in 1787. It was and is a remarkable document, an ingenious balancing act between competing political forces and philosophies and intended, above all, to prevent just that sort of concentration of unrepresented authority that had led up to the revolution and the break away from the English. But the careful tripartite structuring of the government and the cautious balance between state and federal power did not satisfactorily answer one of the driving concerns of the revolution, the protection of individual rights from both state and federal government. Accordingly, again after much debate and heated argument, the Congress proposed and the various states affirmed a "Bill of Rights," ten amendments to the completed Constitution, whose whole purpose was to spell out and guarantee (some of) the specific natural rights that had only been suggested in the Declaration of Independence. (Article 9 insists on the incompleteness of the list.) Again, the right to property as such is not itself the subject of any of the amendments, but it is clearly presupposed and "self-evident" in them (e.g. in numbers three and four, which prohibit the quartering of troops in any house without the consent of the owner, and guarantee "the right of people to be secure in their houses, papers and effect," and in number five, where private property in general is twice guaranteed.) So, too, it has been argued that the right to privacy itself, though not spelled out as such in any of these amendments, is presupposed and guaranteed by virtually all of them. Some of the amendments have been generously interpreted, notoriously the second, which guarantees, for the sake of "a well-regulated militia," "the right of the people to keep and bear arms." The right to freedom of speech and the *qualified* separation of religion and government ("Congress shall pass no law respecting . . . or prohibiting . . .") are perennial sources of debate, of course, as is the "cruel and unusual punishment" prohibition in Article 8, for example, with reference to inhumanly overcrowded and violent prisons and, in particular, the death penalty. (See Part IV for some of this discussion.) But beneath these debates it is all too easy for Americans to miss the remarkable fact—so evident in contrast to civilized countries that do not have any such formal document (e.g., Australia)—that we do take personal rights to be the very bedrock of our culture, the ultimate justification for having a government in the first place and the absolute limit of government interference.

The original ten amendments or "Bill of Rights" went into effect on December 15, 1791. Since then, the moral and political development of the United States has forced Congress to debate, formulate and ratify over a dozen further amendments. Some of these have to do with the specific workings of the government (e.g., the amendment limiting a president to two terms in office), but others, like the Bill of Rights itself, are profound moral statements about what we as a people take to be basic rights. The most morally important of these, no doubt, is the belated recognition that slavery however conceived has no place whatever in a society of free individuals and, second, that the basic rights guaranteed by the Constitution should apply to everyone, regardless of race or sex. These articles are included below, along with the proposed "Equal Rights Amendment," proposed March 22, 1972, but still not ratified as of this writing.

When in the Course of human events, it becomes necessary for one people to dissolve the political bonds which have connected them with another, and to assume among the powers of the earth, the separate and equal station to which the Laws of Nature and of Nature's God entitle them, a decent respect to the opinions of mankind requires that they should declare the causes which impel them to the separation.— We hold these truths to be self-evident, that all men are created equal, that they are endowed by their Creator with certain unalienable Rights, that among these are Life, Liberty and the pursuit of Happiness.—That to secure these rights, Governments are instituted among Men, deriving their just powers from the consent of the governed.—That whenever any Form of Government becomes destructive of these ends, it is the Right of the People to alter or to abolish it, and to institute new Government, laying its foundation on such principles and organizing its powers in such form, as to them shall seem most likely to effect their Safety and Happiness. Prudence, indeed, will dictate that Governments long established should not be changed for light and transient causes; and accordingly all experience hath shown, that mankind are more disposed to suffer, while evils are sufferable, than to right themselves by abolishing the forms to which they are accustomed. But when a long train of abuses and usurpations, pursuing invariably the same Object evinces a design to reduce them under absolute Despotism, it is their right, it is their duty, to throw off such Government, and to provide new Guards for their future security.—

* * *

A Prince, whose character is thus marked by every act which may define a Tyrant, is unfit to be the ruler of a free people. Nor have We been wanting in attentions to our British brethren. We have warned them from time to time of attempts by their legislature to extend an unwarrantable jurisdiction over us. We have reminded them of the circumstances of our emigration and settlement here. We have appealed to their native justice and magnanimity, and we have conjured them by the ties of our common kindred to disavow these usurpations, which would inevitably interrupt our connections and correspondence. They too have been deaf to the voice of justice and of consanguinity. We must, therefore, acquiesce in the necessity, which denounces our Separation, and hold them, as we hold the rest of mankind, Enemies in War, in Peace Friends.—

We, Therefore, the Representatives of the United States of America, in General Congress, Assembled, appealing to the Supreme Judge of the world for the rectitude of our intentions, do, in the Name, and by Authority of the good People of these Colonies, solemnly publish and declare, That these United Colonies are, and of Right ought to be Free and Independent States; that they are Absolved from all Allegiance to the British Crown, and that all political connection between them and the State of Great Britain, is and ought to be totally dissolved; and that as Free and Independent States, they have full Power to levy War, conclude Peace, contract Alliances, establish Commerce, and to do all other Acts and Things which Independent States may of right do.—And for the support of this Declaration, with a firm reliance on the protection of Divine Providence, we mutually pledge to each other our Lives, our Fortunes and our sacred Honor.

Amendments to the Constitution

Articles in addition to, and Amendment of the Constitution of the United States of America, proposed by Congress, and ratified by the Legislatures of the several States, pursuant to the fifth Article of the original Constitution.

Article I.

Congress shall make no law respecting an establishment of religion, or prohibiting the free exercise thereof; or abridging the freedom of speech, or of the press; or the right of the people peaceably to assemble, and to petition the Government for a redress of grievances.

Article II.

A well regulated Militia, being necessary to the security of a free State, the right of the people to keep and bear Arms, shall not be infringed.

Article III.

No Soldier shall, in time of peace be quartered in any house, without the consent of the Owner, nor in time of war, but in a manner to be prescribed by law.

Article IV.

The right of the people to be secure in their persons, houses, papers, and effects, against unreasonable searches and seizures, shall not be violated, and no Warrants shall issue, but upon probable cause, supported by Oath or affirmation, and particularly describing the place to be searched, and the persons or things to be seized.

Article V.

No person shall be held to answer for a capital, or otherwise infamous crime, unless on a presentment or indictment of a Grand Jury, except in cases arising in the land or naval forces, or in the Militia, when in actual service in time of War or public danger; nor shall any person be subject for the same offence to be twice put in jeopardy of life or limb; nor shall be compelled in any criminal case to be a witness against himself, nor be deprived of life, liberty, or property, without due process of law; nor shall private property be taken for public use, without just compensation.

Article VI.

In all criminal prosecutions, the accused shall enjoy the right to a speedy and public trial, by an impartial jury of the State and district wherein the crime shall have been committed, which district shall have been previously ascertained by law, and to be informed of the nature and cause of the accusation; to be confronted with the witnesses against him; to have compulsory process for obtaining witnesses in his favor, and to have the Assistance of Counsel for his defence.

Article VII.

In Suits at common law, where the value in controversy shall exceed twenty dollars, the right of trial by jury shall be preserved, and no fact tried by a jury, shall be otherwise re-examined in any Court of the United States, than according to the rules of the common law.

Article VIII.

Excessive bail shall not be required, or excessive fines imposed, nor cruel and unusual punishments inflicted.

Article IX.

The enumeration in the Constitution, of certain rights, shall not be construed to deny or disparage others retained by the people.

Article X.

The powers not delegated to the United States by the Constitution, nor prohibited by it to the States, are reserved to the States respectively, or to the people.

Article XIII.

Neither slavery nor involuntary servitude, except as a punishment for crime whereof the party shall have been duly convicted, shall exist within the United States, or any place subject to their jurisdiction.

Article XIV.

All persons born or naturalized in the United States, and subject to the jurisdiction thereof, are citizens of the United States and of the State wherein they reside. No State shall make or enforce any law which shall abridge the privileges or immunities of citizens of the United States; nor shall any State deprive any person of life, liberty, or property, without due process of law; nor deny to any person within its jurisdiction the equal protection of the laws.

Article XV.

The right of citizens of the United States to vote shall not be denied or abridged by the United States or by any State on account of race, color, or previous condition of servitude.

Article XIX.

The right of citizens of the United States to vote shall not be denied or abridged by the United States or by any State on account of sex.

EQUAL RIGHTS FOR WOMEN

Equality of rights under the law shall not be denied or abridged by the United States or by any State on account of sex.

John Locke, "Of Property," from *The Second Treatise on Government* (1632–1704)

John Locke's suggestion that private property is to be counted as a natural right, as appealing and obvious as that sounds to us, was something of a revolutionary act in itself, "a landmark in the history of thought" (according to O. Giercke in *Natural Law and the Theory of Society 1550 to 1800* [Cambridge 1950, p. 103, quoted by Vlastos p. 34]). The idea that "the earth belongs to men" (as a gift of God) seemed obvious enough to the philosophers and politicians of Locke's times, but that men should have the right, as individuals, to possess bits and pieces of the whole he took to be something of a problem. He begins modestly and persuasively enough: every one (and even every animal) has an essential and in an obvious sense "inalienable" piece of property in his, her, or its own body. But Locke moves quickly and without argument to the subsequent claim that whatever one does or makes with this body is also one's own and whatever objects one "mixes one's labor with" are also one's own. Since labor is "the unquestionable property of the laborer," whatever one produces through one's labor is also one's property. This essential argument has a proviso, and that is "where there is enough and as good left in common for others"—by itself enough to halt the many developers and latter-day robber barons who leave in the wake of their labors so much of nature destroyed and devastated communities. Locke also makes it quite clear that greed and gluttony are not justified or excused by his defense of property as a natural right. "Nothing was made by God for man to spoil or destroy," he says, and, indeed, the whole argument is ultimately framed within the understanding that, by taking one's own, one thereby does not injure anyone else. Thus the right to private property is not at all absolute, and its legitimacy is based upon and bounded by utility, by our ability to "improve" the earth "for the benefit of life." Indeed, Locke's argument against the need for consent from other people (that is, their permission to take what is one's own) is not that "it is none of their business" but rather that it would be too impractical and unwieldy: "children or servants could not cut the meat which their father or master had provided for them in common without assigning to everyone his particular part." There is no doubt that the argument for property as a right is fundamental in Locke, but it should not be thought that its status as such is completely independent of and prior to any considerations of utility and the public good. Indeed, it is quite clear that Locke saw the former as a means to the latter and certainly not as incompatible.

Of Property

25. Whether we consider natural reason, which tells us that men, being once born, have a right to their preservation, and consequently to meat and drink and such other things as nature affords for their subsistence; or revelation, which gives us an account of those grants God made of the world to Adam, and to Noah and his sons; it is very clear that God, as King David says (Psalm cxv. 16), "has given the earth to the children of men," given it to mankind in common. But this being supposed, it seems to some a very great difficulty how any one should ever come to have a property in anything. I will not content myself to answer that if it be difficult to make out property upon a supposition that God gave the world to Adam and his posterity in common, it is impossible that any man but one universal monarch should have any property upon a supposition that God gave the world to Adam and his heirs in succession, exclusive of all the rest of his posterity. But I shall endeavor to show how men might come to have a property in several parts of that which God gave to mankind in common, and that without any express compact of all the commoners.

26. God, who has given the world to men in common, has also given them reason to make use of it to the best advantage of life and convenience. The earth and all that is therein is given to men for the support and comfort of their being. And though all the fruits it naturally produces and beasts it feeds belong to mankind in common, as they are produced by the spontaneous hand of nature; and nobody has originally a private dominion exclusive of the rest of mankind in any of them, as they are thus in their natural state; yet, being given for the use of men, there must of necessity be a means to appropriate them some way or other before they can be of any use or at all beneficial to any particular man. The fruit or venison which nourishes the wild Indian, who knows no enclosure and is still a tenant in common, must be his, and so his, i.e., a part of him, that another can no longer have any right to it before it can do him any good for the support of his life.

27. Though the earth and all inferior creatures be common to all men, yet every man has a property in his own person; this nobody has any right to but himself. The labor of his body and the work of his hands, we may say, are properly his. Whatsoever then he removes out of the state that nature has provided and left it in, he has mixed his labor with, and joined to it something that is his own, and thereby makes it his property. It being by him removed from the common state nature has placed it in, it has by this labor something annexed to it that excludes the common right of other men. For this labor being the unquestionable property of the laborer, no man but he can have a right to what that is once joined to, at least where there is enough and as good left in common for others.

28. He that is nourished by the acorns he picked up under an oak, or the apples he gathered from the trees in the wood, has certainly appropriated them to himself. Nobody can deny but the nourishment is his. I ask, then, When did they begin to be his? When he digested or when he ate or when he boiled or when he brought them home? Or when he picked them up? And it is plain, if the first gathering made them not his, nothing else could. That labor put a distinction between them and common; that added something to them more than nature, the common mother of all, had done; and so they became his private right. And will anyone say he had no right to those acorns or apples he thus appropriated because he had not the consent of all mankind to make them his? Was it a robbery thus to assume to himself what

belonged to all in common? If such a consent as that was necessary, man had starved, notwithstanding the plenty God had given him. We see in commons, which remain so by compact, that it is the taking any part of what is common and removing it out of the state nature leaves it in which begins the property, without which the common is of no use. And the taking of this or that part does not depend on the express consent of all the commoners. Thus the grass my horse has bit, the turfs my servant has cut, and the ore I have digged in any place where I have a right to them in common with others, become my property without the assignation or consent of anybody. The labor that was mine, removing them out of that common state they were in, has fixed my property in them.

29. By making an explicit consent of every commoner necessary to any one's appropriating to himself any part of what is given in common, children or servants could not cut the meat which their father or master had provided for them in common without assigning to every one his peculiar part. Though the water running in the fountain be every one's, yet who can doubt but that in the pitcher is his only who drew it out? His labor has taken it out of the hands of nature where it was common and belonged equally to all her children, and has thereby appropriated it to himself.

30. Thus this law of reason makes the deer that Indian's who has killed it; it is allowed to be his goods who has bestowed his labor upon it, though before it was the common right of every one. And amongst those who are counted the civilized part of mankind, who have made and multiplied positive laws to determine property, this original law of nature, for the beginning of property in what was before common, still takes place; and by virtue thereof what fish any one catches in the ocean, that great and still remaining common of mankind, or what ambergris any one takes up here, is, by the labor that removes it out of that common state nature left it in, made his property who takes that pains about it. And even amongst us, the hare that anyone is hunting is thought his who pursues her during the chase; for, being a beast that is still looked upon as common and no man's private possession, whoever has employed so much labor about any of that kind as to find and pursue her has thereby removed her from the state of nature wherein she was common, and has begun a property.

31. It will perhaps be objected to this that "if gathering the acorns, or other fruits of the earth, etc., makes a right to them, then any one may engross as much as he will." To which I answer: not so. The same law of nature that does by this means give us property does also bound that property, too. "God has given us all things richly" (1 Tim. vi. 17), is the voice of reason confirmed by inspiration. But how far has he given it us? To enjoy. As much as any one can make use of to any advantage of life before it spoils, so much he may by his labor fix a property in; whatever is beyond this is more than his share and belongs to others. Nothing was made by God for man to spoil or destroy. And thus considering the plenty of natural provisions there was a long time in the world, and the few spenders, and to how small a part of that provision the industry of one man could extend itself and engross it to the prejudice of others, especially keeping within the bounds set by reason of what might serve for his use, there could be then little room for quarrels or contentions about property so established.

32. But the chief matter of property being now not the fruits of the earth and the beasts that subsist on it, but the earth itself, as that which takes in and carries

with it all the rest, I think it is plain that property in that, too, is acquired as the former. As much land as a man tills, plants, improves, cultivates, and can use the product of, so much is his property. He by his labor does, as it were, enclose it from the common. Nor will it invalidate his right to say everybody else has an equal title to it, and therefore he cannot appropriate, he cannot enclose, without the consent of all his fellow commoners—all mankind. God, when he gave the world in common to all mankind, commanded man also to labor, and the penury of his condition required it of him. God and his reason commanded him to subdue the earth, i.e., improve it for the benefit of life, and therein lay out something upon it that was his own, his labor. He that in obedience to this command of God subdued, tilled, and sowed any part of it, thereby annexed to it something that was his property, which another had no title to, nor could without injury take from him.

33. Nor was this appropriation of any parcel of land by improving it any prejudice to any other man, since there was still enough and as good left, and more than the yet unprovided could use. So that, in effect, there was never the less left for others because of his enclosure for himself; for he that leaves as much as another can make use of does as good as take nothing at all. Nobody could think himself injured by the drinking of another man, though he took a good draught, who had a whole river of the same water left him to quench his thirst; and the case of land and water, where there is enough for both, is perfectly the same.

David Hume, "Justice as Utility" from
An Enquiry Concerning the Principles
of Morals and *Treatise of Human Nature*
(1711–1776)

David Hume was a Scotsman, gone south to London to fulfill an ambition he described as "literary fame." Though steeped in the Scottish enlightenment he was anxious to make a name for himself in English letters, and his philosophy is a curious mix of Scots's "common sense" and English conservatism, plus, of course, his own unique blend of Paganism and skepticism. But there is little of that famed skepticism in Hume's writings in social philosophy. He is concerned, like Locke before him, to defend the idea of private property, and to defend a conception of justice which is, first and foremost, supportive of a public good conceived in terms of peace and security. In the passage that follows, he considers the limited circumstances in which justice would be conceivable at all; within these he argues that public utility is not just *a* but "the sole" justification of justice. In a world filled with abundance, there would be no need for justice, and in a world devoured by law, there would be no place for justice, either. In either case, there would be no *use* for justice. Thus justice is, according to Hume, an "artificial" rather than a "natural" virtue, invented by men in circumstances of moderate scarcity and relative peace to assure social tranquility and public utility, and in particular to protect private property—which is the basis of society.

―――――――

That Justice is useful to society, and consequently that *part* of its merit, at least, must arise from that consideration, it would be a superfluous undertaking to prove. That public utility is the *sole* origin of justice, and that reflections on the beneficial consequences of this virtue are the *sole* foundation of its merit; this proposition, being more curious and important, will better deserve our examination and enquiry.

Let us suppose that nature has bestowed on the human race such profuse *abundance* of all *external* conveniencies, that, without any uncertainty in the event, without any care or industry on our part, every individual finds himself fully provided with whatever his most voracious appetites can want, or luxurious imagination wish or desire. His natural beauty, we shall suppose, surpasses all acquired ornaments; the perpetual clemency of the seasons renders useless all clothes or covering; the raw herbage affords him the most delicious fare; the clear fountain, the richest beverage. No laborious occupation required: no tillage: no navigation. Music, poetry, and contemplation form his sole business: conversation, mirth, and friendship his sole amusement.

It seems evident that, in such a happy state, every other social virtue would flourish, and receive tenfold increase; but the cautious, jealous virtue of justice would never once have been dreamed of. For what purpose make a partition of goods, where every one has already more than enough? Why give rise to property, where there cannot possibly be any injury? Why call this object *mine*, when upon the seizing of it by another, I need but stretch out my hand to possess myself to what is equally valuable? Justice, in that case, being totally useless, would be an idle ceremonial, and could never possibly have place in the catalogue of virtues.

We see, even in the present necessitous condition of mankind, that, wherever any benefit is bestowed by nature in an unlimited abundance, we leave it always in common among the whole human race, and make no subdivisions of right and property. Water and air, though the most necessary of all objects, are not challenged as the property of individuals; nor can any man commit injustice by the most lavish use and enjoyment of these blessings. In fertile extensive countries, with few inhabitants, land is regarded on the same footing. And no topic is so much insisted on by those, who defend the liberty of the seas, as the unexhausted use of them in navigation. Were the advantages, procured by navigation, as inexhaustible, these reasoners had never had any adversaries to refute; nor had any claims ever been advanced of a separate, exclusive dominion over the ocean.

It may happen, in some countries, at some periods, that there be established a property in water, none in land[1]; if the latter be in greater abundance than can be used by the inhabitants, and the former be found, with difficulty, and in very small quantities.

Again; suppose, that, though the necessities of human race continue the same as at present, yet the mind is so enlarged, and so replete with friendship and generosity, that every man has the utmost tenderness for every man, and feels no more concern for his own interest than for that of his fellows; it seems evident, that the use of justice would, in this case, be suspended by such an extensive benevolence, nor would the divisions and barriers of property and obligation have ever been thought of. Why should I bind another, by a deed or promise, to do me any good office, when I know that he is already prompted, by the strongest inclination, to seek my happiness, and would, of himself, perform the desired service; except the hurt, he thereby receives, be greater than the benefit accruing to me? in which case, he knows, that, from my innate humanity and friendship, I should be the first to oppose myself to his imprudent generosity. Why raise land-marks between my neighbour's field and mine, when my heart has made no division between our interests; but shares all his joys and sorrows with the same force and vivacity as if originally my own? Every man, upon this supposition, being a second self to another, would trust all his interests to the discretion of every man; without jealousy, without partition, without distinction. And the whole human race would form only one family; where all would lie in common, and be used freely, without regard to property; but cautiously too, with as entire regard to the necessities of each individual, as if our own interests were most intimately concerned.

[1]Genesis, chaps. xiii and xxi.

In the present disposition of the human heart, it would, perhaps, be difficult to find complete instances of such enlarged affections; but still we may observe, that the case of families approaches towards it; and the stronger the mutual benevolence is among the individuals, the nearer it approaches; till all distinction of property be, in a great measure, lost and confounded among them. Between married persons, the cement of friendship is by the laws supposed so strong as to abolish all division of possessions; and has often, in reality, the force ascribed to it. And it is observable, that, during the ardour of new enthusiasms, when every principle is inflamed into extravagance, the community of goods has frequently been attempted; and nothing but experience of its inconveniencies, from the returning or disguised selfishness of men, could make the imprudent fanatics adopt anew the ideas of justice and of separate property. So true is it, that this virtue derives its existence entirely from its necessary *use* to the intercourse and social state of mankind.

To make this truth more evident, let us reverse the foregoing suppositions; and carrying everything to the opposite extreme, consider what would be the effect of these new situations. Suppose a society to fall into such want of all common necessaries, that the utmost frugality and industry cannot preserve the greater number from perishing, and the whole from extreme misery; it will readily, I believe, be admitted, that the strict laws of justice are suspended, in such a pressing emergence, and give place to the stronger motives of necessity and self-preservation. Is it any crime, after a shipwreck, to seize whatever means or instrument of safety one can lay hold of, without regard to former limitations of property? Or if a city besieged were perishing with hunger; can we imagine, that men will see any means of preservation before them, and lose their lives, from a scrupulous regard to what, in other situations, would be the rules of equity and justice? The use and tendency of that virtue is to procure happiness and security, by preserving order in society: but where the society is ready to perish from extreme necessity, no greater evil can be dreaded from violence and injustice; and every man may now provide for himself by all the means, which prudence can dictate, or humanity permit. The public, even in less urgent necessities, opens granaries, without the consent of proprietors; as justly supposing, that the authority of magistracy may, consistent with equity, extend so far: but were any number of men to assemble, without the tie of laws or civil jurisdiction; would an equal partition of bread in a famine, though effected by power and even violence, be regarded as criminal or injurious?

Suppose likewise, that it should be a virtuous man's fate to fall into the society of ruffians, remote from the protection of laws and government; what conduct must he embrace in that melancholy situation? He sees such a desperate rapaciousness prevail; such a disregard to equity, such contempt of order, such stupid blindness to future consequences, as must immediately have the most tragical conclusion, and must terminate in destruction to the greater number, and in a total dissolution of society to the rest. He, meanwhile, can have no other expedient than to arm himself, to whomever the sword he seizes, or the buckler, may belong: To make provision of all means of defence and security: And his particular regard to justice being no longer of use to his own safety or that of others, he must consult the dictates of self-preservation alone, without concern for those who no longer merit his care and attention.

When any man, even in political society, renders himself by his crimes, obnoxious to the public, he is punished by the laws in his goods and person; that is, the ordinary rules of justice are, with regard to him, suspended for a moment, and it becomes equitable to inflict on him, for the *benefit* of society, what otherwise he could not suffer without wrong or injury.

The rage and violence of public war; what is it but a suspension of justice among the warring parties, who perceive, that this virtue is now no longer of any *use* or advantage to them? The laws of war, which then succeed to those of equity and justice, are rules calculated for the *advantage* and *utility* of that particular state, in which men are now placed. And were a civilized nation engaged with barbarians, who observed no rules even of war, the former must also suspend their observance of them, where they no longer serve to any purpose; and must render every action or rencounter as bloody and pernicious as possible to the first aggressors.

Thus, the rules of equity or justice depend entirely on the particular state and condition in which men are placed, and owe their origin and existence to that utility, which results to the public from their strict and regular observance. Reverse, in any considerable circumstance, the condition of men: Produce extreme abundance or extreme necessity: Implant in the human breast perfect moderation and humanity, or perfect rapaciousness and malice: By rendering justice totally *useless*, you thereby totally destroy its essence, and suspend its obligation upon mankind.

* * *

It must, indeed, be confessed, that nature is so liberal to mankind, that, were all her presents equally divided among the species, and improved by art and industry, every individual would enjoy all the necessaries, and even most of the comforts of life; nor would ever be liable to any ills, but such as might accidentally arise from the sickly frame and constitution of his body. It must also be confessed, that, wherever we depart from this equality, we rob the poor of more satisfaction than we add to the rich, and that the slight gratification of a frivolous vanity, in one individual, frequently costs more than bread to many families, and even provinces. It may appear withal, that the rule of equality, as it would be highly *useful*, is not altogether *impracticable*; but has taken place, at least in an imperfect degree, in some republics; particularly that of Sparta; where it was attended, it is said, with the most beneficial consequences. Not to mention that the Agrarian laws, so frequently claimed in Rome, and carried into execution in many Greek cities, proceeded, all of them, from a general idea of the utility of this principle.

But historians, and even common sense, may inform us, that, however specious these ideas of *perfect* equality may seem, they are really, at bottom, *impracticable*; and were they not so, would be extremely *pernicious* to human society. Render possessions ever so equal, men's different degrees of art, care, and industry will immediately break that equality. Or if you check these virtues, you reduce society to the most extreme indigence; and instead of preventing want and beggary in a few, render it unavoidable to the whole community. The most rigorous inquisition too is requisite to watch every inequality on its first appearance; and the most severe jurisdiction, to punish and redress it. But besides, that so much authority must soon degenerate into tyranny, and be exerted with great partialities; who can possibly be possessed of it, in such a situation as is here supposed? Perfect equality of possessions, destroying all subordination, weakens extremely the authority of magistracy, and must reduce all power nearly to a level, as well as property.

We may conclude, therefore, that, in order to establish laws for the regulation of property, we must be acquainted with the nature and situation of man; must reject appearances, which may be false, though specious; and must search for those rules, which are, on the whole, most *useful* and *beneficial*. Vulgar sense and slight experience are sufficient for this purpose; where men give not way to too selfish avidity, or too extensive enthusiasm.

Who sees not, for instance, that whatever is produced or improved by a man's art or industry ought, for ever, to be secured to him, in order to give encouragement to such *useful* habits and accomplishments? That the property ought also to descend to children and relations, for the same *useful* purpose? That it may be alienated by consent, in order to beget that commerce and intercourse, which is so *beneficial* to human society? And that all contracts and promises ought carefully to be fulfilled, in order to secure mutual trust and confidence, by which the general *interest* of mankind is so much promoted?

Examine the writers on the laws of nature; and you will always find, that, whatever principles they set out with, they are sure to terminate here at last, and to assign, as the ultimate reason for every role which they establish, the convenience and necessities of mankind.

* * *

No one can doubt that the convention for the distinction of property and for the stability of possession is of all circumstances the most necessary to the establishment of human society, and that after the agreement for the fixing and observing of this rule there remains little or nothing to be done towards settling a perfect harmony and concord. All the other passions, beside this of interest, are either easily restrained, or are not of such pernicious consequence when indulged. *Vanity* is rather to be esteemed a social passion and a bond of union among men. *Pity* and *love* are to be considered in the same light. And as to *envy* and *revenge*, though pernicious, they operate only by intervals, and are directed against particular persons whom we consider as our superiors or enemies. This avidity alone of acquiring goods and possessions for ourselves and our nearest friends is insatiable, perpetual, universal, and directly destructive of society. There scarce is any one who is not actuated by it; and there is no one who has not reason to fear from it, when it acts without any restraint and gives way to its first and most natural movements. So that, upon the whole, we are to esteem the difficulties in the establishment of society to be greater or less, according to those we encounter in regulating and restraining this passion.

It is certain that no affection of the human mind has both a sufficient force and a proper direction to counterbalance the love of gain, and render men fit members of society by making them abstain from the possessions of others. Benevolence to strangers is too weak for this purpose; and as to the other passions, they rather inflame this avidity, when we observe that the larger our possessions are, the more ability we have of gratifying all our appetites. There is no passion, therefore, capable of controlling the interested affection but the very affection itself, by an alteration of its direction. Now, this alteration must necessarily take place upon the least reflection; since it is evident that the passion is much better satisfied by its restraint than by its liberty, and that, in preserving society, we make much greater advances in the acquiring possessions than in the solitary and forlorn condition which must follow upon violence and an universal licence. The question, therefore, concerning the wickedness or goodness of human nature enters not in the least into that other question concerning the origin of society; nor is there anything to be considered but the degrees of men's sagacity or folly. For whether the passion of self-interest be esteemed vicious or virtuous, it is all a case, since itself alone restrains it; so that if it be virtuous, men become social by their virtue; if vicious, their vice has the same effect.

* * *

Here then is a proposition which, I think, may be regarded as certain, *that it is only from the selfishness and confined generosity of man, along with the scanty provision nature has made for his wants that justice derives its origin.* If we look backward we shall find that this proposition bestows an additional force on some of those observations which we have already made on this subject.

First, we may conclude from it that a regard to public interest, or a strong extensive benevolence, is not our first and original motive for the observation of the rules of justice, since it is allowed that if men were endowed with such a benevolence, these rules would never have been dreamed of.

Secondly, we may conclude from the same principle that the sense of justice is not founded on reason, or on the discovery of certain connections and relations of ideas which are eternal, immutable, and universally obligatory. For since it is confessed that such an alteration as that above mentioned, in the temper and circumstances of mankind, would entirely alter our duties and obligations, it is necessary upon the common system *that the sense of virtue is derived from reason,* to show the change which this must produce in the relations and ideas. But it is evident that the only cause why the extensive generosity of man and the perfect abundance of everything would destroy the very idea of justice is because they render it useless; and that, on the other hand, his confined benevolence and his necessitous condition give rise to that virtue only by making it requisite to the public interest and to that of every individual. It was therefore a concern for our own and the public interest which made us establish the laws of justice; and nothing can be more certain than that it is not any relation of ideas which gives us this concern, but our impressions and sentiments, without which everything in nature is perfectly indifferent to us, and can never in the least affect us. The sense of justice, therefore, is not founded on our ideas but on our impressions.

Thirdly, we may further confirm the foregoing proposition *that those impressions, which give rise to this sense of justice, are not natural to the mind of man, but arise from artifice and human conventions.* For since any considerable alienation of temper and circumstances destroys equally justice and injustice, and since such an alteration has an effect only by changing our own and the public interest, it follows that the first establishment of the rules of justice depends on these different interests. But if men pursued the public interest naturally, and with a hearty affection, they would have never dreamed of restraining each other by these rules; and if they pursued their own interest, without any precaution, they would run headlong into every kind of injustice and violence. These rules, therefore, are artificial and seek their end in an oblique and indirect manner; nor is the interest which gives rise to them of a kind that could be pursued by the natural and inartificial passions of men.

To make this more evident, consider that, though the rules of justice are established merely by interest, their connection with interest is somewhat singular, and is different from what may be observed on other occasions. A single act of justice is frequently contrary to *public interest*; and were it to stand alone, without being followed by other acts, may in itself be very prejudicial to society. When a man of merit, of a beneficent disposition, restores a great fortune to a miser or a seditious bigot, he has acted justly and laudably; but the public is a real sufferer. Nor is every single act of justice, considered apart, more conducive to private interest than to public; and it is easily conceived how a man may impoverish himself by a single instance of integrity, and have reason to wish that, with regard to that single act, the laws of justice were for a moment suspended in the universe. But however single acts of justice may be contrary either to public or private interest, it is certain that the whole plan or scheme is highly conducive, or indeed absolutely requisite, both to the support of society and the well-being of every individual. It is impossible to separate the good from the ill. Property must be stable, and must be fixed by general rules.

Adam Smith, on Justice as a Moral Sentiment and on the Virtues of the Free Market, from *A Theory of the Moral Sentiments* and *The Wealth of Nations* (1723–1790)

Economists and defenders of the free market should no longer be surprised by this, but Adam Smith wrote *two* great books in philosophy, not only that "Bible" of capitalism, *The Wealth of Nations*. The other, which preceded it, was *A Theory of the Moral Sentiments*, and it was a book that argued that greed—or what later would be called (by capitalism's critics) "the profit motive"—was *not* the motive that moved modern civilization. Smith never doubted the existence or the power of self-interest, but he always insisted that it was balanced in us by a number of *moral sentiments*, especially sympathy for our fellow citizens. In sympathy, we in some sense share or at least "feel for" the sufferings of others, and such feelings hold in check our temptations to hurt others as well. Foremost among those prohibitive feelings that prevent us from doing harm to others is the feeling of *justice*, which Smith, here going against his Edinburgh friend David Hume, takes to be a "natural" and not merely an "artificial" sentiment. Furthermore, Smith (like Hume) takes one of the primary motives of our behavior to be *the approval of others*, thus blending and (to stubborn egoists) confusing motives of self-interest and altruism. Thus justice has threefold support: the force of the sentiment itself, a sense of sympathy, which renders us incapable of being indifferent to the sufferings of others, and the near unanimous approval of our fellow citizens. (Thrasymachus, presumably, is not among them.) But justice, above all, is ultimately *useful* and this is what ultimately justifies it; in this Smith wholly agrees with his friend Hume. Smith distinguishes between what he calls "beneficence" ("wanting to do good") and justice, insisting that both are desirable but only the latter is essential, for while a society can survive (but not happily) without citizens who are helpful and generous with one another it could not survive at all in the presence of general mutual harm.

In *The Wealth of Nations*, Adam Smith famously argues for the virtues of the free market system. That system presupposes, first of all, the institution of private property, for what could one buy or sell (and what would one bother to produce) if the right to ownership were not assumed from the outset. But the institution of private property and the workings of the free market virtually assure some great inequalities between the few rich and the relatively many poor, and the inevitable enmity between them, according to Smith—as for Locke and

Hume—demands the existence of civil government. But it is not the right to private property that justifies the market itself, according to Smith, and here he differs from some of his more illustrious philosophical followers (e.g. Robert Nozick). Smith's defense of the market is ultimately utilitarian, based on the straightforward empirical argument that the market alone will assure greater prosperity for all citizens. But it should be said that Smith did not have much to say about the very poor, those who do not benefit from any rise in prosperity, no matter how far down the socioeconomic ladder it might trickle. and it should also be said that Smith never thought that the market could similarly take care of all social problems or that the government should be "minimal" and limited to little more than the protection of private property and other individual rights. To the contrary, the strong sense of community defended in *A Theory of the Moral Sentiments* is presupposed throughout *The Wealth of Nations*, and there can be no sense of justice based on prosperity alone.

There is, however, another virtue, of which the observance is not left to the freedom of our own wills, which may be extorted by force, and of which the violation exposes to resentment, and consequently to punishment. This virtue is justice: the violation of justice is injury: it does real and positive hurt to some particular persons, from motives which are naturally disapproved of. It is, therefore, the proper object of resentment, and of punishment, which is the natural consequence of resentment. As mankind go along with, and approve of the violence employed to avenge the hurt which is done by injustice, so they much more go along with, and approve of, that which is employed to prevent and beat off the injury, and to restrain the offender from hurting his neighbours. The person himself who mediates an injustice is sensible of this, and feels that force may, with the utmost propriety, be made use of, both by the person whom he is about to injure, and by others, either to obstruct the execution of his crime, or to punish him when he has executed it. And upon this is founded that remarkable distinction between justice and all the other social virtues, which has of late been particularly insisted upon by an author of very great and original genius, that we feel ourselves to be under a strict obligation to act according to justice, than agreeably to friendship, charity, or generosity; that the practise of these last mentioned virtues seems to be left in some measure to our own choice, but that, somehow or other, we feel ourselves to be in a peculiar manner tied, bound, and obligated to the observation of justice. We feel, that is to say, that force may, with the utmost propriety, and with the approbation of all mankind, be made use of to constrain us to observe the rules of the one, but not to follow the precepts of the other. . . .

Chapter II

Of the sense of Justice, of Remorse, and of the consciousness of Merit

There can be no proper motive for hurting our neighbour, there can be no incitement to do evil to another, which mankind will go along with, except just indignation for evil which that other has done to us. To disturb his happiness merely because it stands in the way of our own, to take from him what is of real use to him merely because it may be of equal or of more use to us, or to indulge, in this manner, at the expence of other people, the natural preference which every man has for his own happiness above that of other people, is what no impartial spectator can go along with. Every man is, no doubt, by nature, first and principally recommended to his own care; and as he is fitter to take care of himself than of any other person, it is fit and right that it should be so. Every man, therefore, is much more deeply interested in whatever immediately concerns himself, than in what concerns any other man: and to hear, perhaps, of the death of another person, with whom we have no particular connection, will give us less concern, will spoil our stomach, or break our rest much less than a very insignificant disaster which has befallen ourselves.

But though the ruin of our neighbour may affect us much less than a very small misfortune of our own, we must not ruin him to prevent that small misfortune, nor even to prevent our own ruin. We must, here, as in all other cases, view ourselves not so much according to that light in which we may naturally appear to ourselves, as according to that in which we naturally appear to others. Though every man may, according to the proverb, be the whole world to himself, to the rest of mankind he is a most insignificant part of it. Though his own happiness may be of more importance to him than that of all the world besides, to every other person it is of no more consequence than that of any other man. Though it may be true, therefore, that every individual, in his own breast, naturally prefers himself to all mankind, yet he dares not look mankind in the face, and avow that he acts according to this principle. He feels that in this preference they can never go along with him, and that how natural soever it may be to him, it must always appear excessive and extravagant to them.

When he views himself in the light in which he is conscious that others will view him, he sees that to them he is but one of the multitude in no respect better than any other in it. If he would act so as that the impartial spectator may enter into the principles of his conduct, which is what of all things he has the greatest desire to do, he must, upon this, as upon all other occasions, humble the arrogance of his self-love, and bring it down to something which other men can go along with. They will indulge it so far as to allow him to be more anxious about, and to pursue with more earnest assiduity, his own happiness than that of any other person. Thus far, whenever they place themselves in his situation, they will readily go along with him. In the race for wealth, and honours, and preferments, he may run as hard as he can, and strain every nerve and every muscle, in order to outstrip all his competitors. But if he should justle, or throw down any of them, the indulgence of the spectators is entirely at an end. It is a violation of fair play, which they cannot admit of. This man is to them, in every respect, as good as he: they do not enter into that self-love by which he prefers himself so much to this other, and cannot go along with the motive from which he hurt him. They readily, therefore, sympathize with the natural resentment of the injured, and the offender becomes the object of their hatred and indignation. He is sensible that he becomes so, and feels that those sentiments are ready to burst out from all sides against him.

As the greater and more irreparable the evil that is done, the resentment of the sufferer runs naturally the higher; so does likewise the sympathetic indignation of the spectator, as well as the sense of guilt in the agent. Death is the greatest evil which one man can inflict upon another, and excites the highest degree of resentment in those who are immediately connected with the slain. Murder, therefore, is the most atrocious of all crimes which affect individuals only, in the sight both of mankind, and of the person who has committed it. To be deprived of that which we are possessed of, is a greater evil than to be disappointed of what we have only the expectation. Breach of property, therefore, theft and robbery, which take from us what we are possessed of, are greater crimes than breach of contract, which only disappoints us of what we expected. The most sacred laws of justice, therefore, those whose violation seems to call loudest for vengeance and punishment, are the laws which guard the life and person of our neighbour; the next are those which guard his property and possessions, and last of all come those which guard what are called his personal rights, or what is due to him from the promises of others.

The violator of the more sacred laws of justice can never reflect on the sentiments which mankind must entertain with regard to him, without feeling all the agonies of shame, and horror, and consternation. When his passion is gratified, and he begins coolly to reflect on his past conduct, he can enter into none of the motives which influenced it. They appear now as detestable to him as they did always to other people. By sympathizing with the hatred and abhorrence which other men must entertain for him, he becomes in some measure the object of his own hatred and abhorrence. The situation of the person, who suffered by his injustice, now calls upon his pity. He is grieved at the thought of it; regrets the unhappy effects of his own conduct, and feels at the same time that they have rendered him the proper object of the resentment and indignation of mankind, and of what is the natural consequence of resentment, vengeance and punishment. The thought of this perpetually haunts him, and fills him with terror and amazement. He dares no longer look society in the face, but imagines himself as it were rejected, and thrown out from the affections of all mankind. He cannot hope for the consolation of sympathy in this his greatest and most dreadful distress. The remembrance of his crimes has shut out all fellow-feeling with him from the hearts of his fellow-creatures. The sentiments which they entertain with regard to him, are the very thing which he is most afraid of. Every thing seems hostile, and he would be glad to fly to some inhospitable desert, where he might never more behold the face of a human creature, nor read in the countenance of mankind the condemnation of his crimes.

But solitude is still more dreadful than society. His own thoughts can present him with nothing but what is black, unfortunate, and disastrous, the melancholy forebodings of incomprehensible misery and ruin. The horror of solitude drives him back into society, and he comes again into the presence of mankind, astonished to appear before them, loaded with shame and distracted with fear, in order to supplicate some little protection from the countenance of those very judges, who he knows have already all unanimously condemned him. Such is the nature of that sentiment, which is properly called remorse; of all the sentiments which can enter the human breast the most dreadful. It is made up of shame from the sense of the impropriety of past conduct; of grief for the effects of it; of pity for those who suffer by it; and of the dread and terror of punishment from the consciousness of the justly provoked resentment of all rational creatures.

The opposite behaviour naturally inspires the opposite sentiment. The man who, not from frivolous fancy, but from proper motive, has performed a generous action, when he looks forward to those whom he has served, feels himself to be the natural object of their love and gratitude, and, by sympathy with them, of the esteem and approbation of all mankind. And when he looks backward to the motive from which he acted, and surveys it in the light in which the indifferent spectator will survey it, he still continues to enter into it, and applauds himself by sympathy with the approbation of this supposed impartial judge. In both these points of view his own conduct appears to him every way agreeable. His mind, at the thought of it, is filled with cheerfulness, serenity, and composure. He is in friendship and harmony with all mankind, and looks upon his fellow creatures with confidence and benevolent satisfaction, secure that he has rendered himself worthy of their most favourable regards. In the combination of all these sentiments consists the consciousness of merit, or of deserved reward.

Chapter III

Of the utility of this constitution of Nature

It is thus that man, who can subsist only in society, was fitted by nature to that situation for which he was made. All the members of human society stand in need of each others assistance, and are likewise exposed to mutual injuries. Where necessary assistance is reciprocally afforded from love, from gratitude, from friendship, and esteem, the society flourishes and is happy. All the different members of it are bound together by the agreeable bands of love and affection, and are, as it were, drawn to one common centre of mutual good offices.

But though the necessary assistance should not be afforded from such generous and disinterested motives, though among the different members of the society there should be no mutual love and affection, the society, though less happy and agreeable, will not necessarily be dissolved. Society may subsist among different men, as among different merchants, from a sense of its utility, without any mutual love or affection; and though no man in it should owe any obligation, or be bound in gratitude to any other, it may still be upheld by a mercenary exchange of good offices according to an agreed valuation.

Society, however, cannot subsist among those who are at all times ready to hurt and injure one another. The moment that injury begins, the moment that mutual resentment and animosity take place, all the bonds of it are broke asunder, and the different members of which it consisted are, as it were, dissipated and scattered abroad by the violence and opposition of their discordant affections. If there is any society among robbers and murderers, they must at least, according to the trite observation, abstrain from robbing and murdering one another. Beneficence, therefore, is less essential to the existence of society than justice. Society may subsist, though not in the most comfortable state, without beneficence; but the prevalence of injustice must utterly destroy it.

* * *

Among nations of hunters, as there is scarce any property, or at least none that exceeds the value of two or three days labour; so there is seldom any established magistrate or any regular administration of justice. Men who have no property can injure one another only in their persons or reputations. But when one man kills, wounds, beats, or defame another, though he to whom the injury is done suffers, he who does it receives no benefit. It is otherwise with the injuries to property. The benefit of the person who does the injury is often equal to the loss of him who suffers it. Envy, malice, or resentment, are the only passions which can prompt one man to injure another in his person or reputation. But the greater part of men are not very frequently under the influence of those passions; and the very worst men are so only occasionally. As their gratification too, how agreeable soever it may be to certain characters, is not attended with any real or permanent advantage, it is in the greater part of men commonly restrained by prudential considerations. Men may live together in society with some tolerable degree of security, though there is no civil magistrate to protect them from the injustice of those passions. But avarice and ambition in the rich, in the poor the hatred of labour and the love of present ease and enjoyment, are the passions which prompt to invade property, passions much more steady in their operation, and much more universal in their influence.

Wherever there is great property, there is great inequality. For one very rich man, there must be at least five hundred poor, and the affluence of the few supposes the indigence of the many. The affluence of the rich excites the indignation of the poor, who are often both driven by want, and prompted by envy, to invade his possessions. It is only under the shelter of the civil magistrate that the owner of that valuable property, which is acquired by the labour of many years, or perhaps of many successive generations, can sleep a single night in security. He is at all times surrounded by unknown enemies, whom, though he never provoked, he can never appease, and from whose injustice he can be protected only by the powerful arm of the civil magistrate continually held up to chastise it. The acquisition of valuable and extensive property, therefore, necessarily requires the establishment of civil government. Where there is no property, or at least none that exceeds the value of two or three days labour, civil government is not so necessary.

On the Division of Labor

It is the great multiplication of the productions of all the different arts, in consequence of the division of labour, which occasions, in a well-governed society, that universal opulence which extends itself to the lowest ranks of the people. Every workman has a great quantity of his own work to dispose of beyond what he himself has occasion for; and every other workman being exactly in the same situation, he is enabled to exchange a great quantity of his own goods for a great quantity, or, what comes to the same thing, for the price of a great quantity of theirs. He supplies them abundantly with what they have occasion for, and a general plenty diffuses itself through all the different ranks of the society.

Observe the accommodation of the most common artificer or day-labourer in a civilised and thriving country, and you will perceive that the number of people of whose industry a part, though but a small part, has been employed in procuring him this accommodation, exceeds all computation. The woollen coat, for example, which covers the day-labourer, as coarse and rough as it may appear, is the produce of the joint labour of a great multitude of workmen. The shepherd, the sorter of the wool, the wool-comber or carder, the dryer, the scribbler, the spinner, the weaver, the fuller, the dresser, with many others, must all join their different arts in order to complete even this homely production. How many merchants and carriers, besides, must have been employed in transporting the materials from some of those workmen to others who often live in a very distant part of the country! How much commerce and navigation in particular, how many shipbuilders, sailors, sailmakers, rope-makers, must have been employed in order to bring together the different drugs made use of by the dyer, which often come from the remotest corners of the world! What a variety of labour too is necessary in order to produce the tools of the meanest of those workmen! To say nothing of such complicated machines as the ship of the sailor, the mill of the fuller, or even the loom of the weaver, let us consider only what a variety of labour is requisite in order to form that very simple machine, the shears with which the shepherd clips the wool. The miner, the builder of the furnace for smelting the ore, the feller of the timber, the burner of the charcoal to be made use of in the smelting house, the brickmaker, the bricklayer, the workmen who attend the furnace, the millwright, the forger, the smith, must all of them join their different arts in order to produce them.

Were we to examine, in the same manner, all the different parts of his dress and household furniture, the coarse linen shirt which he wears next his skin, the shoes which cover his feet, the bed which he lies on, and all the different parts which compose it, the kitchen grate at which he prepares his victuals, the coals which he makes use of for that purpose, dug from the bowels of the earth, and brought to him perhaps by a long sea and a long land carriage, all the other utensils of his kitchen, all the furniture of his table, the knives and forks, the earthen or pewter plates upon which he serves up and divides his victuals, the different hands employed in prepar-ing his bread and his beer, the glass window which lets in the heat and the light, and keeps out the wind and the rain, with all the knowledge and art requisite for preparing that beautiful and happy invention, without which these northern parts of the world could scarce have afforded a very comfortable habitation, together with the tools of all the different workmen employed in producing those different conve-niences; if we examine, I say, all these things, and consider what a variety of labour is employed about each of them, we shall be sensible that without the assistance and cooperation of many thousands, the very meanest person in a civilised country could not be provided, even according to, what we very falsely imagine, the easy and simple manner in which he is commonly accommodated. Compared, indeed, with the more extravagant luxury of the great, his accommodation must no doubt appear extremely simple and easy; and yet it may be true, perhaps, that the accommodation of an European prince does not always so much exceed that of an industrious and frugal peasant, as the accommodation of the latter exceeds that of many an African king, the absolute master of the lives and liberties of ten thousand naked savages.

This division of labour, from which so many advantages are derived, is not originally the effect of any human wisdom, which foresees and intends that general opulence to which it gives occasion. It is the necessary, though very slow and gradual consequence of a certain propensity in human nature which has in view no such extensive utility; the propensity to truck, barter, and exchange one thing for another.

Whether this propensity be one of those original principles in human nature, of which no further account can be given; or whether, as seems more probable, it be the necessary consequence of the faculties of reason and speech, it belongs not to our present subject to enquire. It is common to all men, and to be found in no other race of animals, which seem to know neither this nor any other species of contracts. Two greyhounds, in running down the same hare, have sometimes the appearance of acting in some sort of concert. Each turns her towards his companion, or endeavours to intercept her when his companion turns her towards himself.

This, however, is not the effect of any contract, but of the accidental concurrence of their passions in the same object at that particular time. Nobody ever saw a dog make a fair and deliberate exchange of one bone for another with another dog. Nobody ever saw one animal by its gestures and natural cries signify to another, this is mine, that yours; I am willing to give this for that. When an animal wants to obtain something either of a man or of another animal, it has no other means of persuasion but to gain the favour of those whose service it requires. A puppy fawns upon its dam, and a spaniel endeavours by a thousand attractions to engage the attention of its master who is at dinner, when it wants to be fed by him. Man sometimes uses the same arts with his brethren, and when he has no other means of engaging them to act according to his inclinations, endeavours by every servile and fawning attention to obtain their good will. He has not time, however, to do this upon every occasion. In civilised society he stands at all times in need of the cooperation and assistance of great multitudes, while his whole life is scarce sufficient to gain the friendship of a few persons.

In almost every other race of animals each individual, when it is grown up to maturity, is entirely independent, and in its natural state has occasion for the assistance of no other living creature. But man has almost constant occasion for the help of his brethren, and it is in vain for him to expect it from their benevolence only. He will be more likely to prevail if he can interest their self-love in his favour, and show them that it is for their own advantage to do for him what he requires of them. Whoever offers to another a bargain of any kind, proposes to do this. Give me that which I want, and you shall have this which you want, is the meaning of every such offer; and it is in this manner that we obtain from one another the far greater part of those good offices which we stand in need of. It is not from the benevolence of the butcher, the brewer, or the baker, that we expect our dinner, but from their regard to their own interest. We address ourselves, not to their humanity but to their self-love, and never talk to them of our own necessities but of their advantages. Nobody but a beggar chooses to depend chiefly upon the benevolence of his fellow-citizens. Even a beggar does not depend upon it entirely. The charity of well-disposed people, indeed, supplies him with the whole fund of his subsistence. But though this principle ultimately provides him with all the necessaries of life which he has occasion for, it neither does nor can provide him with them as he has occasion for them. The greater part of his occasional wants are supplied in the same manner as those of other people, by treaty, by barter, and by purchase. With the money which one man gives him he purchases food. The old clothes which another bestows upon him he exchanges for other old clothes which suit him better, or for lodging, or for food, or for money, with which he can buy either food, clothes, or lodging, as he has occasion.

* * *

And thus the certainty of being able to exchange all that surplus part of the produce of his own labour, which is over and above his own consumption, for such parts of the produce of other men's labour as he may have occasion for, encourages every man to apply himself to a particular occupation, and to cultivate and bring to perfection whatever talent or genius he may possess for that particular species of business.

The difference of natural talents in different men is, in reality, much less than we are aware of, and the very different genius which appears to distinguish men of different professions, when grown up to maturity, is not upon many occasions so much the cause, as the effect of the division of labour. The difference between the most dissimilar characters, between a philosopher and a common street porter, for example, seems to arise not so much from nature, as from habit, custom, and education. When they came into the world, and for the first six or eight years of their existence, they were, perhaps, very much alike, and neither their parents nor play-fellows could perceive any remarkable difference. About that age, or soon after, they come to be employed in very different occupations. The difference of talents comes then to be taken notice of, and widens by degrees, till at last the vanity of the philosopher is willing to acknowledge scarce any resemblance. But without the disposition to truck, barter, and exchange, every man must have procured to himself every necessary and convenience of life which he wanted. All must have had the same duties to perform, and the same work to do, and there could have been no such difference of employment as could alone give occasion to any great difference of talents.

As it is this disposition which forms that difference of talents, so remarkable among men of different professions, so it is this same disposition which renders that difference useful. Many tribes of animals acknowledged to be all of the same species, derive from nature a much more remarkable distinction of genius, than what, antecedent to custom and education, appears to take place among men. By nature a philosopher is not in genius and disposition half so different from a street porter, as a mastiff is from a greyhound, or a greyhound from a spaniel, or this last from a shepherd's dog. Those different tribes of animals, however, though all of the same species, are of scarce any use to one another. The strength of the mastiff is not, in the least, supported either by the swiftness of the greyhound, or by the sagacity of the spaniel, or by the docility of the shepherd's dog. The effects of those different geniuses and talents, for want of the power or disposition to barter and exchange, cannot be brought into a common stock, and do not in the least contribute to the better accommodation and convenience of the species. Each animal is still obliged to support and defend itself, separately and independently, and derives no sort of advantage from that variety of talents with which nature has distinguished its fellows. Among men, on the contrary, the most dissimilar geniuses are of use to one another; the different produces of their respective talents, by the general disposition to truck, barter, and exchange, being brought, as it were, into a common stock, where every man may purchase whatever part of the produce of other men's talents he has occasion for.

Is this improvement in the circumstances of the lower ranks of the people to be regarded as an advantage or as an inconvenience to the society? The answer seems at first sight abundantly plain. Servants, labourers, and workmen of different kinds, make up the far greater part of every great political society. But what improves the circumstances of the greater part can never be regarded as an inconvenience to the whole. No society can surely be flourishing and happy, of which the far greater part of the members are poor and miserable. It is but equity, besides, that they who feed, clothe, and lodge the whole body of the people, should have such a share of the produce of their own labour as to be themselves tolerably well fed, clothed, and lodged.

... the annual revenue of every society is always precisely equal to the exchangeable value of the whole annual produce of its industry, or rather is precisely the same thing with that exchangeable value. As every individual, therefore, endeavours as much as he can both to employ his capital in the support of domestic industry, and so to direct that industry that its produce may be of the greatest value; every individual necessarily labours to render the annual revenue of the society as great as he can. He generally, indeed, neither intends to promote the public interest, nor knows how much he is promoting it. By preferring the support of domestic to that of foreign industry, he intends only his own security and by directing that industry in such a manner as its produce may be of the greatest value, he intends only his own gain, and he is in this, as in many other cases, led by an invisible hand to promote an end which was no part of his intention. Nor is it always the worse for the society that it was no part of it. By pursuing his own interest he frequently promotes that of the society more effectually than when he really intends to promote it. I have never known much good done by those who affected to trade for the public good. It is an affectation, indeed, not very common among merchants, and very few words need be employed in dissuading them from it.

Jean-Jacques Rousseau, on Property, from
The Social Contract (1712–1778)

In the following brief excerpt, Rousseau takes issue with John Locke's notion of a "natural right" to private property, shifting the emphasis from respect for property as such to recognition of the *limits* of one's own rights.

———

Each member of the community gives himself to it at the instant of its constitution, just as he actually is, himself and all his forces, including all the goods in his possession. This is not to say that by this act possession changes its nature as it changes hands and becomes property in the hands of the sovereign. Rather, since the forces of the city are incomparably greater than those of a private individual, public possession is by that very fact stronger and more irrevocable, without being more legitimate, at least to strangers. For with regard to its members, the state is master of all their goods in virtue of the social contract, which serves in the state as the basis of all rights. But with regard to other powers, the state is master only in virtue of the right of the first occupant, which it derives from private individuals.

The right of first occupant, though more real than the right of the strongest, does not become a true right until after the establishment of the right of property. Every man by nature has a right to everything he needs; however, the positive act whereby he becomes a proprietor of some goods excludes him from all the rest. Once his lot has been determined, he should limit himself thereto, no longer having any right against the community. This is the reason why the right of the first occupant, so weak in the state of nature, is able to command the respect of every man living in the civil state. In this right, one respects not so much what belongs to others as what does not belong to oneself.

In general, the following rules must obtain in order to authorize the right of the first occupant on any land. First, this land may not already be occupied by anyone. Second, no one may occupy more than the amount needed to subsist. Third, one is to take possession of it not by an empty ceremony, but by working and cultivating it—the only sign of property that ought, in the absence of legal titles, to be respected by others.

Immanuel Kant, from *The Philosophy of Law*

Kant utterly rejected the utility-minded and sentiment-based ethics of his illustrious Scottish predecessors Hume and Smith, and for him the notion of "justice" or "rights"—including the right to private property or "possession"—required a firm *rational* basis. Kant distinguishes between merely "sensible" possession of an object—that is, my physically possessing it—and the more philosophically significant notion of "intelligible" possession, which means that I have a *right* to the thing in question even when I am not using it, and it is not in my physical possession at all. Kant argues that ownership is not something external to one's self or "Will" but essential to it. The violation of a person's property, in other words, is not just a cause of inconvenience but a violation of the person's freedom, that is, his or her most basic sense of self. In developing this view, Kant asks a typically Kantian question, "how is possession possible?" He distinguishes, as he does in his other works, between an "analytic" (or trivially true) and a "synthetic" proposition, where the latter (but not the former) provides us with some substantial information about the subject in question, in this case the idea that one can rightfully possess or own an object not physically in his or her possession. That is, according to Kant, a matter of *necessity* and not merely "empirical." Property law thus protects what is already a law of a more basic kind, and one can see in Kant some close affinity to Locke, except that Locke's "natural" right to property becomes a strictly "rational" (or "*de jure*") right for Kant, part of the definition of one's self as well as (as in Locke) the definition of the object in question.

Anything is *'Mine' by Right*, or is rightfully Mine, when I am so connected with it, that if any other Person should make use of it without my consent, he would do me a lesion or injury. The subjective condition of the use of anything is *Possession* of it.

An *external* thing, however, as such could only be mine, if I may assume it to be possible that I can be wronged by the use which another might make of it *when it is not actually in my possession*. Hence it would be a contradiction to have anything External as one's own, were not the conception of Possession capable of two different meanings, as *sensible* Possession that is perceivable by the senses, and *rational* Possession that is perceivable only by the Intellect. By the former is to be understood a *physical* Possession, and by the latter, a purely *juridical* Possession of the same object.

The description of an Object as '*external* to me' may signify either that it is merely 'different and distinct from me as a Subject,' or that it is also 'a thing placed

174

outside of me, and to be found elsewhere in space or time.' Taken in the first sense, the term Possession signifies 'rational Possession'; and, in the second sense, it must mean 'Empirical Possession.' A rational or *intelligible* Possession, if such be possible, is Possession *viewed apart from physical holding or detention (detentio).*

* * *

It is possible to have any external object of my Will as Mine. In other words, a Maxim to this effect—were it to become law—that any object on which the Will can be exerted must remain objectively in itself *without an owner*, as 'res nullius,' is contrary to the Principle of Right.

For an object of any act of my Will, is something that it would be *physically* within my power to use. Now, suppose there were things that *by right* should absolutely not be in our power, or, in other words, that it would be wrong or inconsistent with the freedom of all, according to universal Law, to make use of them. On this supposition, Freedom would so far be depriving itself of the use of its voluntary activity, in thus putting *useable* objects out of all possibility of *use.* In practical relations, this would be to annihilate them, by making them *res nullius*, notwithstanding the fact that acts of Will in relation to such things would formally harmonize, in the actual use of them, with the external freedom of all according to universal Laws.

* * *

Any one who would assert the Right to a thing as his, must be in possession of it as an object. Were he not its actual possessor or owner, he could not be wronged or injured by the use which another might make of it without his consent. For, should anything external to him, and in no way connected with him by Right, affect this object, it could not affect himself as a Subject, nor do him any wrong, unless he stood in a relation of Ownership to it.

* * *

Definitions are *nominal* or *real.* A nominal Definition is sufficient merely to *distinguish* the object defined from all other objects, and it springs out of a complete and definite *exposition* of its conception. A real Definition further suffices for a *Deduction* of the conception defined, so as to furnish a knowledge of the reality of the object.—The *nominal Definition* of the external 'Mine' would thus be: 'The external Mine is anything outside of myself, such that any hindrance of my use of it at will, would be doing me an injury or wrong as an infringement of that Freedom of mine which may coexist with the freedom of all others according to a universal Law.' The *real Definition* of this conception may be put thus: 'The external Mine is anything outside of myself, such that any prevention of my use of it would be a wrong, *although I may not be in possession of it* so as to be actually holding it as an object.'—I must be in some kind of possession of an external object, if the object is to be regarded as *mine*; for, otherwise, any one interfering with this object would not, in doing so, affect me; nor, consequently, would he thereby do me any wrong. Hence, a *rational Possession (possession noumenon)* must be assumed as possible, if there is to be rightly an external 'Mine and Thine.' Empirical Possession is thus only phenomenal possession of holding (detention) of the object in the sphere of sensible *appearance (possessio phenomenon).*

* * *

The question, 'How is an *external Mine and Thine* possible?' resolves itself into this
other question, 'How is a *merely juridical* or *rational* Possession possible?'

* * *

All Propositions of Right—as juridical propositions—are Propositions *a priori*, for
they are practical Laws of Reason. But the juridical Proposition *a priori* respecting
empirical Possession is *analytical*; for it says nothing more than what follows by the
principle of Contradiction, from the conception of such possession; namely, that if I
am the holder of a thing in the way of being physically connected with it, any one
interfering with it without my consent—as, for instance, in wrenching an apple out
of my hand—affects and detracts from my freedom as that which is internally Mine;
and consequently the maxim of his action is in direct contradiction to the Axiom of
Right. The proposition expressing the principle of an empirical rightful Possession,
does not therefore go beyond the Right of a Person in reference to himself.

On the other hand, the Proposition expressing the possibility of the Possession
of a thing external to me, after abstraction of all the conditions of empirical
possession in space and time—consequently presenting the assumption of the possibility of a *Possessio Noumenon*—goes beyond these limiting conditions; and because
this Proposition asserts a possession even without physical holding, as necessary to
the conception of the external Mine and Thine, it is *synthetical*. And thus it becomes
a problem for Reason to show how such a Proposition, extending its range beyond
the conception of empirical possession, is possible *a priori*.

In this manner, for instance, the act of taking possession of a particular portion
of the soil, is a mode exercising the private free-will without being an act of
usurpation. The possessor founds upon the innate Right of *common possession* of
the surface of the earth, and upon the universal Will corresponding *a priori* to it,
which allows a *private Possession* of the soil; because what are mere things would be
otherwise made in themselves and by a Law, into inappropriable objects. Thus a first
appropriator acquires originally by primary possession a particular portion of the
ground; and by Right (*jure*) he resists every other person who would hinder him in
the private use of it, although while the 'state of Nature' continues, this cannot be
done by juridical means (*de jure*), because a public Law does not yet exist.

* * *

Simple physical Possession, or holding of the soil, involves already certain relations
of Right to the thing, although it is certainly not sufficient to enable me to regard it
as Mine. Relative to others, so far as they know, it appears as a first possession in
harmony with the law of external freedom; and, at the same time, it is embraced in
the universal original possession which contains *a priori* the fundamental principle
of the possibility of a private possession. Hence to disturb the first occupier or holder
of a portion of the soil in his use of it, is a lesion or wrong done to him. The first
taking of Possession has therefore a Title of Right in its favour, which is simply the
principle of the original common possession; and the saying that 'It is well for those
who are in possession,' when one is not bound to authenticate his possession, is a
principle of Natural Right that establishes the juridical act of taking possession, as a
ground of acquisition upon which every first possessor may found.

* * *

Now it is just an abstraction from physical possession of the object of my free-will in the sphere of sense, that the Practical Reason wills that a rational possession of it shall be thought, according to intellectual conceptions which are not empirical, but contain *a priori* the conditions of rational possession. Hence it is in this fact, that we found the ground of the validity of such a rational conception of possession (*possessio noumenon*) as the principle of a universally valid *Legislation*. For such a Legislation is implied and contained in the expression, 'This external object is *mine*,' because an Obligation is thereby imposed upon all others in respect of it, who would otherwise not have been obliged to abstain from the use of this object.

The mode, then, of having something External to myself as Mine, consists in a specially juridical connection of the Will of the Subject with that object, independently of the empirical relations to it in Space and Time, and in accordance with the conception of a rational possession.

G. W. F. Hegel, from *The Philosophy of Right* (1770–1831)

Hegel published his *Philosophy of Right* in 1821, less than 15 years after *The Phenomenology* but a virtual millennium in terms of social and political change. The earlier book had been composed during the exhilarating days of Napoleon's greatest successes and the promise of liberal reform throughout Europe. The later book was written during "the Reaction," when conservative forces clamped down on all liberal and revolutionary ideas and sentiments, and Hegel, now a distinguished professor in Berlin, had given up his sense of a new and better world. He now tries to come to grips with what's already there, announces that "the rational is actual and the actual is rational" (p. 10) and attacks those social philosophies that would try to imagine a merely ideal society. Philosophy should not tell the world what it ought to be; it is enough to understand it. (One can and should contrast this book, which young Karl Marx knew well, with Marx's own declaration, years later, that "philosophers have only tried to understand the world, . . . the point, however, is to change it.")

In *The Philosophy of Right*, Hegel attempts to lay out a very different picture of society and justice than the one(s) we find in the British philosophers. Hegel, like most Germans, found the British emphasis on "utility" simply vulgar. The British emphasis on the individual and the neglect of community they found utterly wrongheaded, and Hegel, more than most, defended a conception of society as primary and "right"—not rights—as first of all a function of the whole—or what he opaquely calls "free infinite personality"—rather than the individual. Hegel's writing here is notoriously difficult, but the theme is easily stated. What we call the individual is the product of a certain way of thinking, not "immediate" but "mediated" by society. The ultimate "will" is not the individual but what Rousseau called "the general will," or "universal will," which has "all particular individuality absorbed within it." Free will, accordingly, is not a metaphysical feature of an individual but a larger notion, inseparable from a society and its history. And Hegel consequently distinguishes what Kant called "morality"—as a function of purely individual autonomy—and "ethical life," or "*Sittlichkeit*," which he takes to be the primary "natural" unit of social life and (though he does not use any term other than "*Rechts*" or "right") justice.

Ethical life or *Sittlichkeit* begins (as in Rousseau) with the family and one's primal sense of belonging, as existing first of all as a family member and not as an individual. This limited and clearly natural unit becomes incorporated into the larger notion of civil society, conceived as an association of "self-subsistent"

From Hegel, *Philosophy of Right*, translated by T. M. Knox and published by Oxford University Press. Reprinted by permission of the publisher.

members who see themselves as associated only because of their individual needs and protected by an "external organization" (think of Nozick's "protective associations"). But this is no true community, although the very notion of the individual in such an association is, of course, ultimately dependent upon the association itself, for the members do not recognize themselves collectively *as* a community. Here, essentially, is Hegel's charge against the British conception of the state as an external force created by individuals to protect their self-interests. Such a conception gets things backwards. What is needed is explicit and formal recognition of the community *as* a community, and this is the function of a constitution, which renders the community a state. The constitution, for Hegel, does not "set up" the state but rather recognizes the legitimacy of what has been there all along. We should also note that "the state" here refers not to the government but to the entire community, as formalized by the constitution.

Hegel goes into some detail about the nature of property and its role in society. But we can anticipate that Hegel's discussion will have an anti-utilitarian twist to it and will not take the concept of individual rights as primary. Property, for Hegel, is not just the "mixing of one's labor" with nature but rather the *expression* of oneself in nature. The difference might seem slight but it is culturally and historically enormous. Property for Hegel is not a right but rather an expression of one's very self and, accordingly, it need not be an individual self. Hegel's sense of property is much more qualified than that found in British philosophy. We might note that he discusses at some length the notion of *alienation* of property, a concept that would have enormous importance 2 decades later in the writings of young Karl Marx. What Hegel means by this, however, is simply the fact that one can give away or sell the products of one's labor, and it is because we can do this that the institution of contracts becomes essential to the concept of private property. But Hegel also makes the point that, while one can give away or sell the products of one's labor, one cannot give away or sell the source of that labor, one's self (one's talents, etc.). I can do this for a period of time, of course, for example, if I work for someone for wages, but if I were to so "alienate" all of myself, I would make myself into another's property and cease to be a human being. For Marx, of course, this is just the problem, that a wage system that ruthlessly exploits the needs of workers reduces them to the subhuman and makes virtual slaves of them.

. . . the truth about Right, Ethics, and the state is as old as its public recognition and formulation in the law of the land, in the morality of everyday life, and in religion. What more does this truth require—since the thinking mind is not content to possess it in this ready fashion? It requires to be grasped in thought as well; the content which is already rational in principle must win the *form* of rationality and so appear well-founded to untrammelled thinking. Such thinking does not remain stationary at the given, whether the given be upheld by the external positive authority of the state or the *consensus hominum*, or by the authority of inward feeling and emotion and by the 'witness of the spirit' which directly concurs with it. On the contrary, thought which is free starts out from itself and thereupon claims to know itself as united in its innermost being with the truth.

* * *

It is just this placing of philosophy in the actual world which meets with misunderstandings, and so I revert to what I have said before, namely that, since philosophy is the exploration of the rational, it is for that very reason the apprehension of the present and the actual, not the erection of a beyond, supposed to exist, God knows where, or rather which exists, and we can perfectly well say where, namely in the error of a one-sided, empty, ratiocination. In the course of this book, I have remarked that even Plato's *Republic*, which passes proverbially as an empty ideal, is in essence nothing but an interpretation of the nature of Greek ethical life. Plato was conscious that there was breaking into that life in his own time a deeper principle which could appear in it directly only as a longing still unsatisfied, and so only as something corruptive. To combat it, he needs must have sought aid from that very longing itself. But this aid had to come from on High and all that Plato could do was to seek it in the first place in a particular external form of that same Greek ethical life. By that means he thought to master this corruptive invader, and thereby he did fatal injury to the deeper impulse which underlay it, namely free infinite personality. Still, his genius is proved by the fact that the principle on which the distinctive character of his Idea of the state turns is precisely the pivot on which the impending world revolution turned at that time.

What is rational is actual and what is actual is rational. On this conviction the plain man like the philosopher takes his stand, and from it philosophy starts in its study of the universe of mind as well as the universe of nature. If reflection, feeling, or whatever form subjective consciousness may take, looks upon the present as something vacuous and looks beyond it with the eyes of superior wisdom, it finds itself in a vacuum, and because it is actual only in the present, it is itself mere vacuity. If on the other hand the Idea passes for 'only an Idea,' for something represented in an opinion, philosophy rejects such a view and shows that nothing is actual except the Idea. Once that is granted, the great thing is to apprehend in the show of the temporal and transient the substance which is immanent and the eternal which is present.

* * *

To comprehend what is, this is the task of philosophy, because what is, is reason. Whatever happens, every individual is a child of his time; so philosophy too is its own time apprehended in thoughts. It is just as absurd to fancy that a philosophy can transcend its contemporary world as it is to fancy that an individual can overleap his own age.

* * *

One word more about giving instruction as to what the world ought to be. Philosophy in any case always comes on the scene too late to give it. As the thought of the world, it appears only when actuality is already there cut and dried after its process of formation has been completed. The teaching of the concept, which is also history's inescapable lesson, is that it is only when actuality is mature that the ideal first appears over against the real and that the ideal apprehends this same real world in its substance and builds it up for itself into the shape of an intellectual realm. When philosophy paints its grey in grey, then has a shape of life grown old. By philosophy's grey in grey it cannot be rejuvenated but only understood. The owl of Minerva spreads its wings only with the falling of the dusk.

* * *

1. The subject-matter of the philosophical science of right is the Idea of right, i.e., the concept of right together with the actualization of that concept.

* * *

3. Right is positive in general (*a*) when it has the *form* of being valid in a particular state, and this legal authority is the guiding principle for the knowledge of right in this positive form, i.e. for the science of positive law. (*b*) Right in this positive form acquires a positive element in its *content*

(α) through the particular national character of a people, its stage of historical development, and the whole complex of relations connected with the necessities of nature;

(β) because a system of positive law must necessarily involve the application of the universal concept to particular, externally given, characteristics of objects and cases. This application lies outside speculative thought and the development of the concept, and is the subsumption by the Understanding [of the particular under the universal];

(γ) through the finally detailed provisions requisite for actually pronouncing judgement in court.

If inclination, caprice, and the sentiments of the heart are set up in opposition to positive right and the laws, philosophy at least cannot recognize authorities of that sort.—That force and tyranny may be an element in law is accidental to law and has nothing to do with its nature.

* * *

Natural law, or law from the philosophical point of view, is distinct from positive law; but to pervert their difference into an opposition and a contradiction would be a gross misunderstanding.

* * *

4. The basis of right is, in general, mind; its precise place and point of origin is the will. The will is free, so that freedom is both the substance of right and its goal, while the system of right is the realm of freedom made actual, the world of mind brought forth out of itself like a second nature.

* * *

5. The will contains (α) the element of pure indeterminacy or that pure reflection of the ego into itself which involves the dissipation of every restriction and every content either immediately presented by nature, by needs, desires, and impulses, or given and determined by any means whatever. This is the unrestricted infinity of absolute abstraction or universality, the pure thought of oneself.

* * *

Division of the Subject

33. In correspondence with the stages in the development of the Idea of the absolutely free will, the will is

> A. immediate; its concept therefore is abstract, namely personality, and its embodiment is an immediate external thing—the sphere of *Abstract* or *Formal Right*;
>
> B. reflected from its external embodiment into itself—it is then characterized as subjective individuality in opposition to the universal. The universal here is characterized as something inward, the good, and also as something outward, a world presented to the will; both these sides of the Idea are here mediated only by each other. This is the Idea in its division or in its existence as particular; and here we have the right of the subjective will in relation to the right of the world and the right of the Idea, though only the Idea implicit—the sphere of *Morality*;
>
> C. the unity and truth of both these abstract moments—the Idea of the good not only apprehended in thought but so realized both in the will reflected into itself and in the external world that freedom exists as substance, as actuality and necessity, no less than as subjective will; this is the Idea in its absolutely universal existence—*Ethical Life*.
>
> But on the same principle the ethical substance is
>
> (*a*) natural mind, the *Family*;
>
> (*b*) on its division and appearance, *Civil Society*;
>
> (*c*) the *State* as freedom, freedom universal and objective even in the free self-subsistence of the particular will. This actual and organic mind (α) of a single nation (β) reveals and actualizes itself through the inter-relation of the particular national minds until (γ) in the process of world-history it reveals and actualizes itself as the universal world-mind whose right is supreme.

* * *

'Morality' and 'ethical life,' which perhaps usually pass current as synonyms, are taken here in essentially different senses. Yet even commonplace thinking seems to be distinguishing them; Kant generally prefers to use the word 'morality' and, since the principles of action in his philosophy are always limited to this conception, they make the standpoint of ethical life completely impossible, in fact they explicitly nullify and spurn it.

* * *

41. A person must translate his freedom into an external sphere in order to exist as Idea. Personality is the first, still wholly abstract, determination of the absolute and infinite will, and therefore this sphere distinct from the person, the sphere capable of embodying his freedom, is likewise determined as what is immediately different and separable from him. [A.]

42. What is immediately different from free mind is that which, both for mind and in itself, is the external pure and simple, a thing, something not free, not personal, without rights.

* * *

43. As the concept in its *immediacy*, and so as in essence a unit, a person has a *natural* existence partly within himself and partly of such a kind that he is related to

it as to an external world.—It is only these things in their immediacy as things, not what they are capable of becoming through the mediation of the will, i.e. things with determinate characteristics, which are in question here where the topic under discussion is personality, itself at this point still in its most elementary immediacy.

Mental aptitudes, erudition, artistic skill, even things ecclesiastical (like sermons, masses, prayers, consecration of votive objects), inventions, and so forth, become subjects of a contract, brought on to a parity, through being bought and sold, with things recognized as things. It may be asked whether the artist, scholar, &c., is from the legal point of view in possession of his art, erudition, ability to preach a sermon, sing a mass, &c., that is, whether such attainments are 'things.' We may hesitate to call such abilities, attainments, aptitudes, &c., 'things,' for while possession of these may be the subject of business dealings and contracts, as if they were things, there is also something inward and mental about it, and for this reason the Understanding may be in perplexity about how to describe such possession in legal terms, because its field of vision is as limited to the dilemma that this is 'either a thing or not a thing' as to the dilemma 'either finite or infinite.' Attainments, erudition, talents, and so forth are, of course, owned by free mind and are something internal and not external to it, but even so, by expressing them it may embody them in something external and alienate them (see below), and in this way they are put into the category of 'things.' Therefore they are not immediate at the start but only acquire this character through the mediation of mind which reduces its inner possessions to immediacy and externality.

It was an unjustifiable and unethical proviso of Roman law that children were from their father's point of view 'things.' Hence he was legally the owner of his children, although, of course, he still also stood to them in the ethical relation of love (though this relation must have been much weakened by the injustice of his legal position). Here, then, the two qualities 'being a thing' and 'not being a thing' were united, though quite wrongly.

In the sphere of abstract light, we are concerned only with the person as person, and therefore with the particular (which is indispensable if the person's freedom is to have scope and reality) only in so far as it is something separable from the person and immediately different from him, no matter whether this separability constitutes the essential nature of the particular, or whether the particular receives it only through the mediation of the subjective will. Hence in this sphere we are concerned with mental aptitudes, erudition, &c., only in so far as they are possessions in a legal sense; we have not to treat here the possession of our body and mind which we can achieve through education, study, habit, &c., and which exists as an *inward* property of mind. But it is not until we come to deal with alienation that we need begin to speak of the *transition* of such mental property into the external world where it falls under the category of property in the legal sense.

44. A person has as his substantive end the right of putting his will into any and every thing and thereby making it his, because it has no such end in itself and derives its destiny and soul from his will. This is the absolute right of appropriation which man has over all 'things.'

* * *

45. To have power over a thing *ab extra* constitutes possession. The particular aspect of the matter, the fact that I make something my own as a result of my natural

need, impulse, and caprice, is the particular interest satisfied by possession. But I as free will am an object to myself in what I possess and thereby also for the first time am an actual will, and this is the aspect which constitutes the category of *property*, the true and right factor in possession.

If emphasis is placed on my needs, then the possession of property appears as a means to their satisfaction, but the true position is that, from the standpoint of freedom, property is the first embodiment of freedom and so is in itself a substantive end.

46. Since my will, as the will of a person, and so as a single will, becomes objective to me in property, property acquires the character of private property; and common property of such a nature that it may be owned by separate persons acquires the character of an inherently dissoluble partnership in which the retention of my share is explicitly a matter of my arbitrary preference.

* * *

53. Property has its modifications determined in the course of the will's relation to the thing. This relation is

    (A) *taking possession* of the thing directly (here it is in the thing *qua* something positive that the will has its embodiment);

    (B) *use* (the thing is negative in contrast with the will and so it is in the thing as something to be negated that the will has its embodiment);

    (C) *alienation*, the reflection of the will back from the thing into itself.

A. Taking Possession

54. We take possession of a thing (α) by directly grasping it physically, (β) by forming it, and (φ) by merely marking it as ours. [A.]

* * *

59. By being taken into possession, the thing acquires the predicate 'mine' and my will is related to it positively. Within this identity, the thing is equally established as something negative, and my will in this situation is a particular will, i.e. need, inclination, and so forth. Yet my need, as the particular aspect of a single will, is the positive element which finds satisfaction, and the thing, as something negative in itself, exists only for my need and is at its service.—The use of the thing is my need being externally realized through the change, destruction, and consumption of the thing. The thing thereby stands revealed as naturally self-less and so fulfils its destiny.

* * *

63. A thing in use is a single thing determined quantitatively and qualitatively and related to a specific need. But its specific utility, being quantitatively determinate, is at the same time comparable with [the specific utility of] other things of like utility. Similarly, the specific need which it satisfies is at the same time need in general and thus is comparable on its particular side with other needs, while the thing in virtue of the same considerations is comparable with things meeting other needs. This, the thing's universality, whose simple determinate character arises from

the particularity of the thing, so that it is *eo ipso* abstracted from the thing's specific quality, is the thing's *value*, wherein its genuine substantiality becomes determinate and an object of consciousness. As full owner of the thing, I am *eo ipso* owner of its value as well as of its use.

* * *

C. Alienation of Property

65. The reason I can alienate my property is that it is mine only in so far as I put my will into it. Hence I may abandon as a *res nullius* anything that I have or yield it to the will of another and so into his possession, provided always that the thing in question is a thing external by nature.

66. Therefore those goods, or rather substantive characteristics, which constitute my own private personality and the universal essence of my self-consciousness are inalienable and my right to them is imprescriptible. Such characteristics are my personality as such, my universal freedom of will, my ethical life, my religion.

* * *

67. Single products of my particular physical and mental skill and of my power to act I can alienate to someone else and I can give him the use of my abilities for a restricted period, because, on the strength of this restriction, my abilities acquire an external relation to the totality and universality of my being. By alienating the whole of my time, as crystallized in my work, and everything I produced, I would be making into another's property the substance of my being, my universal activity and actuality, my personality.

* * *

68. What is peculiarly mine in a product of my mind may, owing to the method whereby it is expressed, turn at once into something external like a 'thing' which *eo ipso* may then be produced by other people. The result is that by taking possession of a thing of this kind, its new owner may make his own the thoughts communicated in it or the mechanical invention which it contains, and it is ability to do this which sometimes (i.e. in the case of books) constitutes the value of these things and the only purpose of possessing them. But besides this, the new owner at the same time comes into possession of the univeral methods of so expressing himself and producing numerous other things of the same sort.

* * *

Transition from Property to Contract

71. Existence as determinate being is in essence being for another. One aspect of property is that it is an existent as an external thing, and in this respect property exists for other external things and is connected with their necessity and contingency. But it is also an existent as an embodiment of the will, and from this point of view the 'other' for which it exists can only be the will of another person. This relation of will to will is the true and proper ground in which freedom is existent.—

The sphere of contract is made up of this mediation whereby I hold property not merely by means of a thing and my subjective will, but by means of another person's will as well and so hold it in virtue of my participation in a common will.

Reason makes it just as necessary for men to enter into contractual relation-ships—gift, exchange, trade, &c.—as to possess property. While all they are con-scious of is that they are led to make contracts by need in general, by benevolence, advantage, &c., the fact remains that they are led to do this by reason implicit within them, i.e. by the Idea of the real existence of free personality, 'real' here meaning 'present in the will alone.'

Contract presupposes that the parties entering it recognize each other as persons and property owners. It is a relationship at the level of mind objective, and so contains and presupposes from the start the moment of recognition.

Contract

72. Contract brings into existence the property whose external side, its side as an existent, is no longer a mere 'thing' but contains the moment of a will (and conse-quently the will of a second person also). Contract is the process in which there is revealed and mediated the contradiction that I am and remain the independent owner of something from which I exclude the will of another only in so far as in identifying my will with the will of another I cease to be an owner.

* * *

74. This contractual relationship, therefore, is the means whereby one identical will can persist within the absolute difference between independent property owners. It implies that each, in accordance with the common will of both, ceases to be an owner and yet is and remains one. It is the mediation of the will to give up a property, a single property, and the will to take up another, i.e. another belonging to someone else; and this mediation takes place when the two wills are associated in an identity in the sense that one of them comes to its decision only in the presence of the other.

* * *

76. Contract is *formal* when the double consent whereby the common will is brought into existence is apportioned between the two contracting parties so that one of them has the negative moment—the alienation of a thing—and the other the positive moment—the appropriation of the thing. Such a contract is *gift*. But contract may be called *real* when each of the two contracting wills is the sum of these mediating moments and therefore in such a contract becomes a property owner and remains so. This is a contract of *exchange*. [A.]

* * *

142. Ethical life is the Idea of freedom in that on the one hand it is the good become alive—the good endowed in self-consciousness with knowing and willing and actualized by self-conscious action—while on the other hand self-consciousness has in the ethical realm its absolute foundation and the end which actuates its effort. Thus ethical life is the concept of freedom developed into the existing world and the nature of self-consciousness.

* * *

144. The objective ethical order, which comes on the scene in place of good in the abstract, is substance made concrete by subjectivity as infinite form. Hence it posits within itself distinctions whose specific character is thereby determined by the concept, and which endow the ethical order with a stable content independently necessary and subsistent in exaltation above subjective opinion and caprice. These distinctions are absolutely valid laws and institutions.

145. It is the fact that the ethical order is the system of these specific determinations of the Idea which constitutes its rationality. Hence the ethical order is freedom or the absolute will as what is objective, a circle of necessity whose moments are the ethical powers which regulate the life of individuals. To these powers individuals are related as accidents to substance, and it is in individuals that these powers are represented, have the shape of appearance, and become actualized.

* * *

151. But when individuals are simply identified with the actual order, ethical life (*das Sittliche*) appears as their general mode of conduct, i.e. as custom (*Sitte*), while the habitual practice of ethical living appears as a second nature which, put in the place of the initial, purely natural will, is the soul of custom permeating it through and through, the significance and the actuality of its existence. It is mind living and present as a world, and the substance of mind thus exists now for the first time as mind. [A.]

152. In this way the ethical substantial order has attained its right, and its right its validity. That is to say, the self-will of the individual has vanished together with his private conscience which had claimed independence and opposed itself to the ethical substance. For, when his character is ethical, he recognizes as the end which moves him to act the universal which is itself unmoved but is disclosed in its specific determinations as rationality actualized. He knows that his own dignity and the whole stability of his particular ends are grounded in this same universal, and it is therein that he actually attains these. Subjectivity is itself the absolute form and existent actuality of the substantial order, and the distinction between subject on the one hand and substance on the other, as the object, end, and controlling power of the subject, is the same as, and has vanished directly along with, the distinction between them in form.

* * *

153. The right of individuals to be subjectively destined to freedom is fulfilled when they belong to an actual ethical order, because their conviction of their freedom finds its truth in such an objective order, and it is in an ethical order that they are actually in possession of their own essence or their own inner universality.

* * *

154. The right of individuals to their *particular* satisfaction is also contained in the ethical substantial order, since particularity is the outward appearance of the ethical order—a mode in which that order is existent.

155. Hence in this identity of the universal will with the particular will, right and duty coalesce, and by being in the ethical order a man has rights in so far as he

has duties, and duties in so far as he has rights. In the sphere of abstract right, I have the right and another has the corresponding duty. In the moral sphere, the right of my private judgment and will, as well as of my happiness, has not, but only ought to have, coalesced with duties and become objective.

156. The ethical substance, as containing independent self-consciousness united with its concept, is the actual mind of a family and a nation.

157. The concept of this Idea has being only as mind, as something knowing itself and actual, because it is the objectification of itself, the movement running through the form of its moments. It is therefore

(A) ethical mind in its natural or immediate phase—the *Family*. This substantiality loses its unity, passes over into division, and into the phase of relation, i.e. into

(B) *Civil Society*—an association of members as self-subsistent individuals in a universality which, because of their self-subsistence, is only absract. Their association is brought about by their needs, by the legal system—the means to security of person and property—and by an external organization for attaining their particular and common interests. This external stage

(C) is brought back to and welded into unity in the *Constitution of the State* which is the end and actuality of both the substantial universal order and the public life devoted thereto.

The Family

158. The family, as the immediate substantiality of mind, is specifically characterized by love, which is mind's feeling of its own unity. Hence in a family, one's frame of mind is to have self-consciousness of one's individuality within this unity as the absolute essence of oneself, with the result that one is in it not as an independent person but as a member. [A.]

159. The right which the individual enjoys on the strength of the family unity and which is in the first place simply the individual's life within this unity, takes on the *form* of right (as the abstract moment of determinate individuality) only when the family begins to dissolve. At that point those who should be family-members both in their inclination and in actuality begin to be self-subsistent persons, and whereas they formerly constituted one specific moment within the whole, they now receive their share separately and so only in an external fashion by way of money, food, educational expenses, and the like. [A.]

160. The family is completed in these three phases:

(*a*) *Marriage*, the form assumed by the concept of the family in its immediate phase;

(*b*) *Family Property and Capital* (the external embodiment of the concept) and attention to these;

(*c*) *The Education of Children and the Dissolution of the Family.*

* * *

Transition of the Family into Civil Society

181. The family disintegrates (both essentially, through the working of the principle of personality, and also in the course of nature) into a plurality of families, each of

which conducts itself as in principle a self-subsistent concrete person and therefore as externally related to its neighbours. In other words, the moments bound together in the unity of the family, since the family is the ethical Idea still in its concept, must be released from the concept to self-subsistent objective reality. This is the stage of difference. This gives us, to use abstract language in the first place, the determination of particularity which is related to universality but in such a way that universality is its basic principle, though still only an inward principle; for that reason, the universal merely shows in the particular as its form. Hence this relation of reflection prima facie portrays the disappearance of ethical life or, since this life as the essence necessarily shows itself, this relation constitutes the world of ethical appearance—civil society.

* * *

Civil Society

182. The concrete person who is himself the subject of his particular aims, is, as a totality of wants and a mixture of caprice and physical necessity, one principle of civil society. But the particular person is essentially so related to other particular persons that each establishes himself and finds satisfaction by means of the others, and at the same time purely and simply by means of the form of universality, the second principle here. [A.]

183. In the course of the actual attainment of selfish ends—an attainment conditioned in this way by universality—there is formed a system of complete interdependence, wherein the livelihood, happiness, and legal status of one man is interwoven with the livelihood, happiness, and rights of all. On this system, individual happiness, &c., depend, and only in this connected system are they actualized and secured. This system may be prima facie regarded as the external state, the state based on need, the state as the Understanding envisages it.

* * *

188. Civil society contains three moments:

(A) The mediation of need and one-man's satisfaction through his work and the satisfaction of the needs of all others—the *System of Needs*.

(B) The actuality of the universal principle of freedom therein contained—the protection of property through the *Administration of Justice*.

(C) Provision against contingencies still lurking in systems (A) and (B), and care for particular interests as a common interest, by means of the *Police* and the *Corporation*.

* * *

199. When men are thus dependent on one another and reciprocally related to one another in their work and the satisfaction of their needs, subjective self-seeking turns into a contribution to the satisfaction of the needs of everyone else. That is to say, by a dialectical advance, subjective self-seeking turns into the mediation of the particular through the universal, with the result that each man in earning, producing, and enjoying on his own account is *eo ipso* producing and earning for the enjoyment

of everyone else. The compulsion which brings this about is rooted in the complex interdependence of each on all, and it now presents itself to each as the universal permanent capital which gives each the opportunity, by the exercise of his education and skill, to draw a share from it and so be assured of his livelihood, while what he thus earns by means of his work maintains and increases the general capital.

200. A particular man's resources, or in other words his opportunity of sharing in the general resources, are conditioned, however, partly by his own unearned principal (his capital), and partly by his skill; this in turn is itself dependent not only on his capital, but also on accidental circumstances whose multiplicity introduces differences in the development of natural, bodily, and mental characteristics, which were already in themselves dissimilar. In this sphere of particularity, these differences are conspicuous in every direction and on every level, and, together with the arbitrariness and accident which this sphere contains as well, they have as their inevitable consequence disparities of individual resources and ability.

* * *

201. The infinitely complex, criss-cross, movements of reciprocal production and exchange, and the equally infinite multiplicity of means therein employed, become crystallized, owing to the universality inherent in their content, and distinguished into general groups. As a result, the entire complex is built up into particular systems of needs, means, and types of work relative to these needs, modes of satisfaction and of theoretical and practical education, i.e. into systems, to one or other of which individuals are assigned—in other words, into class-divisions.

* * *

208. As the private particularity of knowing and willing, the principle of this system of needs contains absolute universality, the universality of freedom, only abstractly and therefore as the right of property. At this point, however, this right is no longer merely implicit but has attained its recognized actuality as the protection of property through the administration of justice.

B. The Administration of Justice

209. The relatedness arising from the reciprocal bearing on one another of needs and work to satisfy these is first of all reflected into itself as infinite personality, as abstract right. But it is this very sphere of relatedness—a sphere of education—which gives abstract right the determinate existence of being something universally recognized, known, and willed, and having a validity and an objective actuality mediated by this known and willed character.

It is part of education, of thinking as the consciousness of the single in the form of universality, that the ego comes to be apprehended as a universal person in which all are identical. A man counts as a man in virtue of his manhood alone, not because he is a Jew, Catholic, Protestant, German, Italian, &c. This is an assertion which thinking ratifies and to be conscious of it is of infinite importance. It is defective only when it is crystallized, e.g. as a cosmopolitanism in opposition to the concrete life of the state. [A.]

210. The objective actuality of the right consists, first, in its existence for consciousness, in its being known in some way or other; secondly, in its possessing

the power which the actual possesses, in its being valid, and so also in its becoming known as universally valid.

(a) RIGHT AS LAW

211. The principle of rightness becomes the law (*Gesetz*) when, in its objective existence, it is posited (*gesetzt*), i.e. when thinking makes it determinate for consciousness and makes it known as what is right and valid; and in acquiring this determinate character, the right becomes positive law in general.

* * *

In becoming law, what is right acquires for the first time not only the form proper to its universality, but also its true determinacy. Hence making a law is not to be represented as merely the expression of a rule of behaviour valid for everyone, though that is one moment in legislation; the more important moment, the inner essence of the matter, is knowledge of the content of the law in its determinate universality.

Since it is only animals which have their law as instinct, while it is man alone who has law as custom, even systems of customary law contain the moment of being thoughts and being known. Their difference from positive law consists solely in this, that they are known only in a subjective and accidental way, with the result that in themselves they are less determinate and the universality of thought is less clear in them. (And apart from this, knowledge of a system of law either in general or in its details, is the accidental possession of a few.) The supposition that it is customary law, on the strength of its character as custom, which possesses the privilege of having become part of life is a delusion, since the valid laws of a nation do not cease to be its customs by being written and codified—and besides, it is as a rule precisely those versed in the deadest of topics and the deadest of thoughts who talk nowadays of 'life' and of 'becoming part of life.' When a nation begins to acquire even a little culture, its customary law must soon come to be collected and put together. Such a collection is a legal code, but one which, as a mere collection, is markedly formless, indeterminate, and fragmentary. The main difference between it and a code properly so-called is that in the latter the principles of jurisprudence in their universality, and so in their determinacy, have been apprehended in terms of thought and expressed. English national law or municipal law is contained, as is well known, in statutes (written laws) and in so-called 'unwritten' laws. This unwritten law, however, is as good as written, and knowledge of it may, and indeed must, be acquired simply by reading the numerous quartos which it fills. The monstrous confusion, however, which prevails both in English law and its administration is graphically portrayed by those acquainted with the matter. In particular, they comment on the fact that, since this unwritten law is contained in court verdicts and judgements, the judges are continually legislators. The authority of precedent is binding on them, since their predecessors have done nothing but give expression to the unwritten law; and yet they are just as much exempt from its authority, because they are themselves repositories of the unwritten law and so have the right to criticize previous judgements and pronounce whether they accorded with the unwritten law or not.

* * *

No greater insult could be offered to a civilized people or to its lawyers than to deny them ability to codify their law; for such ability cannot be that of constructing a legal system with a novel content, but only that of apprehending, i.e. grasping in thought, the content of existing laws in its determinate universality and then applying them to particular cases.

* * *

(c) THE COURT OF JUSTICE

219. By taking the form of law, right steps into a determinate mode of being. It is then something on its own account, and in contrast with particular willing and opining of the right, it is self-subsistent and has to vindicate itself as something universal. This is achieved by recognizing it and making it actual in a particular case without the subjective feeling of private interest; and this is the business of a public authority—the court of justice.

John Stuart Mill, from *Utilitarianism* (Chapter V) (1806–1873)

Mill was not unaware that one of the strongest objections to his theory of utilitarianism was the charge that the principle of utility took no account of rights and justice. Hume, of course, had argued a full century before that the "sole" justification for justice was public utility and Adam Smith had repeated a similar argument after him, but the point was not yet convincing (nor is it still, considering the repetition of much the same charges by such contemporary philosophers as John Rawls). Mill begins by shifting the burden, questioning whether the ideal of justice would make any sense or have any appeal at all if it were not also expedient, and much of Mill's argument (following a much more belligerent argument of a similar sort by Jeremy Bentham many years before) consists in undermining the appearance of "justice" as a clear and immediately comprehensible ideal. He runs through the notion of rights, in particular, and distinguishes legal and moral rights and the notion of desert. He briefly discusses punishment and makes the point (wholly consistent with his own insistence on equality in utilitarianism) that justice requires impartiality. He then discusses equality as such, pointing out that "so many diverse applications" of the term "justice" surely ought to be an "embarrassment." Mill goes on to soothe this embarrassment (but in fact to further aggravate it) by pointing out the etymology of the word, meaning *law* in various languages, and concluding that what the word really expresses, in all of its various usages, is the recognition that some things ought and others ought not to be done coupled with the desire that there should always be some enforcement or sanction for these, whether by actual law (often inexpedient) or not. The result of this somewhat sneaky argument is to make the point that justice is not a particular ideal that can therefore be contrasted with utility, but rather a very general moral concern that can be easily reconciled (if not identified) with social utility.

So, too, Mill considers rights as independent ideals embodied in persons and argues instead that a right should rather be considered as a *claim* that a person can legitimately make on society. Why should society acknowledge or defend me in my rights? Mill's answer, predictably, is for the reason of general utility. The opposition to this answer, Mill somewhat unfairly argues, is it can only be because of the natural "thirst for retaliation"—or revenge—that is not satisfied with the security and well-being of all. Needless to say, Mill has virtually the whole of modern thought on his side here as well as the particular history of utilitarianism, which began (in part) as an attack on the barbarity of the penal system and an insistence on effective deterrence rather than retribution and revenge. But clearly the sort of insistence that goes into the juxtaposition of justice and utility does not turn on the much more limited notion of retribution alone. And whether or not the long-term effects of justice are indeed conducive to general utility (a hard claim to prove, one way or the other), it is clear that *particular* acts of injustice, especially if undetected, may substantially improve public utility while particular acts of justice may well not.

Chapter V. On the Connection Between Justice and Utility

In all ages of speculation one of the strongest obstacles to the reception of the doctrine that utility or happiness is the criterion of right and wrong has been drawn from the idea of justice. The powerful sentiment and apparently clear perception which that word recalls with a rapidity and certainty resembling an instinct have seemed to the majority of thinkers to point to an inherent quality in things; to show that the just must have an existence in nature as something absolute, generically distinct from every variety of the expedient and, in idea, opposed to it, though (as is commonly acknowledged) never, in the long run, disjoined from it in fact.

In the case of this, as of our other moral sentiments, there is no necessary connection between the question of its origin and that of its binding force. That a feeling is bestowed on us by nature does not necessarily legitimate all its promptings. The feeling of justice might be a peculiar instinct, and might yet require, like our other instincts, to be controlled and enlightened by a higher reason. If we have intellectual instincts leading us to act in a particular way, as well as animal instincts that prompt us to act in a particular way, there is no necessity that the former should be more infallible in their sphere than the latter in theirs; it may as well happen that wrong judgments are occasionally suggested by those, as wrong actions by these. But though it is one thing to believe that we have natural feelings of justice, and another to acknowledge them as an ultimate criterion of conduct, these two opinions are very closely connected in point of fact. Mankind are always predisposed to believe that any subjective feeling, not otherwise accounted for, is a revelation of some objective reality. Our present object is to determine whether the reality to which the feeling of justice corresponds is one which needs any such special revelation, whether the justice or injustice of an action is a thing intrinsically peculiar and distinct from all its other qualities or only a combination of certain of those qualities presented under a peculiar aspect. For the purpose of this inquiry it is practically important to consider whether the feeling itself, of justice and injustice, is *sui generis* like our sensations of color and taste or a derivative feeling formed by a combination of others. And this it is the more essential to examine, as people are in general willing enough to allow that objectively the dictates of justice coincide with a part of the field of general expediency; but inasmuch as the subjective mental feeling of justice is different from that which commonly attaches to simple expediency, and, except in the extreme cases of the latter, is far more imperative in its demands, people find it difficult to see in justice only a particular kind or branch of general utility, and think that its superior binding force requires a totally different origin.

To throw light upon this question, it is necessary to attempt to ascertain what is the distinguishing character of justice, or of injustice; what is the quality, or whether there is any quality, attributed in common to all modes of conduct designated as unjust (for justice, like many other moral attributes, is best defined by its opposite), and distinguishing them from such modes of conduct as are disapproved, but without having that particular epithet of disapprobation applied to them. If in everything which men are accustomed to characterize as just or unjust some one common attribute or collection of attributes is always present, we may judge whether this particular attribute or combination of attributes would be capable of gathering round it a sentiment of that peculiar character and intensity of virtue of the general laws of our emotional constitution, or whether the sentiment is inexplicable and requires to be regarded as a special provision of nature. If we find the former to be the case, we shall, in resolving this question, have resolved also the main problem; if the latter, we shall have to seek for some other mode of investigating it:

To find the common attributes of a variety of objects, it is necessary to begin by surveying the objects themselves in the concrete. Let us therefore advert successively to the various modes of action and arrangements of human affairs which are classed, by universal or widely spread opinion, as just or as unjust. The things well known to excite the sentiments associated with those names are of a very multifarious character. I shall pass them rapidly in review, without studying any particular arrangement.

In the first place, it is mostly considered unjust to deprive anyone of his personal liberty, his property, or any other thing which belongs to him by law. Here, therefore, is one instance of the application of the terms "just" and "unjust" in a perfectly definite sense, namely, that it is just to respect, unjust to violate, the *legal rights* of anyone. But this judgment admits of several exceptions, arising from the other forms in which the notions of justice and injustice present themselves. For example, the person who suffers the deprivation may (as the phrase is) have *forfeited* the rights which he is so deprived of—a case to which we shall return presently. But also—

Secondly, the legal rights of which he is deprived may be rights which *ought* not to have belonged to him; in other words, the law which confers on him these rights may be a bad law. When it is so or when (which is the same thing for our purpose) it is supposed to be so, opinions will differ as to the justice or injustice of infringing it. Some maintain that no law, however bad, ought to be disobeyed by an individual citizen; that his opposition to it, if shown at all, should only be shown in endeavoring to get it altered by competent authority. This opinion (which condemns many of the most illustrious benefactors of mankind, and would often protect pernicious institutions against the only weapons which, in the state of things existing at the time, have any chance of succeeding against them) is defended by those who hold it on grounds of expediency, principally on that of the importance to the common interest of mankind, of maintaining inviolate the sentiment of submission to law. Other persons, again, hold the directly contrary opinion that any law, judged to be bad, may blamelessly be disobeyed, even though it be not judged to be unjust but only expedient, while others would confine the license of disobedience to the case of unjust laws; but, again, some say that all laws which are inexpedient are unjust, since every law imposes some restriction on the natural liberty of mankind, which restriction is an injustice unless legitimated by tending to their good. Among these diversities of opinion it seems to be universally admitted that there may be unjust laws, and that law, consequently, is not the ultimate criterion of justice, but may give to one person a benefit, or impose on another an evil, which justice condemns. When, however, a law is thought to be unjust, it seems always to be regarded as being so in the same way in which a breach of law is unjust, namely, by infringing somebody's right, which, as it cannot in this case be a legal right, receives a different appellation and is called a moral right. We may say, therefore, that a second case of injustice consists in taking or withholding from any person that to which he has a *moral right*.

Thirdly, it is universally considered just that each person should obtain that (whether good or evil) which he *deserves*, and unjust that he should obtain a good or be made to undergo an evil which he does not deserve. This is, perhaps, the clearest and most emphatic form in which the idea of justice is conceived by the general mind. As it involves the notion of desert, the question arises what constitutes desert? Speaking in a general way, a person is understood to deserve good if he does right, evil if he does wrong; and in a more particular sense, to deserve good from those to whom he does or has done good, and evil from those to whom he does or has done evil. The precept of returning good for evil has never been regarded as a case of the fulfillment of justice, but as one in which the claims of justice are waived, in obedience to other considerations.

Fourthly, it is confessedly unjust to *break* faith with anyone: to violate an engagement, either express or implied, or disappoint expectations raised by our own conduct, at least if we have raised those expectations knowingly and voluntarily. Like the other obligations of justice already spoken of, this one is not regarded as absolute, but as capable of being overruled by a stronger obligation of justice on the other side, or by such conduct on the part of the person concerned as is deemed to absolve us from our obligation to him and to constitute a *forfeiture* of the benefit which he has been led to expect.

Fifthly, it is, by universal admission, inconsistent with justice to be *partial*—to show favor or preference to one person over another in matters to which favor and preference do not properly apply. Impartiality, however, does not seem to be regarded as a duty in itself, but rather as instrumental to some other duty; for it is admitted that favor and preference are not always censurable, and, indeed, the cases in which they are condemned are rather the exception than the rule. A person would be more likely to be blamed than applauded for giving his family or friends no superiority in good offices over strangers when he could do so without violating any other duty; and no one thinks it unjust to seek one person in preference to to another as a friend, connection, or companion. Impartiality where rights are concerned is of course obligatory, but this is involved in the more general obligation of giving to everyone his right. A tribunal, for example, must be impartial because it is bound to award, without regard to any other consideration, a disputed object to the one of two parties who has the right to it. There are other cases in which impartiality means being solely influenced by desert, as with those who, in the capacity of judges, preceptors, or parents, administer reward and punishment as such. There are cases, again, in which it means being solely influenced by consideration for the public interest, as in making a selection among candidates for a government employment. Impartiality, in short, as an obligation of justice, may be said to mean being exclusively influenced by the considerations which it is supposed ought to influence the particular case in hand, and resisting solicitation of any motives which prompt to conduct different from what those considerations would dictate.

Nearly allied to the idea of impartiality is that of *equality*, which often enters as a component part both into the conception of justice and into the practice of it, and, in the eyes of many persons, constitutes its essence. But in this, still more than in any other case, the notion of justice varies in different persons, and always conforms in its variations to their notion of utility. Each person maintains that equality is the dictate of justice, except where he thinks that expediency requires inequality. The justice of giving equal protection to the rights of all is maintained by those who support the most outrageous inequality in the rights themselves. Even in slave countries it is theoretically admitted that the rights of the slave, such as they are, ought to be as sacred as those of the master, and that a tribunal which fails to enforce them with equal strictness is wanting in justice; while, at the same time, institutions which leave to the slave scarcely any rights to enforce are not deemed unjust because they are not deemed inexpedient. Those who think that utility requires distinctions of rank do not consider it unjust that riches and social privileges should be unequally dispensed; but those who think this inequality inexpedient think it unjust also. Whoever thinks that government is necessary sees no injustice in as much inequality as is constituted by giving to the magistrate powers not granted to other people. Even among those who hold leveling doctrines, there are differences of opinion about expediency. Some communists consider it unjust that the produce of labor of the community should be shared on any other principle than that of exact equality; others think it just that those should receive most whose wants are greatest; while others hold that those who work harder, or who produce more, or whose services are more valuable to the community, may justly claim a larger quota in the division of the produce. And the sense of natural justice may be plausibly appealed to in behalf of every one of these opinions.

Among so many diverse applications of the term "justice," which yet is not regarded as ambiguous, it is a matter of some difficulty to seize the mental link which holds them together, and on which the moral sentiment adhering to the term essentially depends. Perhaps, in this embarrassment, some help may be derived from the history of the word, as indicated by its etymology.

In most if not in all languages, the etymology of the word which corresponds to "just" points distinctly to an origin connected with the ordinances of law. *Justum* is a form of *jussum*, that which has been ordered. *Dikaion* comes directly from *dike*, a suit at law. *Recht*, from which came *right* and *righteous*, is synonymous with law. The courts of justice, the administration of justice, are the courts and the administration of law. *La justice*, in French, is the established term for judicature. I am not committing the fallacy, imputed with some show of truth to Horne Tooke, of assuming that a word must still continue to mean what it originally meant. Etymology is slight evidence of what the idea now signified is, but the very best evidence of how it sprang up. There can, I think, be no doubt that the *idée mère*, the primitive element, in the formation of the notion of justice was conformity to law. It constituted the entire idea among the Hebrews, up to the birth of Christianity; as might be expected in the case of a people whose laws attempted to embrace all subjects on which precepts were required, and who believed those laws to be a direct emanation from the Supreme Being. But other nations, and in particular the Greeks and Romans, who knew that their laws had been made originally, and still continued to be made, by men, were not afraid to admit that those men might make bad laws; might do, by law, the same things, and from the same motives, which if done by individuals without the sanction of law would be called unjust. And hence the sentiment of injustice came to be attached, not to all violations of law, but only to violations of such laws as *ought* to exist, including such as ought to exist but do not, and to laws themselves if supposed to be contrary to what ought to be law. In this manner the idea of law and of its injunctions was still predominant in the notion of justice, even when the laws actually in force ceased to be accepted as the standard of it.

It is true that mankind consider the idea of justice and its obligations as applicable to many things which neither are, nor is it desired that they should be, regulated by law. Nobody desires that laws should interfere with the whole detail of private life; yet everyone allows that in all daily conduct a person may and does show himself to be either just or unjust. But even here, the idea of the breach of what ought to be law still lingers in a modified shape. It would always give us pleasure, and chime in with our feelings of fitness, that acts which we deem unjust should be punished, though we do not always think it expedient that this should be done by the tribunals. We forego that gratification on account of incidental inconveniences. We should be glad to see just conduct enforced and injustice repressed, even in the minutest details, if we were not, with reason, afraid of trusting the magistrate with so unlimited an amount of power over individuals. When we think that a person is bound in justice to do a thing, it is an ordinary form of language to say that he ought to be compelled to do it. We should be gratified to see the obligation enforced by anybody who had the power. If we see that its enforcement by law would be inexpedient, we lament the impossibility, we consider the impunity given to injustice as an evil, and strive to make amends for it by bringing a strong expression of our own and the public disapprobation to bear upon the offender. Thus the idea of legal constraint is still the generating idea of the notion of justice, though undergoing several transformations before that notion as it exists in an advanced state of society becomes complete.

* * *

I have, throughout, treated the idea of a *right* residing in the injured person and violated by the injury, not as a separate element in the composition of the idea and sentiment, but as one of the forms in which the other two elements clothe themselves. These elements are a hurt to some assignable person or persons, on the one hand, and a demand for punishment, on the other. An examination of our own minds, I think, will show that these two things include all that we mean when we speak of violation of a right. When we call anything a person's right, we mean that he has a valid claim on society to protect him in the possession of it, either by the force of law or by that of education and opinion. If he has what we consider a sufficient claim, on whatever account, to have something guaranteed to him by society, we say that he has a right to it. If we desire to prove that anything does not belong to him by right, we think this done as soon as it is admitted that society ought not to take measures for securing it to him, but should leave him to change or to his own exertions. Thus a person is said to have a right to what he can earn in fair professional competition, because society ought not to allow any other person to hinder him from endeavoring to earn in that manner as much as he can. But he has not a right to three hundred a year, though he may happen to be earning it; because society is not called on to provide that he shall earn that sum. On the contrary, if he owns ten thousand pounds three-per-cent stock, he *has* a right to three hundred a year because society has come under an obligation to provide him with an income of that amount.

To have a right, then, is, I conceive, to have something which society ought to defend me in the possession of. If the objector goes on to ask why it ought, I can give him no other reason than general utility. If that expression does not seem to convey a sufficient feeling of the strength of the obligation, nor to account for the peculiar energy of the feeling, it is because there goes to the composition of the sentiment, not a rational only but also an animal element—the thirst for retaliation; and this thirst derives its intensity, as well as its moral justification, from the extraordinarily important and impressive kind of utility which is concerned. The interest involved is that of security, to everyone's feelings the most vital of all interests. All other earthly benefits are needed by one person, not needed by another; and many of them can, if necessary, be cheerfully foregone or replaced by something else; but security no human being can possibly do without; on it we depend for all our immunity from evil and for the whole value of all and every good, beyond the passing moment, since nothing but the gratification of the instant could be of any worth to us if we could be deprived of everything the next instant by whoever was momentarily stronger than ourselves. Now this most indispensable of all necessaries, after physical nutriment, cannot be had unless the machinery for providing it is kept unintermittedly in active play. Our notion, therefore, of the claim we have on our fellow creatures to join in making safe for us the very groundwork of our existence gathers feelings around it so much more intense than those concerned in any of the more common cases of utility that the difference in degree (as is often the case in psychology) becomes a real difference in kind. The claim assumes that character of absoluteness, that apparent infinity and incommensurability with all other considerations which constitute the distinction between the feeling of right and wrong and that of ordinary expediency and inexpediency. The feelings concerned are so powerful, and we count so positively on finding a responsive feeling in others (all being alike interested) that *ought* and *should* grow into *must*, and recognized indispensability becomes a moral necessity, analogous to physical, and often not inferior to it in binding force.

If the preceding analysis, or something resembling it, be not the correct account of the notion of justice—if justice be totally independent of utility, and be a standard *per se*, which the mind can recognize by simple introspection of itself—it is hard to understand why that internal oracle is so ambiguous, and why so many things appear either just or unjust, according to the light in which they are regarded.

* * *

Most of the maxims of justice current in the world, and commonly appealed to in its transactions, are simply instrumental to carrying into effect the principles of justice which we have now spoken of. That a person is only responsible for what he has done voluntarily, or could voluntarily have avoided, that it is unjust to condemn any person unheard; that the punishment ought to be proportioned to the offense, and the like, are maxims intended to prevent the just principle of evil for evil from being perverted to the infliction of evil without that justification. The greater part of these common maxims have come into use from the practice of courts of justice, which have been naturally led to a more complete recognition and elaboration than was likely to suggest itself to others, of the rules necessary to enable them to fulfill their double function—of inflicting punishment when due, and of awarding to each person his right.

That first of judicial virtues, impartiality, is an obligation of justice, partly for the reason last mentioned, as being a necessary condition of the fulfillment of other

obligations of justice. But this is not the only source of the exalted rank, among
human obligations, of those maxims of equality and impartiality, which, both in
popular estimation and in that of the most enlightened, are included among the
precepts of justice. In one point of view, they may be considered as corollaries from
the principles already laid down. If it is a duty to do to each according to his deserts,
returning good for good, as well as repressing evil by evil, it necessarily follows that
we should treat all equally well (when no higher duty forbids) who have deserved
equally well of *us*, and that society should treat all equally well who have deserved
equally well of *it*, that is, who have deserved equally well absolutely. This is the
highest abstract standard of social and distributive justice, toward which all institutions and the efforts of all virtuous citizens should be made in the utmost possible
degree to converge. But this great moral duty rests upon a still deeper foundation,
being a direct emanation from the first principle of morals, and not a mere logical
corollary from secondary or derivative doctrines. It is involved in the very meaning
of utility, or the greatest happiness principle. That principle is a mere form of words
without rational signification unless one person's happiness, supposed equal in
degree (with the proper allowance made for kind), is counted for exactly as much as
another's. Those conditions being supplied, Bentham's dictum, "everybody to count
for one, nobody for more than one," might be written under the principle of utility as
an explanatory commentary. The equal claim of everybody to happiness, in the
estimation of the moralist and of the legislator, involves an equal claim to all
the means of happiness except in so far as the inevitable conditions of human life and
the general interest in which that of every individual is included set limits to the
maxim; and those limits ought to be strictly construed. As every other maxim of
justice, so this is by no means applied or held applicable universally; on the contrary,
as I have already remarked, it bends to every person's ideas of social expediency. But
in whatever case it is seemed applicable at all, it is held to be the dictate of justice. All
persons are deemed to have a *right* to equality of treatment, except when some
recognized social expediency requires the reverse. And hence all social inequalities
which have ceased to be considered expedient assume the character, not of simple
inexpediency, but of injustice, and appear so tyrannical that people are apt to
wonder how they ever could have been tolerated—forgetful that they themselves,
perhaps, tolerate other inequalities under an equally mistaken notion of expediency,
the correction of which would make that which they approve seem quite as monstrous as what they have at last learned to condemn. The entire history of social
improvement has been a series of transitions by which one custom or institution after
another, from being a supposed primary necessity of social existence, has passed into
the rank of a universally stigmatized injustice and tyranny. So it has been with the
distinctions of slaves and freemen, nobles and serfs, patricians and plebeians; and so
it will be, and in part already is, with the aristocracies of color, race, and sex.

It appears from what has been said that justice is a name for certain moral
requirements which, regarded collectively, stand higher in the scale of social utility,
and are therefore of more paramount obligation, than any others, though particular
cases may occur in which some other social duty is so important as to overrule any
one of the general maxims of justice. Thus, to save a life, it may not only be
allowable, but a duty, to steal or take by force the necessary food or medicine, or to
kidnap and compel to officiate the only qualified medical practitioner. In such cases,

as we do not call anything justice which is not a virtue, we usually say, not that justice must give way to some other moral principle, but that what is just in ordinary cases is, by reason of that principle, not just in the particular case. By this useful accommodation of language, the character of indefeasibility attributed to justice is kept up, and we are saved from the necessity of maintaining that there can be laudable injustice.

Friedrich Engels, *Principles of Communism* and "against arm-chair justice" (from *Anti-Duhring*) (1820–1895)

Engels met Marx in the 1840s, when the latter was in virtual exile from his German homeland and the two became lifelong friends. Engels was the well-to-do son of a Manchester industrialist, but he spent much of his life formulating and defending the principles of communism and the attack on capitalism, and he supported his financially much worse off colleague for much of his life. Together, the two of them wrote the *Communist* in 1848. Engels's pamphlet, *The Principles of Communism*, was written first, and served as something of a prototype for the *Manifesto*.

It may seem remarkable that, in the voluminous collected works of Marx and Engels—two philosophers who would have seemed more than any others obsessed with questions of justice—they virtually never discussed the question of justice as such. The short excerpt from Engels's polemic, *Anti-Duhring*, gives us an important clue why. They considered the concepts of "justice" and "injustice" nothing but abstract philosophical ideals, useless for the purpose of actually changing the world and making it more just. Moreover, one can easily surmise that these two radical thinkers suspected that the terms "justice" and "injustice" had been so thoroughly co-opted by the very system they opposed. A vocabulary that had been so taken over by discussions of "merit" and "rights" would have no place in the new philosophy of "from each according to [his] abilities, to each according to [his] needs."

Principles of Communism

Question 1. What is communism?

Answer. Communism is the doctrine of the conditions of the liberation of the proletariat.

Question 2. What is the proletariat?

Answer. The proletariat is that class in society which lives entirely from the sale of its labor and does not draw profit from any kind of capital; whose weal and woe, whose life and death, whose whole existence depends on the demand for labor, hence on the changing state of business, on the vagaries of unbridled competition. The proletariat, or the class of proletarians, is, in a word, the working class of the nineteenth century.

Question 3. Proletarians, then, have not always existed?

Answer. No. There have always been poor and working classes; and the working classes have mostly been poor. But there have not always been workers and poor people living under conditions as they are today; in other words, there have not always been proletarians, any more than there has always been free unbridled competition.

Question 4. How did the proletariat originate?

Answer. The proletariat originated in the industrial revolution which took place in England in the last half of the last [eighteenth] century, and which has since then been repeated in all the civilized countries of the world. This industrial revolution was precipitated by the discovery of the steam engine, various spinning machines, the mechanical loom, and a whole series of other mechanical devices. These machines, which were very expensive and hence could be bought only by big capitalists, altered the whole mode of production and displaced the former workers, because the machines turned out cheaper and better commodities than the workers could produce with their inefficient spinning wheels and handlooms. The machines delivered industry wholly into the hands of the big capitalists and rendered entirely worthless the meager property of the workers (tools, looms, etc.). The result was that the capitalists soon had everything in their hands and nothing remained to the workers. This marked the introduction of the factory system into the textile industry.

Once the impulse to the introduction of machinery and the factory system had been given, this system spread quickly to all other branches of industry, especially cloth- and book-printing, pottery, and the metal industries. Labor was more and more divided among the individual workers so that the worker who previously had done a complete piece of work now did only part of that piece. This division of labor made it possible to produce things faster and cheaper. It reduced the activity of the individual worker to simple, endlessly repeated mechanical motions which could be performed not only as well but much better by a machine. In this way, all these industries fell, one after another, under the dominance of steam, machinery, and the factory system, just as spinning and weaving had already done. But at the same time they also fell into the hands of big capitalists, and their workers were deprived of whatever independence remained to them. Gradually, not only genuine manufacture but also handicrafts came within the province of the factory system as big capitalists increasingly displaced the small master craftsmen by setting up huge workshops which saved many expenses and permitted an elaborate division of labor.

This is how it has come about that in civilized countries at the present time nearly all kinds of labor are performed in factories, and in nearly all branches of work handicrafts and manufacture have been superseded. This process has to an ever greater degree ruined the old middle class, especially the small handicraftsmen; it has entirely transformed the condition of the workers; and two new classes have been created which are gradually swallowing up all the others. These are:

> (1) The class of big capitalists, who in all civilized countries are already in almost exclusive possession of all the means of subsistence and of the instruments (machines, factories) and materials necessary for the production of the means of subsistence. This is the bourgeois class, or the bourgeoisie.
>
> (2) The class of the wholly propertyless, who are obliged to sell their labor to the bourgeoisie in order to get in exchange the means of subsistence necessary for their support. This is called the class of proletarians, or the proletariat.

Question 5. Under what conditions does this sale of the labor of the proletarians to the bourgeoisie take place?

Answer. Labor is a commodity like any other and its price is therefore determined by exactly the same laws that apply to other commodities. In a regime of big industry or of free competition—as we shall see, the two come to the same thing—the price of a commodity is on the average always equal to its costs of production. Hence the price of labor is also equal to the costs of production of labor. But the costs of production of labor consist of precisely the quantity of means of subsistence necessary to enable the worker to continue working and to prevent the working class from dying out. The worker will therefore get no more for his labor than is necessary for this purpose; the price of labor or the wage will, in other words, be the lowest, the minimum, required for the maintenance of life. However, since business is sometimes better and sometimes worse, it follows that the worker sometimes gets more and sometimes less, just as the industrialist sometimes gets more and sometimes less for his commodities. But again, just as the industrialist, on the average of good times and bad, gets no more and no less for his commodities than what they cost, similarly on the average the worker gets no more and no less than this minimum. This economic law of wages operates the more strictly the greater the degree to which big industry has taken possession of all branches of production.

Question 6. What working classes were there before the industrial revolution?

Answer. The working classes have always, according to the different stages of development of society, lived in different circumstances and had different relations to the owning and ruling classes. In antiquity, the workers were the *slaves* of the owners, just as they still are in many backward countries and even in the southern part of the United States. In the Middle Ages, they were the *serfs* of the landowning nobility, as they still are in Hungary, Poland, and Russia. In the Middle Ages, and indeed right up to the industrial revolution, there were also journeymen in the cities who worked in the service of petty bourgeois masters. Gradually, as manufacture developed, these journeymen became manufacturing workers who were even then employed by larger capitalists.

Question 7. In what way do proletarians differ from slaves?

Answer. The slave is sold once and for all; the proletarian must sell himself daily and hourly. The individual slave, property of one master, is assured an existence, however miserable it may be, because of the master's interest. The individual proletarian, property as it were of the entire bourgeois class which buys his labor only when someone has need of it, has no secure existence. This existence is assured only to the *class* as a whole. The slave is outside competition; the proletarian is in it and experiences all its vagaries. The slave counts as a thing, not as a member of civil society. Thus the slave can have a better existence than the proletarian, while the proletarian belongs to a higher stage of social development and himself stands on a higher social level than the slave. The slave frees himself when, of all the relations of private property, he abolishes only the relation of slavery and thereby becomes a proletarian; the proletarian can free himself only by abolishing private property in general.

Question 8. In what way do proletarians differ from serfs?

Answer. The serf possesses and uses an instrument of production, a piece of land, in exchange for which he gives up a part of his product or part of the services of his labor. The proletarian works with the instruments of production of another, for the account of this other, in exchange for a part of the product. The serf gives up, the proletarian receives. The serf has an assured existence, the proletarian has not. The serf is outside competition, the proletarian is in it. The serf liberates himself in one of three ways; either he runs away to the city and there becomes a handicraftsman; or, instead of products and services, he gives money to his lord and thereby becomes a free tenant; or he overthrows his feudal lord and himself becomes a property-owner. In short, by one route or another he gets into the owning class and enters into competition. The proletarian liberates himself by abolishing competition, private property, and all class differences.

* * *

Question 11. What were the immediate consequences of the industrial revolution and of the division of society into bourgeoisie and proletariat?

Answer. First, the lower and lower prices of industrial products brought about by machine labor totally destroyed in all countries of the world the old system of manufacture or industry based upon hand labor. In this way, all semi-barbarian countries, which had hitherto been more or less strangers to historical development and whose industry had been based on manufacture, were violently forced out of their isolation. They bought the cheaper commodities of the English and allowed their own manufacturing workers to be ruined. Countries which had known no progress for thousands of years, for example India, were thoroughly revolutionized, and even China is now on the way to a revolution. We have come to the point where a new machine invented in England deprives millions of Chinese workers of their livelihood within a year's time. In this way big industry has brought all the people of the earth into contact with each other, has merged all local markets into one world market, has spread civilization and progress everywhere and has thus ensured that whatever happens in the civilized countries will have repercussions in all other

countries. It follows that if the workers in England or France now liberate themselves, this must set off revolutions in all other countries—revolutions which sooner
or later must accomplish the liberation of their respective working classes.

Second, wherever big industries displaced manufacture, the bourgeoisie developed in wealth and power to the utmost and made itself the first class of the country.
The result was that wherever this happened the bourgeoisie took political power into
its own hands and displaced the hitherto ruling classes, the aristocracy, the guildmasters, and their representative, the absolute monarchy. The bourgeoisie annihi-
lated the power of the aristocracy, the nobility, by abolishing the entailment of
estates, in other words by making landed property subject to purchase and sale, and
by doing away with the special privileges of the nobility. It destroyed the power of
the guildmasters by abolishing guilds and handicraft privileges. In their place, it put
competition, that is, a state of society in which everyone has the right to enter into
any branch of industry, the only obstacle being a lack of the necessary capital. The
introduction of free competition is thus a public declaration that from now on the
members of society are unequal only to the extent that their capitals are unequal,
that capital is the decisive power, and that therefore the capitalists, the bourgeoisie,
have become the first class in society. Free competition is necessary for the establishment of big industry, because it is the only condition of society in which big industry
can make its way. Having destroyed the social power of the nobility and the
guildmasters, the bourgeoisie also destroyed their political power. Having raised
itself to the actual position of the first class in society, it proclaims itself to be also the
dominant political class. This it does through the introduction of the representative
system which rests on bourgeois equality before the law and the recognition of free
competition, and in European countries takes the form of constitutional monarchy.
In these constitutional monarchies, only those who possess a certain capital are
voters, that is to say, only members of the bourgeoisie. These bourgeois voters
choose the deputies, and these bourgeois deputies, by using their right to refuse to
vote taxes, choose a bourgeois government.

Third, everywhere the proletariat develops in step with the bourgeoisie. In
proportion as the bourgeoisie grows in wealth the proletariat grows in numbers. For,
since proletarians can be employed only by capital, and since capital expands only
through employing labor, it follows that the growth of the proletariat proceeds at
precisely the same pace as the growth of capital. Simultaneously, this process draws
members of the bourgeoisie and proletarians together into the great cities where
industry can be carried on most profitably, and by thus throwing great masses in one
spot it gives to the proletarians a consciousness of their own strength. Moreover, the
further this process advances, the more new labor-saving machines are invented, the
greater is the pressure exercised by big industry on wages, which, as we have seen,
sink to their minimum and therewith render the condition of the proletariat increasingly unbearable. The growing dissatisfaction of the proletariat thus joins with its
rising power to prepare a proletarian social revolution.

*Question 12. What were the further consequences of the industrial revolu-
tion?*

Answer. Big industry created in the steam engine and other machines the means of
endlessly expanding industrial production, speeding it up, and cutting its costs. With

production thus facilitated, the free competition which is necessarily bound up with big industry assumed the most extreme forms; a multitude of capitalists invaded industry, and in a short while more was produced than was needed. As a consequence, finished commodities could not be sold, and a so-called commercial crisis broke out. Factories had to be closed, their owners went bankrupt, and the workers were without bread. Deepest misery reigned everywhere. After a time, the superfluous products were sold, the factories began to operate again, wages rose, and gradually business got better than ever. But it was not long before too many commodities were again produced and a new crisis broke out, only to follow the same course as its predecessor. Ever since the beginning of this [nineteenth] century, the condition of industry has constantly fluctuated between periods of prosperity and periods of crisis; nearly every five to seven years a fresh crisis has intervened, always with the greatest hardship for workers, and always accompanied by general revolutionary stirrings and direst peril to the whole existing order of things.

Question 13. What follows from these periodic commercial crises?

Answer. First: That though big industry in its earliest stage created free competition, it has now outgrown free competition; that for big industry competition and generally the individualistic organization of production have become a fetter which it must and will shatter; that so long as big industry remains on its present footing it can be maintained only at the cost of general chaos every seven years, each time threatening the whole of civilization and not only plunging the proletarians into misery but also ruining large sections of the bourgeoisie; hence either that big industry must itself be given up, which is an absolute impossibility, or that it makes unavoidably necessary an entirely new organization of society in which production is no longer directed by mutually competing individual industrialists but rather by the whole society operating according to a definite plan and taking account of the needs of all.

Second: That big industry and the limitless expansion of production which it makes possible bring within the range of feasibility a social order in which so much is produced that every member of society will be in a position to exercise and develop all his powers and faculties in complete freedom. It thus appears that the very qualities of big industry which in our present-day society produce misery and crises are those which in a different form of society will abolish this misery and these catastrophic depressions. We see with the greatest clarity:

(1) That all these evils from now on to be ascribed solely to a social order which no longer corresponds to the requirements of the real situation; and

(2) That it is possible, through a new social order, to do away with these evils altogether.

Question 14. What will this new social order have to be like?

Answer. Above all, it will have to take the control of industry and of all branches of production out of the hands of mutually competing individuals, and instead institute a system in which all these branches of production are operated by society as a whole, that is, for the common account, according to a common plan, and with the participation of all members of society. It will, in other words, abolish competition and replace it with association. Moreover, since the management of industry by individuals necessarily implies private property, and since competition is in reality merely the manner and form in which the control of industry by private property owners expresses itself, it follows that private property cannot be separated from competition and the individual management of industry. Private property must therefore be abolished and in its place must come the common utilization of all instruments of production and the distribution of all products according to common agreement—in a word, what is called the communal ownership of goods. In fact, the abolition of private property is doubtless the shortest and most significant way to characterize the revolution in the whole social order which has been made necessary by the development of industry, and for this reason it is rightly advanced by communists as their main demand.

Question 15. Was not the abolition of private property possible at an earlier time?

Answer. No. Every change in the social order, every revolution in property relations, is the necessary consequence of the creation of new forces of production which no longer fit into the old property relations. Private property itself originated in this way. For private property has not always existed. When, towards the end of the Middle Ages, there arose a new mode of production which could not be carried on under the then existing feudal and guild forms of property, this manufacture, which had outgrown the old property relations, created a new property form, private property. And for manufacture and the earliest stage of development of big industry, private property was the only possible property form; the social order based on it was the only possible social order. So long as it is not possible to produce so much that there is enough for all, with more left over for expanding the social capital and extending the forces of production—so long as this is not possible, there must always be a ruling class directing the use of society's productive forces, and a poor, oppressed class. How these classes are constituted depends on the stage of development. The agrarian Middle Ages gives us the baron and the serf; the cities of the later Middle Ages show us the guildmaster and the journeyman and the day laborer; the seventeenth century has its manufacturers and manufacturing workers; the nineteenth has big factory owners and proletarians. It is clear that up to now the forces of production have never been developed to the point where enough could be produced for all, and that private property has become a fetter and a barrier in relation to the further development of the forces of production. Now, however, the development of big industry has ushered in a new period. Capital and the forces of production have been expanded to an unprecedented extent, and the means are at hand to multiply them without limit in the near future. Moreover, the forces of production have been concentrated in the hands of a few bourgeois, while the great mass of the people are more and more falling into the proletariat, their situation becoming more wretched and intolerable in proportion to the increase of wealth of the bourgeoisie. And finally, these mighty and easily extended forces of production have so far outgrown private property and the bourgeoisie that they threaten at any moment to unleash the most violent disturbances of the social order. Now, under these conditions, the abolition of private property has become not only possible but absolutely necessary.

Question 16. Will the peaceful abolition of private property be possible?

Answer. It would be desirable if this could happen, and the communists would certainly be the last to oppose it. Communists know only too well that all conspiracies are not only useless but even harmful. They know all too well that revolutions are not made intentionally and arbitrarily, but that everywhere and always they have been the necessary consequence of conditions which were wholly independent of the will and direction of individual parties and entire classes. But they also see that the development of the proletariat in nearly all civilized countries has been violently suppressed, and that in this way the opponents of communism have been working toward a revolution with all their strength. If the oppressed proletariat is finally driven to revolution, then we communists will defend the interests of the proletarians with deeds as we now defend them with words.

Question 17. Will it be possible for private property to be abolished at one stroke?

Answer. No, no more than existing forces of production can at one stroke be multiplied to the extent necessary for the creation of a communal society. In all probability, the proletarian revolution will transform existing society gradually and will be able to abolish private property only when the means of production are available in sufficient quantity.

Question 18. What will be the course of this revolution?

Answer. Above all, it will establish a *democratic constitution* and through this the direct or indirect dominance of the proletariat. Direct in England, where the proletarians are already a majority of the people. Indirect in France and Germany, where the majority of the people consists not only of proletarians but also of small peasants and petty bourgeois who are in the process of falling into the proletariat, who are more and more dependent in all their political interests on the proletariat, and who must therefore soon adapt themselves to the demands of the proletariat. Perhaps this will cost a second struggle, but the outcome can only be the victory of the proletariat.

Democracy would be wholly valueless to the proletariat if it were not immediately used as a means for putting through measures directed against private property and ensuring the livelihood of the proletariat. The main measures, emerging as the necessary result of existing relations, are the following:

> (1) Limitation of private property through progressive taxation, heavy inheritance taxes, abolition of inheritance through collateral lines (brothers, nephews, etc.), forced loans, etc.
>
> (2) Gradual expropriation of landowners, industrialists, railroad magnates and shipowners, partly through competition by state industry, partly directly through compensation in the form of bonds.
>
> (3) Confiscation of the possessions of all emigrants and rebels against the majority of the people.
>
> (4) Organization of labor or employment of proletarians on publicly owned land, in factories and workshops, with competition among the workers being abolished and with the factory owners, insofar as they still exist, being obliged to pay the same high wages as those paid by the state.

(5) An equal obligation on all members of society to work until such time as private property has been completely abolished. Formation of industrial armies, especially for agriculture.

(6) Centralization of money and credit in the hands of the state through a national bank operating with state capital, and the suppression of all private banks and bankers.

(7) Expansion of the number of national factories, workshops, railroads, ships; bringing new lands into cultivation and improvement of land already under cultivation—all in proportion to the growth of the capital and labor force at the disposal of the nation.

(8) Education of all children, from the moment they can leave their mothers' care, in national establishments at national cost. Education and production together.

(9) Construction, on public lands, of great palaces as communal dwellings for associated groups of citizens engaged in both industry and agriculture and combining in their way of life the advantages of urban and rural conditions while avoiding the one-sidedness and drawbacks of each.

(10) Destruction of all unhealthy and jerry-built dwellings in urban districts.

(11) Equal inheritance rights for children born in and out of wedlock.

(12) Concentration of all means of transportation in the hands of the nation.

It is impossible, of course, to carry out all these measures at once. But one will always bring others in its wake. Once the first radical attack on private property has been launched, the proletariat will find itself forced to go ever further, to concentrate increasingly in the hands of the state all capital, all agriculture, all transport, all trade. All the foregoing measures are directed to this end; and they will become practicable and feasible, capable of producing their centralizing effects to precisely the degree that the proletariat through its labor multiplies the country's productive forces. Finally, when all capital, all production, all exchange have been brought together in the hands of the nation, private property will disappear of its own accord, money will become superfluous, and production will so expand and man so change that society will be able to slough off whatever of its old economic habits may remain.

Anti-Dühring

If for the impending overthrow of the present mode of distribution of the products of labour, with its crying contrasts of want and luxury, starvation and surfeit, we had no better guarantee than the consciousness that this mode of distribution is unjust, and that justice must eventually triumph, we should be in a pretty bad way, and we might have a long time to wait. The mystics of the Middle Ages who dreamed of the coming millennium were already conscious of the injustice of class antagonisms. On the threshold of modern history, three hundred and fifty years ago, Thomas Münzer proclaimed it to the world. In the English and the French bourgeois revolutions the same call resounded—and died away. And if today the same call for the abolition of class antagonisms and class distinctions, which up to 1830 had left the working and suffering classes cold, if today this call is re-echoed a millionfold, if it takes hold of one country after another in the same order and in the same degree of intensity that modern industry develops in each country, if in one generation it has gained a strength that enables it to defy all the forces combined against it and to be confident of victory in the near future—what is the reason for this? The reason is that modern large-scale industry has called into being on the one hand a proletariat, a class which for the first time in history can demand the abolition, not of this or that particular class organisation, or of this or that particular class privilege, but of classes themselves, and which is in such a position that it must carry through this demand on pain of sinking to the level of the Chinese coolie. On the other hand this same large-scale industry has brought into being, in the bourgeoisie, a class which has the monopoly of all the instruments of production and means of subsistence, but which in each speculative boom period and in each crash that follows it proves that it has become incapable of any longer controlling the productive forces, which have grown beyond its power; a class under whose leadership society is racing to ruin like a locomotive whose jammed safety-valve the driver is too weak to open. In other words, the reason is that both the productive forces created by the modern capitalist mode of production and the system of distribution of goods established by it have come into crying contradiction with that mode of production itself, and in fact to such a degree that, if the whole of modern society is not to perish, a revolution in the mode of production and distribution must take place, a revolution which will put an end to all class distinctions. On this tangible, material fact, which is impressing itself in a more or less clear form, but with insuperable necessity, on the minds of the exploited proletarians—on this fact, and not on the conceptions of justice and injustice held by any armchair philosopher, is modern socialism's confidence in victory founded.

Friedrich von Hayek, from *The Mirage of Social Justice* (1899–)

Friedrich von Hayek has long been one of the foremost defenders of the *laissez-faire* free market system and a harsh critic of all of the schemes—however well-intended—to interfere with the market and impose upon it some predesigned pattern of distribution, including, especially, attempts to distribute the wealth more evenly and equally and efforts to make sure that the market in fact rewards merit. Social justice, he tells us, is an illusion at best, and, in some hands, it is an idea that greatly endangers our personal freedoms. The central argument is that any effort to predetermine distribution paterns, no matter what the criteria, necessarily involves placing more power in the hands of a central authority—presumably the government—and thus takes it away from individual citizens. The attempt to distribute the goods of society more equally, for example, requires taking wealth away from some, who presumably would not like to part with it and perhaps would not do so of their own free will. Thus their liberty is violated and the power required by government to take it from them "must progressively approach nearer and nearer to a totalitarian system." What is less recognized, however, is the fact that the free market, whatever its other virtues, does not dependably reward merit—whether hard work or talent or successful results, and is as much a matter of luck as it is of effort or desert. We are predictably and perhaps naturally upset, von Hayek admits, when we see virtue (particularly our own) go unrewarded, but the cost of correcting such disappointments is, once again, unwarranted and dangerous meddling in the market. Who is to say who deserves what, and what authority are we willing to entrust with such judgments? "Society has simply become the new deity to which we complain and clamour for redress if it does not fulfill the expectations it has created," but there is, von Hayek insists, no one to blame. It is this wholesale dismissal of responsibility, perhaps, that has aroused such indignation in response. E. F. Schumacher, in partial retort, has called the free market system "the institutionalization of irresponsibility." And why, it may be asked, should the market itself be more trustworthy, with no one in control (that is, assuming that it is in fact truly "free"), than some selected commission of judges, or duly elected representatives? But Hayek argues, persuasively for many, that justice cannot be conceived in terms of anything like results or rewards, and the uncertainty of the "economic game" is an essential part of the practice. At best, we can try to make the rules themselves as fair as possible and take some trouble to make sure that people don't cheat, but even then, a "level playing field" (as some say now) cannot guarantee that everyone will have an even starting point, and von Hayek argues—as most conservatives who otherwise share his sense of the market do not—against equal opportunity as just another version of the same government interference, ultimately unlimited in its scope and eventually, he warns, nothing less than a nightmare.

The Conquest of Public Imagination by 'Social Justice'

The appeal to 'social justice' has nevertheless by now become the most widely used and most effective argument in political discussion. Almost every claim for government action on behalf of particular groups is advanced in its name, and if it can be made to appear that a certain measure is demanded by 'social justice', opposition to it will rapidly weaken. People may dispute whether or not the particular measure is required by 'social justice'. But that this is the standard which ought to guide political action, and that the expression has a definite meaning, is hardly ever questioned. In consequence, there are today probably no political movements or politicians who do not readily appeal to 'social justice' in support of the particular measures which they advocate.

It also can scarcely be denied that the demand for 'social justice' has already in a great measure transformed the social order and is continuing to transform it in a direction which those who called for it never foresaw. Though the phrase has undoubtedly helped occasionally to make the law more equal for all, whether the demand for justice in distribution has in any sense made society juster or reduced discontent must remain doubtful.

The expression of course described from the beginning the aspirations which were at the heart of socialism. Although classical socialism has usually been defined by its demand for the socialization of the means of production, this was for it chiefly a means thought to be essential in order to bring about a 'just' distribution of wealth; and since socialists have later discovered that this redistribution could in a great measure, and against less resistance, be brought about by taxation (and government services financed by it), and have in practice often shelved their earlier demands, the realization of 'social justice' has become their chief promise. It might indeed be said that the main difference between the order of society at which classical liberalism aimed and the sort of society into which it is now being transformed is that the former was governed by principles of just individual conduct while the new society is to satisfy the demands for 'social justice'—or, in other words, that the former demanded just action by the individuals while the latter more and more places the duty of justice on authorities with power to command people what to do.

* * *

I believe that 'social justice' will ultimately be recognized as a will-o'-the-wisp which has lured men to abandon many of the values which in the past have inspired the development of civilization—an attempt to satisfy a craving inherited from the traditions of the small group but which is meaningless in the Great Society of free men. Unfortunately, this vague desire which has become one of the strongest bonds spurring people of good will to action, not only is bound to be disappointed. This would be sad enough. But, like most attempts to pursue an unattainable goal, the striving for it will also produce highly undesirable consequences, and in particular lead to the destruction of the indispensable environment in which the traditional moral values alone can flourish, namely personal freedom.

* * *

It is now necessary clearly to distinguish between two wholly different problems which the demand for 'social justice' raises in a market order.

The first is whether within an economic order based on the market the concept of 'social justice' has any meaning or content whatever.

The second is whether it is possible to preserve a market order while imposing upon it (in the name of 'social justice' or any other pretext) some pattern of remuneration based on the assessment of the performance or the needs of different individuals or groups by an authority possessing the power to enforce it.

The answer to each of these questions is a clear no.

Yet it is the general belief in the validity of the concept of 'social justice' which drives all contemporary societies into greater and greater efforts of the second kind and which has a peculiar self-accelerating tendency: the more dependent the position of the individuals or groups is seen to become on the actions of government, the more they will insist that the governments aim at some recognizable scheme of distributive justice; and the more governments try to realize some preconceived pattern of desirable distribution, the more they must subject the position of the different individuals and groups to their control. So long as the belief in 'social justice' governs political action, this process must progressively approach nearer and nearer to a totalitarian system.

* * *

The first insight which should shake this certainty is that we experience the same feelings also with respect to differences in human fates for which clearly no human agency is responsible and which it would therefore clearly be absurd to call injustice. Yet we do cry out against the injustice when a succession of calamities befalls one family while another steadily prospers, when a meritorious effort is frustrated by some unforeseeable accident, and particularly if of many people whose endeavours seem equally great, some succeed brilliantly while others utterly fail. It is certainly tragic to see the failure of the most meritorious efforts of parents to bring up their children, of young men to build a career, or of an explorer or scientist pursuing a brilliant idea. And we will protest against such a fate although we do not know anyone who is to blame for it, or any way in which such disappointments can be prevented.

It is no different with regard to the general feeling of injustice about the distribution of material goods in a society of free men. Though we are in this case less ready to admit it, our complaints about the outcome of the market as unjust do not really assert that somebody has been unjust; and there is no answer to the question of *who* has been unjust. Society has simply become the new deity to which we complain and clamour for redress if it does not fulfil the expectations it has created. There is no individual and no cooperating group of people against which the sufferer would have a just complaint, and there are no conceivable rules of just individual conduct which would at the same time secure a functioning order and prevent such disappointments.

The only blame implicit in those complaints is that we tolerate a system in which each is allowed to choose his occupation and therefore nobody can have the power and the duty to see that the results correspond to our wishes. For in such a system in which each is allowed to use his knowledge for his own purposes the concept of 'social justice' is necessarily empty and meaningless, because in it nobody's will can determine the relative incomes of the different people, or prevent that they be partly dependent on accident. 'Social justice' can be given a meaning only in a directed or 'command' economy (such as an army) in which the individuals are ordered what to do; and any particular conception of 'social justice' could be realized only in such a centrally directed system. It presupposes that people are guided by specific directions and not by rules of just individual conduct. Indeed, no system of rules of just individual conduct, and therefore no free action of the individuals, could produce results satisfying any principle of distributive justice.

The Rationale of the Economic Game in Which Only the Conduct of the Players but not the Result Can Be Just

We have seen earlier that justice is an attribute of human conduct which we have learnt to exact because a certain kind of conduct is required to secure the formation and maintenance of a beneficial order of actions. The attribute of justice may thus be predicated about the intended results of human action but not about circumstances which have not deliberately been brought about by men. Justice requires that in the 'treatment' of another person or persons, i.e. in the intentional actions affecting the well-being of other persons, certain uniform rules of conduct be observed. It clearly has no application to the manner in which the impersonal process of the market allocates command over goods and services to particular people: this can be neither just nor unjust, because the results are not intended or foreseen, and depend on a multitude of circumstances not known in their totality to anybody. The conduct of the individuals in that process may well be just or unjust; but since their wholly just actions will have consequences for others which were neither intended nor foreseen, these effects do not thereby become just or unjust.

The fact is simply that we consent to retain, and agree to enforce, uniform rules for a procedure which has greatly improved the chances of all to have their wants satisfied, but at the price of all individuals and groups incurring the risk of unmerited failure. With the acceptance of this procedure the recompense of different groups and individuals becomes exempt from deliberate control. It is the only procedure yet discovered in which information widely dispersed among millions of men can be effectively utilized for the benefit of all—and used by assuring to all an individual liberty desirable for itself on ethical grounds. It is a procedure which of course has never been 'designed' but which we have learnt gradually to improve after we had discovered how it increased the efficiency of men in the groups who had evolved it.

It is a procedure which, as Adam Smith (and apparently before him the ancient Stoics) understood, in all important respects (except that normally it is not pursued solely as a diversion) is wholly analogous to a game, namely a game partly of skill and partly of chance. It proceeds, like all games, according to rules guiding the actions of individual participants whose aims, skills, and knowledge are different, with the consequence that the outcome will be unpredictable and that there will regularly be winners and losers. And while, as in a game, we are right in insisting that it be fair and that nobody cheat, it would be nonsensical to demand that the results for the different players be just. They will of necessity be determined partly by skill and partly by luck. Some of the circumstances which make the services of a person more or less valuable to his fellows, or which may make it desirable that he change the direction of his efforts, are not of human design or foreseeable by men.

* * *

The long and the short of it all is that men can be allowed to decide what work to do only if the remuneration they can expect to get for it corresponds to the value their services have to those of their fellows who receive them; and that *these values which their services will have to their fellows will often have no relations to their individual merits or needs.* Reward for merit earned and indication of what a person should do, both in his own and in his fellows' interest, are different things. It is not good intentions or needs but doing what in fact most benefits others, irrespective of motive, which will secure the best reward. Among those who try to climb Mount Everest or to reach the Moon, we also honour not those who made the greatest efforts, but those who got there first.

The Alleged Necessity of a Belief in the Justice of Rewards

It has been argued persuasively that people will tolerate major inequalities of the material positions only if they believe that the different individuals get on the whole what they deserve, that they did in fact support the market order only because (and so long as) they thought that the differences of remuneration corresponded roughly to differences of merit, and that in consequence the maintenance of a free society presupposes the belief that some sort of 'social justice' is being done. The market order, however, does not in fact owe its origin to such beliefs, or was originally justified in this manner. This order could develop, after its earlier beginnings had decayed during the middle ages and to some extent been destroyed by the restrictions imposed by authority, when a thousand years of vain efforts to discover substantively just prices or wages were abandoned and the late schoolmen recognized them to be empty formulae and taught instead that the prices determined by just conduct of the parties in the market, i.e. the competitive prices arrived at without fraud, monopoly and violence, was all that justice required. It was from this tradition that John Locke and his contemporaries derived the classical liberal conception of justice for which, as has been rightly said, it was only 'the way in which competition was carried on, not its results,' that could be just or unjust.

It is unquestionably true that, particularly among those who were very successful in the market order, a belief in a much stronger moral justification of individual success developed, and that, long after the basic principles of such an order had been fully elaborated and approved by catholic moral philosophers, it had in the Anglo–Saxon world received strong support from Calvinist teaching. It certainly is important in the market order (or free enterprise society, misleadingly called 'capitalism') that the individuals believe that their well-being depends primarily on their own efforts and decisions. Indeed, few circumstances will do more to make a person energetic and efficient than the belief that it depends chiefly on him whether he will reach the goals he has set himself. For this reason this belief is often encouraged by education and governing opinion—it seems to me, generally much to the benefit of most of the members of the society in which it prevails, who will owe many important material and moral improvements to persons guided by it. But it leads no doubt also to an exaggerated confidence in the truth of this generalization which to those who regard themselves (and perhaps are) equally able but have failed must appear as a bitter irony and severe provocation.

It is probably a misfortune that, especially in the USA, popular writers like Samuel Smiles and Horatio Alger, and later the sociologist W. G. Sumner, have defended free enterprise on the ground that it regularly rewards the deserving, and it bodes ill for the future of the market order that this seems to have become the only defence of it which is understood by the general public. That it has largely become the basis of the self-esteem of the businessman often gives him an air of self-righteousness which does not make him more popular.

It is therefore a real dilemma to what extent we ought to encourage in the young the belief that when they really try they will succeed, or should rather emphasize that inevitably some unworthy will succeed and some worthy fail—whether we ought to allow the views of those groups to prevail with whom the over-confidence in the appropriate reward of the able and industrious is strong and who in consequence will do much that benefits the rest, and whether without such partly erroneous beliefs the large numbers will tolerate actual differences in rewards which will be based only partly on achievement and partly on mere chance.

* * *

'Social Justice' and Equality

The most common attempts to give meaning to the concept of 'social justice' resort to egalitarian considerations and argue that every departure from equality of material benefits enjoyed has to be justified by some recognizable common interest which these differences serve. This is based on a specious analogy with the situation in which some human agency has to distribute rewards, in which case indeed justice would require that these rewards be determined in accordance with some recognizable rule of general applicability. But earnings in a market system, though people tend to regard them as rewards, do not serve such a function. Their rationale (if one may use this term for a role which was not designed but developed because it assisted human endeavour without people understanding how), is rather to indicate to people what they ought to do if the order is to be maintained on which they all rely. The prices which must be paid in a market economy for different kinds of labour and other factors of production if individual efforts are to match, although they will be affected by effort, diligence, skill, need, etc., cannot conform to any one of these magnitudes; and considerations of justice just do not make sense with respect to the determination of a magnitude which does not depend on anyone's will or desire, but on circumstances which nobody knows in their totality.

* * *

The postulate of material equality would be a natural starting point only if it were a necessary circumstance that the shares of the different individuals or groups were in such a manner determined by deliberate human decision. In a society in which this were an unquestioned fact, justice would indeed demand that the allocation of the means for the satisfaction of human needs were effected according to some uniform principle such as merit or need (or some combination of these), and that, where the principle adopted did not justify a difference, the shares of the different individuals should be equal. The prevalent demand for material equality is probably often based on the belief that the existing inequalities are the effect of somebody's decision—a belief which would be wholly mistaken in a genuine market order and has still only very limited validity in the highly interventionist 'mixed' economy existing in most countries today. This now prevalent form of economic order has in fact attained its character largely as a result of governmental measures aiming at what was thought to be required by 'social justice'.

When the choice, however, is between a genuine market order, which does not and cannot achieve a distribution corresponding to any standard of material justice, and a system in which a government uses its powers to put some such standard into effect, the question is not whether government ought to exercise, justly or unjustly, powers it must exercise in any case, but whether government should possess and exercise additional powers which can be used to determine the shares of the different members of society. The demand for 'social justice', in other words, does not merely require government to observe some principle of action according to uniform rules in those actions which it must perform in any case, but demands that it undertake additional activities, and thereby assume new responsibilities—tasks which are not necessary for maintaining law and order and providing for certain collective needs which the market could not satisfy.

The great problem is whether this new demand for equality does not conflict with the equality of the rules of conduct which government must enforce on all in a free society. There is, of course, a great difference between government treating all citizens according to the same rules in all the activities it undertakes for other purposes, and government doing what is required in order to place the different citizens in equal (or less unequal) material positions. Indeed, there may arise a sharp conflict between these two aims. Since people will differ in many attributes which government cannot alter, to secure for them the same material position would require that government treat them very differently. Indeed, to assure the same material position to people who differ greatly in strength, intelligence, skill, knowledge and perseverance as well as in their physical and social environment, government would clearly have to treat them very differently to compensate for those disadvantages and deficiencies it could not directly alter. Strict equality of those benefits which government could provide for all, on the other hand, would clearly lead to the inequality of the material positions.

This, however, is not the only and not even the chief reason why a government aiming to secure for its citizens equal material positions (or any determined pattern of material welfare) would have to treat them very unequally. It would have to do so because under such a system it would have to undertake to tell people what to do. Once the rewards the individual can expect are no longer an appropriate indication of how to direct their efforts to where they are most needed, because these rewards correspond not to the value which their services have for their fellows, but to the moral merit or desert the persons are deemed to have earned, they lose the guiding function they have in the market order and would have to be replaced by the commands of the directing authority. A central planning office would, however, have to decide on the tasks to be allotted to the different groups or individuals wholly on grounds of expediency or efficiency and, in order to achieve its ends, would have to impose upon them very different duties and burdens. The individuals might be treated according to uniform rules so far as their rewards were concerned, but certainly not with respect to the different kinds of work they would have to be made to do. In assigning people to their individual tasks, the central planning authority would have to be guided by considerations of efficiency and expediency and not by principles of justice or equality. No less than in the market order would the individuals in the common interest have to submit to great inequality—only these inequalities would be determined not by the interaction of individual skills in an impersonal process, but by the uncontradictable decision of authority.

As is becoming clear in ever increasing fields of welfare policy, an authority instructed to achieve particular results for the individuals must be given essentially arbitrary powers to make the individuals do what seems necessary to achieve the required result. Full equality for most cannot but mean the equal submission of the great masses under the command of some élite who manages their affairs. While an equality of rights under a limited government is possible and an essential condition of individual freedom, a claim for equality of material position can be met only by a government with totalitarian powers.

'Equality of Opportunity'

It is of course not to be denied that in the existing market order not only the results but also the initial chances of different individuals are often very different; they are affected by circumstances of their physical and social environment which are beyond their control but in many particular respects might be altered by some governmental action. The demand for equality of opportunity or equal starting conditions (*Startgerechtigkeit*) appeals to, and has been supported by, many who in general favour the free market order. So far as this refers to such facilities and opportunities as are of necessity affected by governmental decisions (such as appointments to public office and the like), the demand was indeed one of the central points of classical liberalism, usually expressed by the French phrase 'la carrière ouverte aux talents'. There is also much to be said in favour of the government providing on an equal basis the means for the schooling of minors who are not yet fully responsible citizens, even though there are grave doubts whether we ought to allow government to administer them.

But all this would still be very far from creating real equality of opportunity, even for persons possessing the same abilities. To achieve this government would have to control the whole physical and human environment of all persons, and have to endeavour to provide at least equivalent chances for each; and the more government succeeded in these endeavours, the stronger would become the legitimate demand that, on the same principle, any still remaining handicaps must be removed—or compensated for by putting extra burden on the still relatively favoured. This would have to go on until government literally controlled every circumstance which could affect any person's well-being. Attractive as the phrase of equality of opportunity at first sounds, once the idea is extended beyond the facilities which for other reasons have to be provided by government, it becomes a wholly illusory ideal, and any attempt concretely to realize it apt to produce a nightmare.

Bernard Williams, "On Equality"
(1929–)

Bernard Williams, one of the leading moral philosophers in England until his move to Berkeley, California just a year or so ago, has often argued against modern moralists in favor of the ancient conceptions of virtue defended by Plato and Aristotle (for example, in his fairly recent *Ethics and the Limits of Philosophy*). But on the subject of equality, he sides with the moderns and against the ancients, defending the often confused insistence that all people are equal in a way that is neither trivial nor absurd. He begins by distinguishing between an obviously false factual claim (all people are in fact equal in their various abilities and endowments) and a more political principle that insists that people be *treated* in the same way. He rejects the saving but inadequate interpretation of the former, which turns into "all men are men," and he acknowledges the substantial qualifications that are due the latter, for instance, the recognition that differences as well as similarities are essential considerations for the nature of treatment, whether reward or punishment. He suggests, however, that "all men are men" is not actually as trivial as it looks, and carries with it some important insights regarding the nature of equality. He also takes Kant's insistence on both our universal moral capacities and our (related) status as "ends in themselves" to insist that this, too, is some ground for our considerations of equality. But the bulk of Williams's argument concerns the important qualification that being treated as equals presupposes the recognition that it is in different and often unequal circumstances that we expect to be so treated, and the factual differences between us result in unequal opportunities that make equal treatment difficult and perplexing. Friedrich von Hayek, in the preceding selection, flatly claims the equal opportunity is illusory and impossible. What Williams attempts to give us is some indication how it can be meaningful and, with considerable conceptual as well as political effort, possible.

The idea of equality is used in political discussion both in statements of fact, or what purport to be statements of fact—that men *are* equal—and in statements of political principles or aims—that men *should be* equal, as at present they are not. The two can be, and often are, combined: the aim is then described as that of securing a state of affairs in which men are treated as the equal beings which they in fact already are, but are not already treated as being. In both these uses, the idea of equality notoriously encounters the same difficulty: that on one kind of interpretation the

From *Problems of the Self* by Bernard Williams, published by Cambridge University Press. Copyright © 1976 by Bernard Williams. Reprinted by permission of the publisher and the author.

statements in which it figures are much too strong, and on another kind much too weak, and it is hard to find a satisfactory interpretation that lies between the two.

To take first the supposed statement of fact: it has only too often been pointed out that to say that all men are equal in all those characteristics in respect of which it makes sense to say that men are equal or unequal, is a patent falsehood; and even if some more restricted selection is made of these characteristics, the statement does not look much better. Faced with this obvious objection, the defender of the claim that all men are equal is likely to offer a weaker interpretation. It is not, he may say, in their skill, intelligence, strength, or virtue that men are equal, but merely in their being men: it is their common humanity that constitutes their equality. On this interpretation, we should not seek for some special characteristics in respect of which men are equal, but merely remind ourselves that they are all men. Now to this might be objected that being men is not a respect in which men can strictly speaking be said to be *equal*; but, leaving that aside, there is the more immediate objection that if all that the statement does is to remind us that men are men, it does not do very much, and in particular does less than its proponents in political argument have wanted it to do. What looked like a paradox has turned into a platitude.

I shall suggest in a moment that even in this weak form the statement is not so vacuous as this objection makes it seem; but it must be admitted that when the statement of equality ceases to claim more than is warranted, it rather rapidly reaches the point where it claims less than is interesting. A similar discomfiture tends to overcome the practical maxim of equality. It cannot be the aim of this maxim that all men should be treated alike in all circumstances, or even that they should be treated alike as much as possible. Granted that, however, there is no obvious stopping point before the interpretation which makes the maxim claim only that men should be treated alike in similar circumstances; and since 'circumstances' here must clearly include reference to what a man is, as well as to his purely external situation, this comes very much to saying that for every difference in the way men are treated, some general reason or principle of differentiation must be given. This may well be an important principle; some indeed have seen in it, or in something very like it, an essential element of morality itself. But it can hardly be enough to constitute the principle that was advanced in the name of *equality*. It would be in accordance with this principle, for example, to treat black men differently from others just because they were black, or poor men differently because they were poor, and this cannot accord with anyone's idea of equality.

In what follows I shall try to advance a number of considerations that can help to save the political notion of equality from these extremes of absurdity and triviality. These considerations are in fact often employed in political argument, but are usually bundled together into an unanalysed notion of equality in a manner confusing to the advocates, and encouraging to the enemies, of that ideal. These considerations will not enable us to define a distinct third interpretation of the statements which use the notion of equality; it is rather that they enable us, starting with the weak interpretations, to build up something that in practice can have something of the solidity aspired to by the strong interpretations. In this discussion, it will not be necessary all the time to treat separately the supposedly factual application of the notion of equality, and its application in the maxim of action. Though it is sometimes important to distinguish them, and there are clear grounds for doing so, similar considerations often apply to both. The two go significantly together: on the one hand, the point of the supposedly factual assertion is to back up social ideals and programmes of political action; on the other hand—a rather less obvious point, perhaps—those political proposals have their force because they are regarded not as gratuitously egalitarian, aiming at equal treatment for reasons, for instance, of simplicity or tidiness, but as affirming an equality which is believed in some sense already to exist, and to be obscured or neglected by actual social arrangements.

1. *Common humanity.* The factual statement of men's equality was seen, when pressed, to retreat in the direction of merely asserting the equality of men as men; and this was thought to be trivial. It is certainly insufficient, but not, after all, trivial. That all men are human is, if a tautology, a useful one, serving as a reminder that those who belong anatomically to the species *homo sapiens*, and can speak a language, use tools, live in societies, can interbreed despite racial differences, etc., are also alike in certain other respects more likely to be forgotten. These respects are notably the capacity to feel pain, both from immediate physical causes and from various situations represented in perception and in thought; and the capacity to feel affection for others, and the consequences of this, connected with the frustration of this affection, loss of its objects, etc. The assertion that men are alike in the possession is, while indisputable and (it may be) even necessarily true, not trivial. For it is certain that there are political and social arrangements that systematically neglect these characteristics in the case of others; that is to say, they treat certain men as though they did not possess these characteristics, and neglect moral claims that arise from these characteristics and which would be admitted to arise from them.

* * *

I have discussed this point in connexion with very obvious human characteristics of feeling pain and desiring affection. There are, however, other and less easily definable characteristics universal to humanity, which may all the more be neglected in political and social arrangements. For instance, there seems to be a characteristic which might be called 'a desire for self-respect'; this phrase is perhaps not too happy, in suggesting a particular culturally-limited, bourgeois value, but I mean by it a certain human desire to be identified with what one is doing, to be able to realise purposes of one's own, and not to be the instrument of another's will unless one has willingly accepted such a role. This is a very inadequate and in some ways rather empty specification of a human desire; to a better specification, both philosophical reflexion and the evidences of psychology and anthropology would be relevant. Such investigations enable us to understand more deeply, in respect of the desire I have gestured towards and of similar characteristics, what it is to be human; and of what it is to be human, the apparently trivial statement of men's equality as men can serve as a reminder.

2. *Moral capacities.* So far we have considered respects in which men can be counted as all alike, which respects are, in a sense, negative: they concern the capacity to suffer, and certain needs that men have, and these involve men in moral relations as the recipients of certain kinds of treatment. It has certainly been a part, however, of the thought of those who asserted that men are equal, that there were more positive respects in which men were alike; that they were equal in certain things that they could do or achieve, as well as in things that they needed and could suffer. In respect of a whole range of abilities, from weight-lifting to the calculus, the

assertion is, as was noted at the beginning, not plausible, and has not often been supposed to be. It has been held, however, that there are certain other abilities, both less open to empirical test and more essential in moral connexions, for which it is true that men are equal. These are certain sorts of moral ability or capacity, the capacity for virtue or achievement of the highest kind of moral worth.

* * *

That men should be regarded from the human point of view, and not merely under these sorts of titles, is part of the content that might be attached to Kant's celebrated injunction 'treat each man as an end in himself, and never as a means only.' But I do not think that this is all that should be seen in this injunction, or all that is concerned in the notion of 'respect'. What is involved in the examples just given could be explained by saying that each man is owed an effort at identification: that he should not be regarded as the surface to which a certain label can be applied, but one should try to see the world (including the label) from his point of view. This injunction will be based on, though not of course fully explained by, the notion that men are conscious beings who necessarily have intentions and purposes and see what they are doing in a certain light. But there seem to be further injunctions connected with the Kantian maxim, and with the notion of 'respect', that go beyond these considerations. There are forms of exploiting men or degrading them which would be thought to be excluded by these notions, but which cannot be excluded merely by considering how the exploited or degraded men see the situation. For it is precisely a mark of extreme exploitation or degradation that those who suffer it do *not* see themselves differently from the way they are seen by the exploiters; either they do not see themselves as anything at all, or they acquiesce passively in the rôle for which they have been cast. Here we evidently need something more than the precept that one should respect and try to understand another man's consciousness of his own activities; it is also that one may not suppress or destroy that consciousnesss.

* * *

3. *Equality in unequal circumstances.* The notion of equality is invoked not only in connexions where men are claimed in some sense all to be equal, but in connexions where they are agreed to be unequal, and the question arises of the distribution of, or access to, certain goods to which their inequalities are relevant. It may be objected that the notion of equality is in fact misapplied in these connexions, and that the appropriate ideas are those of fairness or justice, in the sense of what Aristotle called 'distributive justice', where (as Aristotle argued) there is no question of regarding or treating everyone as equal, but solely a question of distributing certain goods in proportion to men's recognised inequalities.

I think it is reasonable to say against this objection that there is some foothold for the notion of equality even in these cases. It is useful here to make a rough distinction between two different types of inequality, inequality of *need* and inequality of *merit*, with a corresponding distinction between goods—on the one hand, goods demanded by the need, and on the other, goods that can be earned by the merit. In the case of needs, such as the need for medical treatment in case of illness, it can be presumed for practical purposes that the persons who have the need actually desire the goods in question, and so the question can indeed be regarded as one of distribution in a simple sense, the satisfaction of an existing desire. In the case of merit, such as for instance the possession of abilities to profit from a university education, there is not the same presumption that everyone who has the merit has the desire for the goods in question, though it may, of course, be the case. Moreover, the good of a university education may be legitimately, even if hopelessly, desired by those who do not possess the merit; while medical treatment or unemployment benefit are either not desired, or not legitimately desired, by those who are not ill or unemployed, i.e. do not have the appropriate need. Hence the distribution of goods in accordance with merit has a competitive aspect lacking in the case of distribution according to need. For these reasons, it is appropriate to speak, in the case of merit, not only of the distribution of the good, but of the distribution of the opportunity of achieving the good. But this, unlike the good itself, can be said to be distributed equally to everybody, and so one does encounter a notion of *general* equality, much vaunted in our society today, the notion of equality of opportunity.

* * *

We may return now to the notion of equality of opportunity; understanding this in the normal political sense of equality of opportunity for *everyone in society* to secure certain goods. This notion is introduced into political discussion when there is question of the access to certain goods which, first, even if they are not desired by everyone in society, are desired by large numbers of people in all sections of society (either for themselves, or, as in the case of education, for their children), or would be desired by people in all sections of society if they knew about the goods in question and thought it possible for them to attain them; second, are goods which people may be said to earn or achieve; and third, are goods which not all the people who desire them can have. This third condition covers at least three different cases, however, which it is worth distinguishing. Some desired goods, like positions of prestige, management, etc., are *by their very nature* limited: whenever there are some people who are in command or prestigious positions, there are necessarily others who are not. Other goods are *contingently* limited, in the sense that there are certain conditions of access to them which in fact not everyone satisfies, but there is no intrinsic limit to the numbers who might gain access to it by satisfying the conditions: university education is usually regarded in this light nowadays, as something which requires certain conditions of admission to it which in fact not everyone satisfies, but which an indefinite proportion of people might satisfy. Third, there are goods which are *fortuitously* limited, in the sense that although everyone or large numbers of people satisfy the conditions of access to them, there is just not enough of them to go round; so some more stringent conditions or system of rationing have to be imposed, to govern access in an imperfect situation. A good can, of course, be both contingently and fortuitously limited at once: when, due to shortage of supply, not even the people who are qualified to have it, limited in numbers though they are, can in every case have it. It is particularly worth distinguishing those kinds of limitation, as there can be significant differences of view about the way in which a certain good is limited. While most would now agree that high education is contingently limited, a Platonic view would regard it as necessarily limited.

Now the notion of equality of opportunity might be said to be the notion that a limited good shall in fact be allocated on grounds which do not *a priori* exclude any section of those that desire it. But this formulation is not really very clear. For suppose grammar school education (a good perhaps contingently, and certainly fortuitously, limited) is allocated on grounds of ability as tested at the age of 11; this would normally be advanced as an example of equality of opportunity, as opposed to a system of allocation on grounds of parents' wealth. But does not the criterion of ability exclude *a priori* a certain section of people, viz. those that are not able—just as the other excludes *a priori* those who are not wealthy? Here it will obviously be said that this was not what was meant by *a priori* exclusion: the present argument just equates this with exclusion of anybody, i.e. with the mere existence of some condition that has to be satisfied. What then is *a priori* exclusion? It must mean exclusion on grounds *other* than those appropriate or rational for the good in question. But this still will not do as it stands. For it would follow from this that so long as those allocating grammar school education on grounds of wealth thought that such grounds were appropriate or rational (as they might in one of the ways discussed above in connexion with private schools), they could sincerely describe their system as one of equality of opportunity—which is absurd.

Hence it seems that the notion of equality of opportunity is more complex than it first appeared. It requires not merely that there should be no exclusion from access on grounds other than those appropriate or rational for the good in question, but that the grounds considered appropriate for the good should themselves be such that people from all sections of society have an equal chance of satisfying them. What now is a 'section of society'? Clearly we cannot include under this term sections of the populace identified just by the characteristics which figure in the grounds for allocating the good—since, once more, any grounds at all must exclude some section of the populace. But what about sections identified by characteristics which are *correlated* with the grounds of exclusion? There are important difficulties here: to illustrate this, it may help first to take an imaginary example.

Suppose that in a certain society great prestige is attached to membership of a warrior class, the duties of which require great physical strength. This class has in the past been recruited from certain wealthy families only; but egalitarian reformers achieve a change in the rules, by which warriors are recruited from all sections of the society, on the results of a suitable competition. The effect of this, however, is that the wealthy families still provide virtually all the warriors, because the rest of the populace is so under-nourished by reason of poverty that their physical strength is inferior to that of the wealthy and well nourished. The reformers protest that equality of opportunity has not really been achieved; the wealthy reply that in fact it has, and that the poor now have the opportunity of becoming warriors—it is just bad luck that their characteristics are such that they do not pass the test. 'We are not,' they might say, 'excluding anyone *for* being poor; we exclude people for being weak, and it is unfortunate that those who are poor are also weak.'

This answer would seem to most people feeble, and even cynical. This is for reasons similar to those discussed before in connexion with equality before the law; that the supposed equality of opportunity is quite empty—indeed, one may say that it does not really exist—unless it is made more effective than this. For one knows that it could be made more effective; one knows that there is a causal connexion between being poor and being undernourished, and between being undernourished and being physically weak. One supposes further that something could be done— subject to whatever economic conditions obtain in the imagined society—to alter the distribution of wealth. All this being so, the appeal by the wealthy to the 'bad luck' of the poor must appear as disingenuous.

It seems then that a system of allocation will fall short of equality of opportunity if the allocation of the good in question in fact works out unequally or disproportionately between different sections of society, if the unsuccessful sections are under a disadvantage which could be removed by further reform or social action. This was very clear in the imaginary example that was given, because the causal connexions involved are simple and well known. In actual fact, however, the situations of this type that arise are more complicated, and it is easier to overlook the causal connexions involved. This is particularly so in the case of educational selection, where such slippery concepts as 'intellectual ability' are involved. It is a known fact that the system of selection for grammar schools by the '11+' examination favours children in direct proportion to their social class, the children of professional homes having proportionately greater success than those from working-class homes. We have every reason to suppose that these results are the product, in good part, of environmental factors; and we further know that imaginative social reform, both of the primary educational system and of living conditions, would favourably affect those environmental factors. In these circumstances, this system of educational selection falls short of equality of opportunity.

This line of thought points to a connexion between the idea of equality of opportunity, and the idea of equality of persons, which is stronger than might at first be suspected. We have seen that one is not really offering equality of opportunity to Smith and Jones if one contents oneself with applying the same criteria to Smith and Jones at, say, the age of 11; what one is doing there is to apply the same criteria to Smith as affected by favourable conditions and to Jones as affected by unfavorable but curable conditions. Here there is a necessary pressure to equal up the conditions: to give Smith and Jones equality of opportunity involves regarding their conditions, where curable, as themselves part of what is done to Smith and Jones, and not part of Smith and Jones themselves. Their identity, for these purposes, does not include their curable environment, which is itself unequal and a contributor of inequality. This abstraction of persons in themselves from unequal environments is a way, if not of regarding them as equal, at least of moving recognisably in that direction; and is itself involved in equality of opportunity.

* * *

When one is faced with the spectacle of the various elements of the idea of equality pulling in these different directions, there is a strong temptation, if one does not abandon the idea altogether, to abandon some of its elements: to claim, for instance, that equality of opportunity is the only ideal that is at all practicable, and equality of respect a vague and perhaps nostalgic illusion; or, alternatively, that equality of respect is genuine equality, and equality of opportunity an inegalitarian betrayal of the ideal—all the more so if it were thoroughly pursued, as now it is not. To succumb to either of these simplifying formulae would, I think, be a mistake. Certainly, a highly rational and efficient application of the ideas of equal opportunity, unmitigated by the other considerations, could lead to a quite inhuman society (if it worked—which, granted a well-known desire of parents to secure a position for their children at least as good as their own, is unlikely). On the other hand, an ideal of equality of respect that made no contact with such things as the economic needs of society for certain skills, and human desire for some sorts of prestige, would be condemned to a futile Utopianism, and to having no rational effect on the distribution of goods, position, and power that would inevitably proceed. If, moreover, as I have suggested, it is not really known how far, by new forms of social structure and of education, these conflicting claims might be reconciled, it is all the more obvious that we should not throw one set of claims out of the window; but should rather seek, in each situation, the best way of eating and having as much cake as possible. It is an uncomfortable situation, but the discomfort is just that of genuine political thought. It is no greater with equality than it is with liberty, or any other noble and substantial political ideal.

David Miller, "On three types of Justice," from *Social Justice*

So far, we have discussed the concept of justice only with reference to a particular kind of society—modern, industrialized, market societies with long histories, a keen sense of progress, and, perhaps, a problematic sense of community and human relationships. It is in such a context that our conceptions of justice bounce back and forth between the notion of individual rights and concerns for public utility, between the concern for private property and larger social considerations. It is in this context, too, that the freedom of the market and the authority of the state come into such apparent conflict, and the rights and desert of the successful rub uncomfortably against the needs of the poor. But if justice can take on such various faces even within our society, how many more appearances must it make if we expand our vision to consider societies very different from our own. For example, a search for the notion of justice (or a rough equivalent) in current-day Japan would be extremely difficult, not because of some mere translation problem, but because the cultural and contextual differences would be so great that our usual concerns, summarized in terms of "merit" and "rights," for instance, would find no comparable social structures. And yet, philosophers since Plato have often talked as if the idea of justice as expressed in their own language represented a universal concept, the same the world over (though some societies approximate it much more than others, of course). To undermine this impression, we have excerpted the skeleton of a much more global argument by David Miller, drawn from his book, *Social Justice*, in which he argues that there are three very different sorts of contexts in which concepts of justice develop, and these in turn are very different indeed. Our central concept of property rights, for example, is all but missing in other conceptions of justice, and the idea that a person may be meaningfully said to "own" goods that he is not actively using and that others desperately need is an idea that some other cultures quite happily live without.

In my inquiry up to this point I have tried to explicate the familiar idea of social justice, first by separating its three distinct, conflicting elements, and then by showing how each element corresponds to a different way of viewing society, represented in my analysis by the political theories of Hume, Spencer, and Kropot-

From *Social Justice* by David Miller. Published by Oxford University Press, 1976, and reprinted by permission of the publisher.

kin. Each of these images of society (the stable order, the competitive market, the solidaristic community) plays a part in the thinking of our contemporaries. Any given person will adhere most closely to one image in particular, and to the corresponding conception of justice. Rather than explore the reasons for these individual differences, however, I want to ask whether ideas of justice do not vary systematically from one social context to the next. Some writers have suggested that men everywhere share a common sense of justice, which can be expressed as a general principle that incorporates more specific conceptions of justice. Although this may hold of the most basic notions of justice (such as the golden rule: treat others as you would like them to treat you under similar circumstances), I shall try to show that substantive ideas of social justice—the principles used to assess the distribution of benefits and burdens among the members of society—take radically different forms in different types of society. To do this, I shall start by comparing the social ideas characteristic of three such types. The types will be referred to as primitive societies, hierarchical societies, and market societies. Our own society will later be presented as a modified form of market society.

This classification is not meant to be an exhaustive social typology: many societies do not belong to any of the three categories. The purpose of the classification is rather to enable us to form a better understanding of the social ideas of market societies by looking at the corresponding ideas of societies with very different structures. In describing each type of society, I shall pick out its most distinctive features, without suggesting that these features will be wholly absent in the other types. Thus primitive societies will display, to a small degree, traits which are characteristic of market societies, and so on. The types are distinguished by the relative prominence or insignificance of features which will be found in societies of every kind.

Let me now outline the three social types to be used in the analysis, before discussing briefly the kind of explanation of social ideas which is to be attempted. Primitive societies, first of all, are small-scale societies in which the basic nexus between man and man is kinship. Men in these societies gain subsistence by hunting and gathering, tending flocks, or simple agriculture. The division of labour is not extensive, and there is no strong, well-defined system of political authority; if such authority exists at all, it will be vested in the village headman or council of elders. Contract, although it may occur occasionally, is not an important feature of these societies.

In hierarchical societies, men are arranged vertically in social strata, each stratum having a definite rank in the hierarchy. There is a strong, though not universal, tendency for economic dominance, social prestige, and political power to combine in the hands of the group at the top of the hierarchy. Each person is assigned to a stratum largely on the basis of his birth, since social mobility is low or non-existent. Stratum membership confers traditionally established rights and obligations, and largely determines the type of work which the individual will perform. Contract is again not an important feature of these societies.

I shall confine my discussion to a particular sub-type of hierarchical society, namely feudal society. In this sub-type the social mobility of individuals is slight rather than nonexistent; a man is not tightly bound within the stratum of his birth (whereas in caste society, for example, there is no individual social mobility and each stratum forms a self-contained social unit with its own rules and authority structure).

Market societies, finally, are typically large, economically developed societies, in which the basic nexus between man and man is contract. The division of labour is extensive, and each man is formally free to decide upon his occupation, to enter into whatever associations he wishes, and generally to choose his place in society. The central institution of these societies is the economic market in which commodities (including human labour) are bought and sold. Kinship is of small importance and authority is largely the creature of contract.

How can this typology be used to explain variations in men's ideas of justice? The fundamental assumption made here is that a man's sense of justice is strongly affected by the nature of the relationships which he enjoys with other men. The social structure of a particular society generates a certain type of interpersonal relationship, which in turn gives rise to a particular way of assessing and evaluating other men, and of judging how benefits and burdens should be distributed. My aim is to show that there is an intelligible connection between the nature of a man's relationship with other men and his conception of justice. The argument, in other words, takes this form: given that a person's relationships with others are predominantly of this type, it is *rational* or *appropriate* for him to adopt the following conception of justice. . . . Such an argument appeals to common notions of rationality and appropriateness; the existence of such notions is an assumption which cannot here be justified. Furthermore, my general methodology—the view that explanation takes the form, first, of demonstrating a correlation between social structures and certain ideas, and, second, of showing an intelligible connection between the two elements—is taken for granted.

Primitive Societies

There is one rather obvious sense in which men in primitive societies do not have our concept of social justice. Whether we place the emphasis on the principle of desert or on the principle of need, we have in either case an understanding of social justice as an ideal which stands in contrast to the existing state of affairs; typically, as I have argued throughout Part II, our ideals of social justice form parts of wider views of society which, to a greater or lesser extent, go beyond society as it now is. Such world-views are not to be found in primitive societies. Primitive men do not conceive of the possibility of transforming society so that it comes to match up to a social ideal. If, therefore, we are to find a concept of social justice in these societies, we must look, not for an abstract ideal, but for an idea which is expressed in actual social practices. We should turn our attention, for instance, to the way in which primitive societies allocate parcels of land to their members, or to the manner in which tribesmen share out the spoils of a hunting expedition. Of course, we must examine not only the physical distribution of goods, but the principles which lie behind this distribution, in so far as we can discover what they are.

My thesis, however, will be that primitive societies do not practise social justice as we understand it. Their treatment of individuals and their manner of distributing benefits are guided by principles which do not correspond to the principles of social justice found in modern societies. There are points of resemblance, but the divergences are more striking. To show this, I shall focus my attention on those elements of social organization which are common to the great majority of primitive societies.

These societies show wide variation in certain respects, but it would be impossible here to give adequate attention to individual differences.

It was commonly held by nineteenth-century investigators of tribal societies that these societies practised communism. No distinction, they said, was made between one man's property and his neighbour's, and each took from the common pool whatever items he needed. (This view was especially strongly maintained by L. H. Morgan, who passed it on to Marx and Engels.) Modern anthropological opinion, however, agrees in regarding this as a serious error. The earlier investigators, failing to find property rights which were the exact analogues of those in market societies, concluded that common ownership must prevail. In fact, private ownership is the norm in primitive societies, but the rights of ownership are different from, and generally more restricted than, the corresponding rights in a society such as our own. At the very least, primitive men own their clothes, items for their personal use, their tools, and weapons. It is also usual for them to own their cattle, their crops, and whatever they have produced, gathered, or killed by their own labour. The owner-ship of land is a rather more complicated matter: with hunters and gatherers, and with primitive pastoralists, the lands generally belong to the tribe as a whole, private ownership having little point in these cases. With agriculturalists, however, private ownership of farming and gardening land is again the norm, though as we shall see the *sense* in which the land is privately owned must be properly understood.

When we speak of a person owning something, our standard meaning is that he has an absolute and exclusive right to the enjoyment of that thing. His right excludes any other person having an equivalent right to the same object, and he is permitted to use the object as he likes within the law, to sell it to whom he pleases, and to destroy it if he wishes. Primitive ownership is rarely of this kind. A man who possesses a personal ornament may be obliged to pass it on to a kinsman at some definite time in the future. A hunter returning with a catch may be under an obligation to distribute some portion of that catch to other persons specified or unspecified. Thus in Samoa 'fishermen on returning to the shore have to give a portion of their catch to anyone they meet in the lagoon or on the shore . . .', and similar practices are to be found elsewhere, for example among several of the Eskimo peoples. The ways in which primitive rights of property in land differ from their equivalents in modern society deserve fuller examination.

* * *

These limitations which primitive people place on rights of ownership furnish us with important clues about their social thinking. We have seen that the strict protection of individual rights which is demanded by our own conception of social justice reflects the view that justice concerns the proper treatment which is due to each individual, irrespective of general social utility. In this perspective, the just act and the socially useful act may on occasion diverge. By contrast, primitive men naturally think of the welfare of the group and subordinate the welfare of particular persons to that general end. While it may be going too far to say with Maine that the subject of rights and duties in primitive societies is the group, not the individual, it is certainly true that rights are granted in these societies when it serves the common interest to do so, and they are limited or rescinded on the same basis. This thinking underlies the restrictions on property rights which we have observed, and is particularly well illustrated by the borrowing of canoes permitted by the people of Tikopia. Of a similar practice which existed among the Chuckchi Eskimos, an observer writes:

> It is contrary to the sense of justice of the natives to allow a good boat to lie idle on shore, when near by are hunters in need of one.

I should say of this practice rather that it shows the *weakness* of the Eskimos' sense of justice, and their willingness to allow considerations of common interest to override individual claims. The same tendency can be seen in primitive thinking on questions of distribution: when John Ladd asked a Navaho Indian how he would distribute 500 sheep among his fellows, his informant replied that they should be given to those best able to look after them.

* * *

The two most fundamental relationships in feudal society were that between lord and vassal, and that between lord and serf. Each took the general form of a contract between superior and inferior, in which the inferior party offered to perform certain specified services for the other in return for protection and the opportunity to make a living. In the case of vassalage, the contract was made between two men who were nominally free (the obligations of a vassal did not pass directly to his descendants), the vassal swearing in general to be his lord's 'man' and in particular to give military service of a stated nature and extent; the lord offering protection and a fief, which generally consisted of a landed estate sufficient to support the vassal. As for serfdom, the contract here was largely tacit since the serf's status was passed on to his descendants without a renewal of the agreement, but the relationship was again a reciprocal one, the serf performing tasks of many kinds for his lord in return for protection and the right to till a share of the estate. The relationship was regulated by the 'custom of the manor'. The difference between these two contracts should not be overstressed since in practice vassalage tended to become heritable, the son of the previous vassal routinely swearing homage to the lord. We may therefore conclude that feudal society was built upon contracts between man and man which differed from the 'free contracts' of market societies in being contracts of allegiance (rather than contracts between equals) whose terms were governed by custom (rather than being chosen by the parties) and which were in practice largely heritable.

An immediate consequence of the feudal relationships was that, as in primitive societies, property rights were divided among a number of men of different status. Apart from a residue of free peasant proprietors (known as 'allodialists'), each piece of land was owned by a number of men, from the tenant who actually worked the land, through the lord to whom he owed service, to the lord of that lord, and so on. Each man had a distinct set of rights over the land, the exercise of which was governed by custom.

Feudal law was likewise fragmented. Each man was subject to the rules of law prevailing in his locality, and little attempt was made to create a uniform legal system. Since a man carried his law with him, there was truth in the remark attributed to the Archbishop of Lyons, that when five men were gathered together it was no surprise if each of them obeyed a different law. Feudal law was also fragmented in another sense: men of different status in the hierachy were tried by different courts for committing the same crime, and were liable to receive different penalties as well. Manorial lords administered the law among their tenants, while they themselves were judged by a group of their peers.

Within the law, custom was dominant. We may conveniently consider this under two aspects. First, as far as the content of the law was concerned, custom took precedence over deliberately enacted law. If two laws were found to be in contradiction, the older of the two was thought to possess superior validity. In fact, the feudal period exhibited a decided preference for unwritten custom, which, as Bloch points out, was by no means irrational since unwritten law could undergo subtle changes to meet new social needs more easily than could codified law. Second, in deciding particular cases, established practice was almost always the clinching factor. Thus if the right to work a piece of land was in dispute between two men, the one who could prove that he (or his ancestors) had ploughed the field in earlier years was certain to succeed in his suit.

Against this background we may understand the conception of justice in medieval Christianity. Justice was seen as an important social virtue, and was thought primarily to consist in the observance or enforcement of the law. In terms of our own analysis, justice was interpreted as the protection of a man's legal rights, these rights deriving by custom from his position in the social hierarchy. We have seen that the law in feudal society was fragmented, and this conception of justice therefore contained no notion of the equal treatment of all men: it was thought rather, as one writer has put it, that 'law and justice have as their true purpose the harmonious preservation of the social *status quo* in the interest of common peace and order'.

Although justice was predominantly identified in the minds of medieval Christian writers with the protection of legally established rights, secondary recognition was given to the claims of need. Here the doctrine that by natural law all property was originally held in common became of practical importance. Because God had in the beginning given the earth to all men to enjoy in common, it followed, first, that no one had a right to take for himself more than he needed, and, second, that a man had only a right to take that of which he made good use. A distinction was drawn between necessities and superfluities, at a level which depended upon a man's station in life. As Tierney puts it, 'medieval canonists took it for granted that different styles of living were appropriate to the different grades in the hierarchy'; superfluities were the goods which a man possessed over and above this accepted standard of living. It was a matter of justice that superfluous goods should be given to those in need, whereas if a man gave away his own necessities to another this was an act of charity or mercy. Aquinas took this doctrine a step further when he argued that a man in need was entitled to avail himself of another man's property;

> It is not theft, properly speaking, to take secretly and use another's property in a case of extreme need; because that which he takes for the support of his life becomes his own property by reason of that need.

What is the connection between the feudal conception of justice and the conditions of life in feudal society? Let us begin by noting that the supply of food and other basic goods was more certain in feudal societies than in most primitive societies. Although poor by comparison to modern market societies, they generally afforded the peasantry a subsistence income and of course allowed privileged groups—the nobility and the clergy—to live off the surplus created by the peasants. In considering primitive societies, I argued that an overriding concern for economic survival militated against the development of a sense of justice; the main preoccupation of feudal society, however, was less economic survival than the preservation of order—what men feared was that society would disintegrate into a congeries of warring platoons, as indeed it did from time to time. To preserve order in the absence of a sovereign power, it was above all necessary that customary rights and obligations should be respected; in such a context it becomes intelligible that justice should become an important value, and that it should be interpreted in a conservative sense, to refer to the protection of established rights. We have repeatedly stressed that in feudal society in the function of justice was to stabilize the social hierarchy in the interests of order.

We may turn now to the social relationships characteristic of feudalism. How did men holding different positions in the hierarchy encounter one another? First, it was clear that each man had a single dominant role, determined mainly by the type of work he performed, and these sole differences were made perfectly manifest in differences of dress, speech, etc. Second, there was at the same time sufficient geographical and social mobility to make it clear that the various estates were not occupied by men of different 'breeds': there was a sense of common humanity and an appreciation that the separation of men into estates was conventional, although of course absolutely necessary. Third, the relationship between superior and inferior was a personal one. The serf was not subordinated to the nobility as a collective group but tied to one particular lord, and equally the vassal swore homage either to a single lord or (through a corruption of the institution) to a small number of lords. Writers on feudalism have observed that the system worked best when this personal relationship was preserved through sustained contact between master and man. Now if we take these three points together, what moral ideas are likely to emerge from such a system of relationships? First, the strength of role divisions is such that more emphasis will be laid on maintaining the various roles than on giving each person his due as an individual. Justice will be thought to consist primarily in treating a man according to his station rather than according to his particular personal qualities. In other words, (as I shall argue later) justice as desert can only emerge as a value when the market has rendered role divisions much more fluid, so that we encounter other men as individuals rather than as the occupants of fixed roles. So long as role distinctions remain strong, the appropriate way of understanding justice is as the protection of a man's rights which derive from his social position.

We have also seen, howwever, that in feudal society the occupants of different roles were linked to one another by ties of personal dependence. Rerlationships of this type naturally created a sense of mutual obligation rather than a sense of mutual indifference. Although there was no question of the lord acknowledging the serf as his equal, he did recognize that they were connected by a personal bond, and that he therefore had a responsibility for the serf's welfare (no doubt in practice such responsibilities were often evaded, but we are here considering the moral responses appropriate to feudal society). This is the basis of the recognition given to the claims of need in the feudal conception of justice. A condition of need gave a man a just claim on the resources of those placed in privileged positions, and especially of course on the master to whom he was personally tied. We should again note the contrast with market society where impersonal market relationships predominate, and men are likely to encounter one another as strangers.

At the beginning of this chapter, I introduced the social typology which I was about to use with the observation that each type of society—primitive, feudal, and market—was distinguished by the prominence of certain features within it, while features characteristic of the other types would be present to a lesser degree. It is particularly important to bear this in mind when discussing market society. I shall describe the early capitalist systems of Western Europe and America as 'market societies', while acknowledging that in real terms these systems superimposed 'non-market' social structures and relationships upon market ones. What are the distinctive social relationships which make up a market society? Its social structure is created out of a series of contracts and exchanges between otherwise free and equal individuals. By contrast to their situation in primitive society, men in market society are not bound by obligations of kinship and traditional status. By contrast to feudalism, they have no fixed place in a hierarchy, and owe no allegiance to any superior. Under a market system, men are equal before the law. They are free to choose their occupations, to join whatever associations they wish, to buy and sell in the market, to make contracts without restriction, to gain wealth and prestige. Any social and economic inequalities arise from these activities. Most of what each man produces is exchanged in the market, rather than consumed directly; this implies a fairly extensive division of labour and a system of mutual interdependence centred on the market.

If we compare this market model with the actual conditions of Western capitalism (at any point in its development) we shall see that, although market relations have always been predominant, they have been combined with other elements of social structure, especially with the existence of vertical class divisions. Capitalist societies are not pure market societies, firstly because they typically contain an aristocratic class whose social position is inherited rather than achieved through production and exchange; and secondly, because they contain a working class whose situation is only partly a market one. The worker enters into market relationships with his employer (he freely exchanges his services for a wage) and with the tradesmen from whom he buys his goods, but his relationship with his fellow workers is not basically a market one. It is rather one of mutual assistance and mutual support for security, and for protection against a common opponent—the employer. Trade unions, the chief institutional form through which this class solidarity has been expressed, have rightly been recognized by theorists of the market as incompatible with a pure market system, rather as the communal compact of medieval burgesses was seen by feudal thinkers as alien to that system. We shall later examine what effect its ambiguous position—half in and half out of the market—has had on the social ideas of the working class.

The fact that in the real world market relations only exist in conjunction with vertical class divisions has the unfortunate consequence that we cannot find a homogeneous market society with an appropriate set of shared social ideas. But it may be possible to overcome this difficulty to a certain degree by finding within market societies a group of people for whom market relations are parcicularly salient. During the earlier period of capitalism, particularly in Britain and the U.S.A., the middle class appear to form such a group. I include among the middle class capitalists, merchants, shopkeepers, and (more arguably) independent artisans and farmers. These men worked for themselves, and their success depended upon their ability to produce and sell in the market. They inherited no fixed position in society, yet the social structure was sufficiently open for them to rise in the world according to their ability and good fortune, at least until they reached the very top of the social scale.

As market relationships became more prominent, and as the middle class began to exert more social influence, a new theory, which we may refer to as individualism, was developed. This theory abandoned any notion of a natural hierarchy in society, and began instead with the idea that men were born free and equal, possessing sets of rights which derived from their inherent natural capacities. Society was seen as the product of the contracts and associations into which these free individuals had entered for their own advantage. The social order was not a constraint on human ambitions, rather whatever order society possessed was the result of human wants and interests, and ideally at least should be readily modifiable if those wants and interests changed. A man's duty was no longer to remain within his station, but instead to take on whatever tasks, and reap whatever rewards, his abilities would allow him. In religious thought, this meant that the doctrine of the 'calling' was transformed from a recipe for economic traditionalism into the demand that each man should seek out the calling most suited to his capacities. The parable of the talents was cited to show that God had granted each man abilities in order that he should put them to the best use possible, and gain his just reward in the process:

An integral part of individualism was a conception of justice as the requital of desert. This criterion is stressed to the exclusion both of the protection of rights and of the fulfilment of needs. Of course, in individualist theory the rights of property are inviolable, and contracts must be enforced, but the ultimate ground for these views is either utilitarian or else resides in the conception of justice referred to. Individualism has no time for rights as such, when these are neither socially useful nor necessary to ensure that desert gets its proper reward; it would not, for instance, acknowledge that the rights traditionally attached to a particular social position were in any way worthy of respect. Anything that a man may justly claim he must earn by exercising his capacities in a socially useful way, in which case the institutions of property, contract, and exchange will ensure that he gets a fair return.

Why should individualism, and the theory of justice as the reward of merit that goes with it, have been produced by market society, and especially by the group in that society for whom market relations were particularly salient? One answer is that individualism is simply a correct description of market society; and that the market really does reward people according to their deserts. But this is manifestly too crude an answer, and neither of the claims contained in it is very plausible. We shall do better to look for an intelligible link between the social position of the group we have singled out for examination and the ideas of justice to which it adhered. The relevant features of the social position of the middle-class entrepreneur or tradesman are as follows: first, the lack of a fixed social status such as would be provided by a hierarchical society, but instead the opportunity to make one's way in the world by one's efforts; second, the absence of a traditional set of rights and obligations— whether to kinsmen, etc., as in primitive society, or to superiors and/or inferiors as in hierarchical society; third, social and geographical mobility, with the consequence that many of one's social encounters are with strangers; fourth, the prominence of market relations themselves—particularly the exchange of goods and services, and free entry into contract.

The fourth of these features is perhaps the most important. The intellectual results of the exchange relationship have been classically analysed by Marx and Bouglé. Both see the market as essentially egalitarian in its implications; it undermines hierarchical social and legal relations because market power is solely a function of the amount of exchange value (goods or money) that a person possesses. From the point of view of the market traditional social distinctions are irrelevant, and there is a contradiction between the equality of the market place and, say, the social inequality of a system of estates. As market relations become more dominant, egalitarian demands are made—for equal human rights, equality before the law, political equality, etc. But this egalitarianism stops short of economic equality in the strong sense, for the market requires equality in the exchange of values rather than the equal treatment of individuals. Its principle of operation is that each man should receive back the exact equivalent in value of what he brings to the market; however, the amount brought to the market will depend upon the productive powers of the individual, which differ of course from person to person. What the market achieves is a distribution according to desert (where desert is measured by the creation of exchange value) rather than economic equality. Bouglé is correct when, in setting out the 'egalitarian ideas' which characterize modern (late nineteenth-century) society, he gives the principle 'to each according to his works' as the criterion of economic distribution, and rejects the principle of equal distribution which he believes to enjoy little favour with his contemporaries.

We can thus see how the experience of the market leads people to adopt a conception of justice as the requital of desert. The paradigmatic exchange relationship takes place between two people who are strangers to one another, who owe neither deference nor services to each other, and whose knowledge is limited to the relative values of the goods or services to be exchanged. This relationship allows no foothold either to the notion of protecting rights or to the notion of satisfying needs. Established rights are irrelevant because neither person stands in any fixed role relationship towards the other, there are no standing obligations, and so on. Needs are irrelevant because each party to the exchange is bent solely on making the most favourable bargain possible, and in any case has no knowledge of the needs of the other party. On the other hand, although an exchange relationship does not have as its *direct* object the requital of desert, it does *indirectly* bring about this result provided that desert is defined in such a way that a person's deserts are embodied in the commodity which he brings to the exchange. At the very least, the definition of justice as desert is the only substantive conception which is consistent with the type of interpersonal relationship characteristic of the market.

The argument of the last chapter, reduced to its simplest terms, is that our concept of social justice has grown out of the specific arrangements of market society, and that while other types of society may embody concepts which are in certain respects analogous to ours, they do not have any single concept with the same range of uses. Furthermore, some of the conflicts which are inherent in our concept can be understood by reference to its historical development under the impact of social change—for instance, the rather uncertain place occupied by the idea of need in our thinking about justice may be explained by reference to the changes in social thought accompanying the transition from free market to organized capitalism. Now if this argument is to be fully successful, it should eventually give a complete sociological account of the structure of the concept which was originally revealed by philosophical analysis. If the philosophical and sociological parts of this inquiry are compared, however, it will be seen that one of the three constituent elements of our concept of justice—the principle of rights—has been omitted from the sociological account of market society; and the reasons for this omission must now be explained.

From our point of view the cardinal feature of market societies, not shared by primitive or hierarchical societies, is that their members characteristically evaluate the social distribution of benefits by reference to certain ideal standards of distribution—namely criteria of desert and need. This is above all what we have in mind when we talk about the idea of social justice. The point I was concerned to make was that although other kinds of society clearly have fixed rules of distribution, etc., they do not have an idea of social justice in the sense just explained. Having said this, and having looked at various differences between market and other societies which may serve to explain it, we may now go on to say that the general criteria of social justice can only be implemented by establishing fixed institutions and rules, and thus by conferring certain positive rights upon individuals. Because such implementation is always imperfect, there must arise the conflicts between established rights and deserts or needs which we analysed in chapter one. Once a right has been established, it gains its own intrinsic value because individuals come to govern their actions by reference to it, and so any interference with rights will affect the security and freedom of action of some people. But in market societies rights are open to criticism as they are not in, say, feudal societies. The question of whether an individual ought to have the rights that he does may always be raised, and if a given distribution of rights is seen to be sufficiently unjust (by a desert or need criterion) we may decide to alter it. Thus what is distinctive about the social thinking of market societies is their assessment of existing rights by ideal standards of social justice, and it is these ideal standards which stand most in need of sociological explanation. This accounts for the direction of the argument in the last chapter.

Finally, we should note that actual market societies still contain some remnants of 'feudal' thinking, in which rights are accepted even though they have no foundation in one of the ideal standards of justice. For instance, a few people will be found to accept the notion of a 'natural aristocracy'—of a class of people with an inherited right to govern their inferiors. This should serve to remind us that our social classification was a classification of ideal types, and that real societies will always contain elements drawn from each type; thus market societies contain some hierarchical elements, and so forth.

PART IV

Punishment is, perhaps, the original meaning of justice and it is, no doubt, one of its most enduring aspects. Whether a society is free market or socialist, whether it is a free-wheeling democracy or a ruthless dictatorship, whether it systematically respects or routinely violates human rights, there are always those who break the law and violate whatever there is of a public trust, and they must be punished. But how do we punish them, and why? One suggestion of long-standing repute was that the breaking of the law—any law, no matter how arguably "minor" or mistaken—deserved severe punishment, on the grounds that it was the disrespect for the law itself (and the authority behind it) that primarily required punishment and only secondarily the crime itself. (Indeed, we still say that a violent crime against a particular person is, nevertheless, "a crime against the state.") One might still adopt some sliding scale of severity for different crimes, and one might argue for or against severity, but at least this conception allows for some more or less uniform sense of what is at stake in punishment. It is breaking the law itself that is at issue, and punishment represents the sanctions that the state applies to uphold its own authority.

But, even so, is the purpose of punishment straightforwardly to *punish* the criminal for his or her crime? Or is it rather itself to uphold and make visible the authority of the state? Is it to frighten other potential wrongdoers from their lawless plans? Or is it to cure the criminal of his or her propensity to disrespect and lawlessness with a therapeutic dose of pain, deprivation, or suffering? If all of the above, in what mixture, or are some of these considerations much more central than others? There is an ancient argument that it would not be punishment—whatever its effectiveness in expressing the authority of the state, deterring others or "curing" the deviance of the criminal—if it were not in some sense "payment for" the crime itself. Indeed, this argument was put forward by no less a modern moral philosopher than Immanuel Kant, who argued uncompromisingly that retribution and retribution alone necessitates and justifies punishment. In fact, Kant even argued that if there were no point in deterring future crime or curing the criminal, if there were no practical purpose to expressing the authority of the state, indeed, if "society as such were to be dissolved tomorrow," it would still be necessary to punish the criminals in our jails, to balance the ledger, and repay crime with punishment. To prevent crime is obviously important, but such measures are not, whatever else they may be, punishment. Indeed, the most effective deterrent might well be a periodic beheading, in the main square or on television, of a randomly chosen, presumably innocent person, just to demonstrate the ruthlessness of the state. But no one would call this justice, and the deterrent could not be confused with punishment.

The idea of punishment is repugnant to us, however, and quite at odds with the conceptions of virtue and mercy that have dominated much of our thinking about justice since Socrates and Jesus, respectively. Punishment as such, the repayment of a debt, strikes us as much too much like mere revenge, a primitive and brutal practice (or is it even an instinct?) that should have been long eclipsed by respect for the law and the justice of the state. Punishment for crimes and the correction of injustices are no longer left in the hands of victims and their families, but are rather the concern of the law. Punishment as such seems archaic and, instead, current debates tend to focus on the effectiveness of punishment (and capital punishment in particular) as a *deterrent* and as a *reformative* agent. But long before the reform movements of the eighteenth century (on which these debates are based), and before punishment was still, more than anything else, a matter of retribution and revenge, there were deep philosophical questions about the nature of punishment. What sort of punishment is appropriate for what sort of crime? How should crimes be considered, as, first of all, the breaking of the law? Or, as first of all, the violation of another citizen's (or citizens') rights? What does it mean to say that the punishment should "fit" the crime? The famous passage from Exodus, "an eye for an eye, a tooth for a tooth . . ." is itself the culmination of a long ancient debate, in which the mark of justice was made out to be the "measure" of retribution, the proper payment of a "debt" to society or to the wronged person or party. But where the "measure" is not so easily determined—how much money should a speeding ticket cost a driver? how many months in prison are "equal" to a burglary?—the appealing clarity of the "eye for an eye" philosophy gets lost. And that in turn renders more appealing the utilitarian insistence on deterrence, for the question—how much punishment does it take to deter a prospective criminal from committing a certain crime?—seems to offer a much more precise answer than the "fit" and "debt" metaphors it is intended to replace.

The most brutal and, from one obvious point of view, ultimate form of punishment is capital punishment—"the death penalty." For good reason, debates about the death penalty have often preoccupied philosophical and public concern with criminal justice, for it seems to bring together and summarize the most difficult questions of all about punishment. For one thing, it is irreversible (but most punishments other than the payment of fines are, too), and (what makes it differ from virtually all other punishments) in the case of a mistake it cannot be compensated. Of course, the relatives of an innocent victim can be paid off and his or her dependents cared for, but this hardly counts as compensation for the victim. Murder cases do seem to fulfill (as few crimes do) the demand that the punishment fit the crime (with the exception that a criminal can only die once for no matter how many murders), but this in turn leads to the obvious objection that, if it was wrong for the murderer to have murdered, why should it be any less wrong for the state to murder the murderer. Furthermore, while capital punishment can be defended as the most dependable possible deterrent regarding any further crimes of this particular criminal, its effectiveness in deterring other criminals is a matter of considerable controversy. In fact, those crimes that most often call for the death penalty—murders of immediate family members—are not at all likely to be repeated in any case, whether because of the usual extreme grief and remorse on the part of the murderer or the scarcity of remaining similar victims. Crime rates in general do not drop in societies that adopt capital punishment, and they do not rise in societies that stop using capital punishment—which makes it all the more evident that the constant clamor in some quarters for the death penalty is indeed a concern for the penalty, a matter of retribution rather than the overall public good. And this in turn makes the death penalty in particular sound more and more like retaliation, like a collective act of vengeance rather than an act of justice. But is revenge in fact opposed to justice, or is this, perhaps, right at the core of the entire debate? Where do revenge and retribution stand in relationship to "retributive" justice, and can we understand justice without them?

"An eye for an eye," from the Bible

We have repeated these two passages, from the Old and New Testaments respectively, because they sum up in a contrast much of the debate that has taken place over the past 2,000 years. On the one hand, there is that basic ("natural"?) urge for retribution, "measure for measure." On the other, there is that elevated sense of being "above" such urges, tempering one's thoughts of vengeance with mercy and appealing to judgment "above"—whether to God or the courts.

Leviticus 24:17–22

17 And he that smiteth any man mortally shall surely be put to death. 18 And he that smiteth a beast mortally shall make it good, life for life. 19 And if a man cause a blemish in his neighbor; as he hath done, so shall it be done to him: 20 breach for breach, eye for eye, tooth for tooth; as he hath caused a blemish in a man, so shall it be rendered unto him.

Matthew 5:38–42

38 Ye have heard that it was said, An eye for an eye, and a tooth for a tooth: 39 but I say unto you, Resist not him that is evil: but whosoever smiteth thee on thy right cheek, turn to him the other also. 40 And if any man would go to law with thee, and take away thy coat, let him have thy cloak also.

Plato, *Crito*

The New Testament virtue of "turning the other cheek" supplanted or in any case supplemented the Old Testament insistence on revenge, but a similar debate had taken shape several centuries before Christ in ancient Athens. Socrates and his student Plato turned against the Homeric conception of vengeance and rejected the idea of punishment altogether as "doing evil for evil." Socrates himself, of course, was executed as punishment for a crime that, in the opinion of history and even in the minds of many of his judges, hardly deserved the severity of the death penalty. Here in *Crito*, Plato argues with his friend that retaliation and injury are never right, even in return for another evil.

———

S: Do we say that one must never in any way do wrong willingly, or must one do wrong in one way and not in another? Is to do wrong never good or admirable, as we have agreed in the past, or have all these former agreements been washed out during the last few days? Have we at our age failed to notice for some time that in our serious discussions we were no different from children? Above all, is the truth such as we used to say it was, whether the majority agree or not, and whether we must still suffer worse things than we do now, or will be treated more gently, that nonetheless, wrongdoing is in every way harmful and shameful to the wrongdoer? Do we say so or not?

C: We do.

S: So one must never do wrong.

C: Certainly not.

S: Nor must one, when wronged, inflict wrong in return, as the majority believe, since one must never do wrong.

C: That seems to be the case.

S: Come now, should one injure anyone or not, Crito?

C: One must never do so.

S: Well then, if one is oneself injured, is it right, as the majority say, to inflict an injury in return, or is it not?

C: It is never right.

S: Injuring people is no different from wrongdoing.

C: That is true.

S: One should never do wrong in return, nor injure any man, whatever injury one has suffered at his hands. And Crito, see that you do not agree to this, contrary to your belief. For I know that only a few people hold this view or will hold it, and there is no common ground between those who hold this view and those who do not, but they inevitably despise each other's views.

From Plato's *Crito*, © Hackett Publishing Co. Reprinted by permission of the publisher.

P. Marongiu and G. Newman, from *Vengeance*

In *Vengeance*, Marongiu and Newman describe a variety of established cultural practices concerning revenge. The idea that vengeance is a primitive emotion or even an instinct neglects the enormous amount of cultural stimulation, support, and structure, as well as some semblance of legitimacy vengeance receives from particular social rules and expectations. In many parts of the world, retaliation and vengeance—not legal retribution—are built into the social system of justice. Vengeance is a personal, family, or tribal matter, not an impersonal function of the state. In the following excerpt, the authors characterize the Sardinian code of vengeance.

The Sardinian Code of Vengeance

> Su sambene no est abba. (Blood is not water)
> —A Sardinian saying.

In Sardinia, the presence of a long established code of vengeance that displays a high degree of formalization and seems, to this day, to regulate the mechanism of the feuds as well as individual vengeance, has been analyzed by Antonio Pigliaru. It has been interpreted as an expression of the wider discrepancy between cultural models of the inner areas of the island and that of Italian society in general. This hypothesis was stated some years ago in the now classic book by Wolfgang and Ferracuti, *The Subculture of Violence*.

Because of its unique history and culture, a particular lifestyle has been maintained in the inner areas of Sardinia that serves to regulate and control violent behavior. The state, traditionally viewed by the inhabitants as a source of "foreign" domination, was always immersed in conflict with the local inhabitants. Historically, in fact, this authority has been sporadic and essentially remote. This condition of conflict, linked to the permanent social organization of extended families and clans that were also in conflict with each other for control of territory and goods, has produced, over the centuries, a climate of isolation and antagonism towards external forces of change. It is not then surprising that the Sardinian pastoral society has developed a system of defining and controlling conflict by avoiding any recourse to a third party—i.e., the state. We can see that this is an attempt to maintain social order without resorting to the obedience model.

The antagonism between the two models of reciprocity and obedience that we have proposed to explain the phenomenon of vengeance is also clear if we consider this customary code, and in general, the conflict between a pastoral subculture and the wider Italian culture. The culture of vengeance that underlies this code is also responsible, at least in part, for the mechanism of identification and support of bandit figures that are opposed to the pressing external cultural forces. In this context, the code seems to express the need of the pastoral society to exercise a limitation and control on the destructive mechanism of vengeful exchange. In order to do this, the code must be able to define offensive and vengeful action in terms of crimes and punishments. It must be able to establish "objective" limitations to vengeful behavior beyond the definitions of offense given by each party or faction.

The Offenses

According to the code of vengeance, "one determinate action is offensive when the event from which depends the existence of such offense is foreseen in order to damage dignity and honor." Property damage in itself is not an offense, and is not sufficient cause for vengeance unless it were done with specific intent to offend the honor of the clan or individual. In this case the offensive will is determined by the basis of objective circumstances or subjective circumstances such as the intensity of malice and condition of the offender, or the relation between the offender and the offended.

In addition, the code of vengeance considers other kinds of offenses of different degrees of gravity. The most serious offense is murder. Other offenses are bearing false witness and spying. The code states clearly that the collective or individual responsibility is the constitutive element of the offense itself. In fact, according to "article 5" of the code, the responsibility for the offensive action is individual or collective according to whether the offense follows the action of a single person or that of a group that is operating as a group. The group is organized according to family structure, and is not responsible for the offenses when they are produced by a single member of the group. Only when the group expresses clear and active solidarity in favor of the culprit, is collective responsibility seen as part of the offense.

Antonio Pigliaru has observed that intentionality is the essential element of the offenses because vengeance, at this stage in its development (i.e. focused on the original offender, rather than unfocused like that of Achilles), must be intentional in order to be directed to the proper source of offense.

Therefore, the offenses can be extinguished only when the "victim/offender" recognizes his responsibility, and takes on himself the charge to repay the "debt" that is "requested" by the offended person, or, when the offender acts in a situation in which he has no other choice (e.g. where he is forced by violence), in which case the author of the violence is responsible for the offense. Once it is stated that the offense is an act set up in order to damage the honor of the other, we must determine exactly what it means to damage the dignity and honor of another.

Pigliaru says that the intent of the offender is either to annihilate or return to a position of weakness the other person or family by hitting them at their weakest point. For example, the use of arson makes clear that there is no material gain from the offense, but ensures that the other faction is placed at a loss. The intent to injure the other party is therefore clear. It is to upset the reciprocity and balance of power between the two factions by taking advantage of the weakness of the other, trying to impose authority with violence, and annihilate the enemy by violent means. The offense is, in this way, similar to, and derivative of, the primitive feeling that we identified at the beginning of this book, of being arbitrarily subjected to a tyrannical power against which one seems powerless to act.

The Punishment

Article 1 of the vengeance code states that vengeance is obligatory. The offenses MUST be avenged. Once the collective or individual responsibility for the offense has been proven, the law of vengeance obliges all persons concerned to take revenge. Obviously, the primary responsibility for implementing vengeance is with the offended person or group. In the Sardinian code, vengeance consists essentially of an offense that is given in response to that which was received. This offense should, in its classic form, match the precipitating offense in kind, but may nevertheless return more damage than was received. The vengeful act should be proportional, but progressive.

We have observed that this tendency of vengeful exchange leads, within the limits of progressive graduation, to the extreme form of retaliation, which is blood vengeance i.e. murder as revenge. The extreme difficulty in reaching a peaceful resolution to the conflict is most evident when, in blood vengeance, one is required to avenge a previous homicide. Predictably, bloody offenses are the only ones for which the code provides no "statute of limitations." Murders are never forgotten. In this way, the limitless "market" for vengeance is perpetuated. The punishment of blood vengeance is inflicted not only for homicides, but also for crimes "against honor," such as breaking a promise of marriage, spying, and bearing false witness.

The progressive tendency of vengeful exchange is clear; the offense that has acted as vengeance in itself constitutes a new offense which should be avenged. Blood vengeance in particular is a capital offense, even though it is given in order to avenge a preceding blood offense. In this instance we see the classic basis of what outsiders are inclined to call "chaos." It is a society in which there is no way to distinguish between crimes and punishments. They are, in fact, interchangeable, equally justifiable, equally criminal.

In recognizing the two elements of proportionality and responsibility (personal and collective) in vengeance, the code seems to be an evolution in comparison with the old conception of vengeance in which the retaliation was automatically and mechanically inflicted on all the members of the offender group. However, the maturation of vengeance is not quite as advanced as it was in Hamlet. That is to say, the societal and cultural form of vengeance is not as "primitive" a form of social control as it seems at first glance. Rather, it seems to be halfway between the primitive and the civilized. On the one hand, it does provide the basic distinctions that most established legal codes provide, such as the analysis of intent, the distinction of individual and collective responsibility; but on the other hand the mechanism of control is unable to go the full way of pronouncing which of the acts is criminal and which is not. To attain this, we need a "civilized" society in which a third party—the state—is able to transcend the opposing factions and pronounce which acts are criminal and which are not. It is apparent in feuding societies that the opposing factions will even join together in order to oppose a third party, almost invariably perceived as foreign and representative of an illegitimate order.

The pastoral society of Sardinia, in its warlike attitude, represents an interesting example, rare in western society, of the problems related to the regulation of social conflict, especially the maintenance of a permanent but unstable "equilibrium," without resort to an external principle authority. According to the Sardinian code, more than what could be done, must be done. The necessity to reap vengeance is also more than a private duty. Rather, it is a public duty in the sense that it must be performed on behalf of a code that is adhered to by both factions. The community is served by each side carrying out its duty to avenge. The avenger will even consider revenge as a necessary and unavoidable destiny from which there is no escape. The existence of rigid and immovable rules is typical of any traditional society. The wonder of vengeance is that, in this context, it provides the moral basis for the social order. Vengeance reveals itself as a kind of violence that eradicates the moral damage that was the consequence of another act of violence—a violence that cancels out each previous act, yet recreates iself. It is just as Marx analyzed capital. It works wonders, creating new commodities, which in turn, through social exchange, recreate capital.

Immanuel Kant, "On Punishment" from
The Philosophy of Law

Kant's moral philosophy centers on the all-important notions of "duty" and "obligation"—an ethics of "right" or "law" (*Recht*) and not merely of human goods and interests. He juxtaposed his philosophy against the various ethics of moral sentiment and of the public good that dominated British philosophy of the period. Against the "moral sentiment theories" of Hume, Smith, and others, Kant insisted that what is right must be a matter of universal, timeless, "a priori" principle, not of a possibly passing sentimentality or mere set of feelings. Against utilitarianism, in particular, Kant argued that mere utility could not explain the uncompromising nature of justice, much less account for those instances in which justice was demonstrably opposed to public utility. Kant's concern with justice, however, is almost exclusively tied up with the necessity of punishment. He has no theory as such of public utility or deterrence, no therapeutic view about rehabilitation or reform. He warns, "woe to him who creeps through the serpent windings of Utilitarianism," for the justification of punishment has nothing to do with its consequences, whether for the criminal or for the community as a whole. Kant gives the example of a criminal condemned to death who is offered the option of participating in a dangerous medical experiment. Kant repudiates such a trade with scorn, for then "Justice would cease to be justice, if it were bartered away for any consideration whatever."

What, then, is the nature and justification of punishment? For Kant, it is straightforward retribution, and nothing but retribution. Retribution is different from vengeance, which is a passion (though hardly a moral sentiment) and personal, insofar as retribution is dictated by reason and is dictated by the law or the court. The principle according to which retribution operates is "the principle of equality," that is, that the punishment inflicted should be the equivalent of the nature and the severity of the crime. "All other standards," he tells us, "are wavering and uncertain." Thus the principle of retribution, like that of "retaliation," insists on matching "like with like." The punishment must not only "fit" but "equal" the crime. In many cases, however, this equivalence is far from obvious—for example, in measuring the equality of pain endured by a rich man and a poor man paying the same fine, or the humiliation suffered by a person of high rank as opposed to one of lower rank. The one crime in which equivalence is not at all in question, however, is murder. There is no equality between the taking of the life and any lesser penalty than death. But note that, in addition to his insistence on "equivalence," Kant also employs what we might call the "erasure" metaphor—a punishment in some sense "erases" (which is not to say "reverses" or "undoes") the crime. It is in this context that Kant makes his infamous demand that, even if society were to be dissolved, it would be necessary to execute the last murderer in prison, "that bloodguiltiness may not remain upon the people."

Kant has especially harsh words to say about the kindlier moral sentiments applied to the problem of punishment. Against Beccaria's recent (1764) defense of the "passionate sentimentality of a humane feeling" (against capital punishment), Kant replies that his arguments are wholesale sophistry "and a perversion of right." In reply to Beccaria's argument, that it is unintelligible that people should agree (in the original "social contract") to their own punishment, Kant insists that a criminal doesn't will punishment, but rather chooses to do some punishable action. But this rather fine point of interpretation yields a rather large conclusion, that the rationality of punishment demands a distinct tribunal of public justice separate from the people over whom it rules. Not only is a crime a violation of the law rather than a mere offense against a person, but judgment against crimes must be made by the law and not by any person or group of persons.

Judicial or Juridical Punishment (*pœna forensis*) is to be distinguished from Natural Punishment (*pœna naturalis*), in which Crime as Vice punishes itself, and does not as such come within the cognizance of the Legislator. Juridical Punishment can never be administered merely as a means for promoting another Good either with regard to the Criminal himself or to Civil Society, but must in all cases be imposed only because the individual on whom it is inflicted *has committed a Crime.* For one man ought never to be dealt with merely as a means subservient to the purpose of another, nor be mixed up with the subjects of Real Right. Against such treatment his Inborn Personality has a Right to protect him, even although he may be condemned to lose his Civil Personality. He must first be found guilty and *punishable*, before there can be any thought of drawing from his Punishment any benefit for himself or his fellow-citizens. The Penal Law is a Categorical Imperative; and woe to him who creeps through the serpent-windings of Utilitarianism to discover some advantage that may discharge him from the Justice of Punishment, or even from the due measure of it, according to the Pharisaic maxim: 'It is better that *one* man should die than that the whole people should perish.' For if Justice and Righteousness perish, human life would no longer have any value in the world.—What, then, is to be said of such a proposal as to keep a Criminal alive who has been condemned to death, on his being given to understand that if he agreed to certain dangerous experiments being performed upon him, he would be allowed to survive if he came happily through them? It is argued that Physicians might thus obtain new information that would be of value to the Commonweal. But a Court of Justice would repudiate with scorn any proposal of this kind if made to it by the Medical Faculty; for Justice would cease to be Justice, if it were bartered away for any consideration whatever.

But what is the mode and measure of Punishment which Public Justice takes as its Principle and Standard? It is just the Principle of Equality, by which the pointer of the Scale of Justice is made to incline no more to the one side than the other. It may be rendered by saying that the undeserved evil which any one commits on another, is to be regarded as perpetrated on himself. Hence it may be said: 'If you slander another, you slander yourself; if you steal from another, you steal from yourself; if you strike another, you strike yourself; if you kill another, you kill yourself.' This is the Right of RETALIATION (*jus talionis*); and properly understood, it is the only Principle which in regulating a Public Court, as distinguished from mere private judgment, can definitely assign both the quality and the quantity of a just penalty. All other standards are wavering and uncertain; and on account of other considerations involved in them, they contain no principle conformable to the sentence of pure and strict Justice. It may appear, however, that difference of social status would not admit the application of the Principle of Retaliation, which is that of 'Like with Like.' But although the application may not in all cases be possible according to the letter, yet as regards the effect it may always be attained in practice, by due regard being given to the disposition and sentiment of the parties in the higher social sphere. Thus a pecuniary penalty on account of a verbal injury, may have no direct proportion to the injustice of slander; for one who is wealthy may be able to indulge himself in this offence for his own gratification. Yet the attack committed on the honour of the party aggrieved may have its equivalent in the pain inflicted upon the pride of the aggressor, especially if he is condemned by the judgment of the Court, not only to retract and apologize, but to submit to some meaner ordeal, as kissing the hand of the injured person. In like manner, if a man of the highest rank has violently assaulted an innocent citizen of the lower orders, he may be condemned not only to apologize but to undergo a solitary and painful imprisonment, whereby, in addition to the discomfort endured, the vanity of the offender would be painfully affected, and the very shame of his position would constitute an adequate Retaliation after the principle of 'Like with Like.' But how then would we render the statement: 'If you *steal* from another, you steal from yourself'? In this way, that whoever steals anything makes the property of all insecure; he therefore robs himself of all security in property, according to the Right of Retaliation. Such a one has nothing, and can acquire nothing, but he has the Will to live; and this is only possible by others supporting him. But as the State should not do this gratuitously, he must for this purpose yield his powers to the State to be used in penal labour; and thus he falls for a time, or it may be for life, into a condition of slavery.—But whoever has committed Murder, must *die*. There is, in this case, no juridical substitute or surrogate, that can be given or taken for the satisfaction of Justice. There is no *Likeness* or proportion between Life, however painful, and Death; and therefore there is no Equality between the crime of Murder and the retaliation of it but what is judicially accomplished by the execution of the Criminal. His death, however, must be kept free from all maltreatment that would make the humanity suffering in his Person loathsome or abominable. Even if a Civil Society resolved to dissolve itself with the consent of all its members—as might be supposed in the case of a People inhabiting an island resolving to separate and scatter themselves throughout the whole world—the last Murderer lying in the prison ought to be executed before the resolution was carried out. This ought to be done in order that every one may realize the desert of his deeds, and that bloodguiltiness may not remain upon the people; for otherwise they might all be regarded as participators in the murder as a public violation of Justice.

The Equalization of Punishment with Crime, is therefore only possible by the cognition of the Judge extending even to the penalty of Death, according to the Right of Retaliation. This is manifest from the fact that it is only thus that a Sentence can be pronounced over all criminals proportionate to their internal *wickedness*; as may be seen by considering the case when the punishment of Death has to be inflicted, not on account of a murder, but on account of a political crime that can only be punished capitally.

* * *

Against these doctrines, the Marquis BECCARIA has given forth a different view. Moved by the compassionate sentimentality of a humane feeling, he has asserted that all Capital Punishment is wrong in itself and unjust. He has put forward this view on the ground that the penalty of death could not be contained in the original Civil Contract; for, in that case, every one of the People would have had to consent to lose his life if he murdered any of his fellow-citizens. But, it is argued, such a consent is impossible, because no one can thus dispose of his own life.—All this is mere sophistry and perversion of Right. No one undergoes Punishment because he has willed to be punished, but because he has willed *a punishable Action*; for it is in fact no Punishment when any one experiences what he wills, and it is impossible for any one to *will* to be punished. To say, 'I *will* to be punished, if I murder any one,' can mean nothing more than, 'I submit myself along with all the other citizens to the Laws'; and if there are any Criminals among the People, these Laws will include Penal Laws. The individual who, as a Co-legislator, enacts *Penal Law*, cannot possibly be the same Person who, as a Subject, is punished according to the Law; for, *quâ* Criminal, he cannot possibly be regarded as having a voice in the Legislation, the Legislator being rationally viewed as just and holy. If any one, then, enact a Penal Law against himself as a Criminal, it must be the pure juridically law-giving Reason (*homo noumenon*), which subjects him as one capable of crime, and consequently as another Person (*homo phenomenon*), along with all the others in the Civil Union, to this Penal Law. In other words, it is not the People taken distributively, but the Tribunal of public Justice, as distinct from the Criminal, that prescribes Capital Punishment; and it is not to be viewed as if the Social Contract contained the Promise of all the individuals to allow themselves to be punished, thus disposing of themselves and their lives. For if the Right to punish must be grounded upon a promise of the wrongdoer, whereby he is to be regarded as being willing to be punished, it ought also to be left to him to find himself deserving of the Punishment; and the Criminal would thus be his own Judge. The chief error of this sophistry consists in regarding the judgment of the Criminal himself, necessarily determined by his Reason, that he is under obligation to undergo the loss of his life, as a judgment that must be grounded on a resolution of his *Will* to take it away himself; and thus the execution of the Right in question is represented as united in one and the same person with the adjudication of the Right.

G. W. F. Hegel, from *The Philosophy of Right*

Hegel begins by distinguishing between mere harm and a violation of society itself, threatening its stability and strength. At the same time, however, the power of society makes the crime of any particular offender all the less dangerous, and this allows for considerable discretion and the possibility of mercy. Like Kant, Hegel distinguishes injury to society as opposed to injury to an individual, and it is the former, not the latter, that requires punishment. (The "subjectively infinite" is the individual, the "universal thing" is society.) Again like Kant, Hegel accepts the idea of retribution and "measurement," in terms of both quantity and quality (how great an injury? what kind of injury?), but the danger to society is the primary concern. This is why, Hegel explains, punishment often does *not* fit the crime, for the seriousness of the injury to the person may have little relationship to the threat to society. Hegel further insists that there is no across-the-board notion of balance or fit; "a penal code is the child of its age and the state of civil society as the time." Again, it is the sanctity of the particular society that is the main concern, not the interests of the various citizens or even the public good, and not, as in Kant, a universal principle of reason as such.

Hegel (in general) has much more appreciation for the passions than Kant did, and so he gives proper recognition to the significance of revenge, but without defending or trying to justify it as such. Revenge, he tells us, is implicit retribution, "implicit right." The motive is the same, but retribution takes on the "objective" form of the law and recognizes itself as not merely personal but as a function of "the injured universal"—the threatened society. Revenge as such is not justified, but when taken up by society and made "universal"—that is, public, objective, a matter of law—it is "the satisfaction of justice" itself. Hegel, too, uses the image of crime as "annulled" by punishment (the "erasure" metaphor). Law is thus "restored." Hegel curiously suggests that punishment also allows for "the reconciliation of the criminal with himself," as if the benefits of punishment accrued not only to the society, but also to the criminal. Taken by itself, this would certainly be a perverse view. But one of Hegel's great virtues is what we might call his "holistic" thinking about these issues. Punishment, for him, does *not* serve retribution, *or* deterrence, *or* rehabilitation, but rather encompasses all of these together. There may be no explicit mention of deterrence or social utility, but these are hardly excluded from consideration. There is no particular celebration of reason or even "right" as such, but these are certainly the auspices under which punishment is justified. And, as in Kant (and before him, Rousseau), Hegel insists that punishment is ultimately not the imposition of society upon the individual so much as the individual imposing the law on himself, facing up to the consequences of his own freely chosen action.

If crime and its annulment (which later will acquire the specific character of punish-ment) are treated as if they were unqualified evils, it must, of course, seem quite unreasonable to will an evil merely because 'another evil is there already.' To give punishment this superficial character of an evil is, amongst the various theories of punishment, the fundamental presupposition of those which regard it as a preven-tive, a deterrent, a threat, as reformative, &c., and what on these theories is supposed to result from punishment is characterized equally superficially as a good. But it is not merely a question of an evil or of this, that, or the other good; the precise point at issue is wrong and the righting of it. If you adopt that superficial attitude to punishment, you brush aside the objective treatment of the righting of wrong, which is the primary and fundamental attitude in considering crime; and the natural consequence is that you take as essential the moral attitude, i.e. the subjective aspect of crime, intermingled with trivial psychological ideas of stimuli, impulses too strong for reason, and psychological factors coercing and working on our ideas (as if freedom were not equally capable of thrusting an idea aside and reducing it to something fortuitous!). The various considerations which are relevant to punishment as a phenomenon and to the bearing it has on the particular consciousness, and which concern its effects (deterrent, reformative, &c.) on the imagination, are an essential topic for examination in their place, especially in connexion with modes of punishment, but all these considerations presuppose as their foundation the fact that punishment is inherently and actually just. In discussing this matter the only impor-tant things are, first, that crime is to be annulled, not because it is the producing of an evil, but because it is an infringement of the right as right, and secondly, the question of what that positive existence is which crime possesses and which must be annulled; it is this existence which is the real evil to be removed, and the essential point is the question of where it lies. So long as the concepts here at issue are not clearly apprehended, confusion must continue to reign in the theory of punishment.

100. The injury [the penalty] which falls on the criminal is not merely *implicitly* just—as just, it is *eo ipso* his implicit will, an embodiment of his freedom, his right; on the contrary, it is also a right *established* within the criminal himself, i.e. in his objectively embodied will, in his action. The reason for this is that his action is the action of a rational being and this implies that it is something universal and that by doing it the criminal has laid down a law which he has explicitly recognized in his action and under which in consequence he should be brought as under his right.

* * *

218. Since property and personality have legal recognition and validity in civil society, wrongdoing now becomes an infringement, not merely of what is subjec-tively infinite, but of the universal thing which is existent with inherent stability and strength. Hence a new attitude arises: the action is seen as a danger to society and thereby the magnitude of the wrongdoing is increased. On the other hand, however, the fact that society has become strong and sure of itself diminishes the external importance of the injury and so leads to a mitigation of its punishment.

The fact that an injury to one member of society is an injury to all others does not alter the conception of wrongdoing, but it does alter it in respect of its outward existence as an injury done, an injury which now affects the mind and consciousness of civil society as a whole, not merely the external embodiment of the person directly injured. In heroic times, as we see in the tragedy of the ancients, the citizens did not feel themselves injured by wrongs which members of the royal houses did to one another.

Implicitly, crime is an infinite injury; but as an existent fact it must be measured in quantity and quality; and since its field of existence here has the essential character of affecting an idea and consciousness of the validity of the laws, its danger to civil society is a determinant of the magnitude of a crime, or even *one* of its qualitative characteristics.

Now this quality of magnitude varies with the state of civil society; and this is the justification for sometimes attaching the penalty of death to a theft of a few pence or a turnip, and at other times a light penalty to a theft of a hundred or more times that amount. If we consider its danger to society, this seems at first sight to aggravate the crime; but in fact it is just this which has been the prime cause of the mitigation of its punishment. A penal code, then, is primarily the child of its age and the state of civil society at the time.

* * *

220. When the right against crime has the form of revenge, it is only right implicit, not right in the form of right, i.e. no *act* of revenge is justified. Instead of the injured party, the injured *universal* now comes on the scene, and this has its proper actuality in the court of law. It takes over the pursuit and the avenging of crime, and this pursuit consequently ceases to be the subjective and contingent retribution of revenge and is transformed into the genuine reconciliation of right with itself, i.e. into punishment. Objectively, this is the reconciliation of the law with itself; by the annulment of the crime, the law is restored and its authority is thereby actualized. Subjectively, it is the reconciliation of the criminal with himself, i.e. with the law known by him as his own and as valid for him and his protection; when this law is executed upon him, he himself finds in this process the satisfaction of justice and nothing save his own act.

John Stuart Mill, from *Utilitarianism*

John Stuart Mill, in marked contrast to both Kant and Hegel, takes social utility to be the whole and only justification for punishment. It is, he speculates, an "outgrowth" of two natural sentiments, self-defense and sympathy. But it is not mere self-defense that emerges in Mill's argument, and we should note from the outset a shifting between two very different claims: self-defense and retaliation. He tells us that "it is natural to resent and to repel and to retaliate" as if these were all varieties of self-defense, but defense is an immediate concern for the future, whereas retaliation is concern for a past and presumably already completed injury or offense. Mill notes that "every animal tries to hurt those who have hurt, or who it thinks are about to hurt," but only the latter is self-defense. The former is not.

Like Kant and Hegel, however, Mill insists that the self-defense that justifies punishment is not merely personal. It is one of the essential facts about us, as opposed to the "lower" animals, that we are able to sympathize on a much larger canvas, even taking into consideration the whole of society. And as in the rest of his moral philosphy, Mill argues that we are not only capable but find it necessary as rational creatures to generalize from our own good to the good of all, in matters of punishment we realize that "any conduct that threatens the security of society generally is threatening to his own." The sentiment of justice is, he concludes, just "the natural feeling of retaliation or vengeance." True, this is a sentiment which (taken as such) "has nothing moral in it"; but with "the exclusive subordination to the social sympathies, so as to wait on and obey their call" it becomes a moral demand. The problem with the natural feeling of vengeance, Mill argues, is that it is indiscriminate—we would resent anything anyone did to us that is disagreeable. But when it is moralized in the direction of the general good, resenting a hurt to society not just oneself, it becomes constrained by the overarching principle that defines all moral action, the public interest or "the greatest good of the greatest number."

Mill readily admits that, when we feel vengeful, we don't always feel as if we are defending society, or for that matter we may not think of society at all. But we do feel, he suggests, as if we are "asserting a rule" that is for the benefit of others as well as ourselves, and (retributive) justice is just our having the interest of mankind collectively in mind. (Note his shift in the burden of the argument to Kant, as he bends Kant to utilitarianism with the claim that any rational being must concern itself with the collective interest. Needless to say, this is not how Kant himself talks, where serving right and justice can even be contrary to collective interests.) Thus Mill on the one hand insists that "the sentiment of justice" is an "animal desire" for punishment coupled with that "enlarged sympathy and human conception of intelligent self-interest." But note again his conflation of the "animal desire to repel or retaliate a hurt or damage to oneself" as if to repel and to retaliate were one and the same. The sense of urgency in self-defense may well explain "its peculiar impressiveness and energy of self-assertion" but it is not clear that the same explanation will do for revenge, or that revenge, so conceived, has anything to do with punishment.

We have seen that the two essential ingredients in the sentiment of justice are the desire to punish a person who has done harm and the knowledge or belief that there is some definite individual or individuals to whom harm has been done.

Now it appears to me that the desire to punish a person who has done harm to some individual is a spontaneous outgrowth from two sentiments, both in the highest degree natural and which either are or resemble instincts: the impulse of self-defense and the feeling of sympathy.

It is natural to resent and to repel or retaliate any harm done or attempted against ourselves or against those with whom we sympathize. The origin of this sentiment is not necessary here to discuss. Whether it be an instinct or a result of intelligence, it is, we know, common to all animal nature; for every animal tries to hurt those who have hurt, or who it thinks are about to hurt, itself or its young. Human beings, on this point, only differ from other animals in two particulars. First, in being capable of sympathizing, not solely with their offspring, or, like some of the more noble animals, with some superior animal who is kind to them, but with all human, and even with all sentient, beings; secondly, in having a more developed intelligence, which gives a wider range to the whole of their sentiments, whether self-regarding or sympathetic. By virtue of his superior intelligence, even apart from his superior range of sympathy, a human being is capable of apprehending a community of interest between himself and the human society of which he forms a part, such that any conduct which threatens the security of the society generally is threatening to his own, and calls forth his instinct (if instinct it be) of self-defense. The same superiority of intelligence, joined to the power of sympathizing with human beings generally, enables him to attach himself to the collective idea of his tribe, his country, or mankind in such a manner that any act hurtful to them raises his instinct of sympathy and urges him to resistance.

The sentiment of justice, in that one of its elements which consists of the desire to punish, is thus, I conceive, the natural feeling of retaliation or vengeance, rendered by intellect and sympathy applicable to those injuries, that is, to those hurts, which wound us through, or in common with, society at large. This sentiment, in itself, has nothing moral in it; what is moral is the exclusive subordination of it to the social sympathies, so as to wait on and obey their call. For the natural feeling would make us resent indiscriminately whatever anyone does that is disagreeable to us; but, when moralized by the social feeling, it only acts in the directions conformable to the general good: just persons resenting a hurt to society, though not otherwise a hurt to themselves, and not resenting a hurt to themselves, however painful, unless it be of the kind which society has a common interest with them in the repression of.

It is no objection against this doctrine to say that, when we feel our sentiment of justice outraged, we are not thinking of society at large or of any collective interest, but only of the individual case. It is common enough, certainly, though the reverse of commendable, to feel resentment merely because we have suffered pain; but a person whose resentment is really a moral feeling, that is, who considers whether an act is blamable before he allows himself to resent it—such a person, though he may not say expressly to himself that he is standing up for the interest of society, certainly does feel that he is asserting a rule which is for the benefit of others as well as for his own. If he is not feeling this, if he is regarding the act solely as it affects him individually, he is not consciously just; he is not concerning himself about the justice of his actions. This is admitted even by anti-utilitarian moralists. When Kant (as before remarked) propounds as the fundamental principle of morals, "So act that thy rule of conduct might be adopted as a law by all rational beings," he virtually acknowledges that the interest of mankind collectively, or at least of mankind indiscriminately, must be in the mind of the agent when conscientiously deciding on the morality of the act. Otherwise he uses words without a meaning; for that a rule even of utter selfishness could not *possibly* be adopted by all rational beings—that there is any insuperable obstacle in the nature of things to its adoption—cannot be even plausibly maintained. To give any meaning to Kant's principle, the sense put upon it must be that we ought to shape our conduct by a rule which all rational beings might adopt *with benefit to their collective interest.*

To recapitulate: the idea of justice supposes two things—a rule of conduct and a sentiment which sanctions the rule. The first must be supposed common to all mankind and intended for their good. The other (the sentiment) is a desire that punishment may be suffered by those who infringe the rule. There is involved, in addition, the conception of some definite person who suffers by the infringement, whose rights (to use the expression appropriated to the case) are violated by it. And the sentiment of justice appears to me to be the animal desire to repel or retaliate a hurt or damage to oneself or to those with whom one sympathizes, widened so as to include all persons, by the human capacity of enlarged sympathy and the human conception of intelligent self-interest. From the latter elements the feeling derives its morality; from the former, its peculiar impressiveness and energy of self-assertion.

Friedrich Nietzsche, from
On the Genealogy of Morals (1844–1900)

Nietzsche's "genealogy" is a combination of history and social psychology with an eye to developing as explanation—not a justification—of our current conceptions of morality and justice. His polemical but now infamous view is that much of what we call morality—or what he calls alternatively "herd" or "slave morality"—is in fact the ingenious expression of that vicious emotion that we know as *resentment*. Morality, in short, is not what it seems—a system of categorical imperatives or the maximization of utility or a set of constraints for the sake of the stability of society; it is rather a grand strategy, motivated by what Nietzsche calls "the will to power," whose aim is to elevate and protect the weak and mediocre against the strong and the noble. Predictably, Nietzsche also attacks the increasingly popular philosophies of democracy and socialism, and many of the standard demands of modern justice, notably the emphasis on equality and fairness, fall target to his displeasure as well. Nietzsche has rightly been cited as the most thorough-going critic of morality, and, in particular, he is recognized (e.g. by Alasdair MacIntyre in his recent *After Virtue*) as the diagnostician of what ails us in both contemporary morals and moral philosophy. But what is too rarely appreciated is that Nietzsche's infamous campaign of destruction has an affirmative side, a positive view of morals as well as a subversive campaign against morality. This is, perhaps, nowhere so evident as in his discussion of justice in the *Genealogy of Morals*. Rather than dismiss justice as one more manifestation of "herd" or "slave morality," as we might expect him to do, Nietzsche defends justice as one of the highest, most noble ideals. It is, in a sense, the virtue of mercy—"forgiving and forgetting"—but not at all from the point of view of meekness, but rather from strength. The idea is that one should be so self-sufficient and satisfied with one's life that one just doesn't worry about the petty injuries inflicted by others. The noble person does not reject revenge or retribution as wrong, but rather as unnecessary, as "beneath" his or her dignity.

As for the nature and origins of punishment, Nietzsche's account is no less striking. The urge to punish is, at least in part, the expression of resentment, but this is very different from justice. Justice is "a stronger power seeking a means of putting an end to the senseless raging of *ressentiment* among the weaker powers . . . —partly but taking the object of *ressentiment* out of the hands of revenge, partly by substituting for revenge the struggle against the enemies of peace and order, partly by devising and . . . imposing settlements . . . but finally the institution of *law*." In conformity with many conservative (and ancient) thinkers, Nietzsche ultimately comes down to an extremely positivistic conception of justice, but not, he hastens to add, as "sovereign and universal," as a means of preventing struggle or treating everyone as equals. That, he suggests, is not justice, but a kind of despair, a dissolute principle that is "hostile to life."

As its power increases, a community ceases to take the individual's transgressions so seriously, because they can no longer be considered as dangerous and destructive to the whole as they were formerly: the malefactor is no longer "set beyond the pale of peace" and thrust out; universal anger may not be vented upon him as unrestrainedly as before—on the contrary, the whole from now on carefully defends the malefactor against this anger, especially that of those he has directly harmed, and takes him under its protection. A compromise with the anger of those directly injured by the criminal; an effort to localize the affair and to prevent it from causing any further, let alone a general, disturbance; attempts to discover equivalents and to settle the whole matter (*compositio*); above all, the increasingly definite will to treat every crime as in some sense *dischargeable*, and thus at least to a certain extent to *isolate* the criminal and his deed from one another—these traits become more and more clearly visible as the penal law evolves. As the power and self-confidence of a community increase, the penal law always becomes more moderate; every weakening or imperiling of the former brings with it a restoration of the harsher forms of the latter. The "creditor" always becomes more humane to the extent that he has grown richer; finally, how much injury he can endure without suffering from it becomes the actual *measure* of his wealth. It is not unthinkable that a society might attain such a *consciousness of power* that it could allow itself the noblest luxury possible to it—letting those who harm it go *unpunished*. "What are my parasites to me?" it might say. "May they live and prosper: I am strong enough for that!"

The justice which began with, "everything is dischargeable, everything must be discharged," ends by winking and letting those incapable of discharging their debt go free: it ends, as does every good thing on earth, by *overcoming itself*. This self-overcoming of justice: one knows the beautiful name it has given itself—*mercy*; it goes without saying that *mercy* remains the privilege of the most powerful man, or better, his—beyond the law.

* * *

Here a word in repudiation of attempts that have lately been made to seek the origin of justice in quite a different sphere—namely in that of *ressentiment*. To the psychologists first of all, presuming they would like to study *ressentiment* close up for once, I would say: this plant blooms best today among anarchists and anti-Semites—where it has always bloomed, in hidden places, like the violet, though with a different odor. And as like must always produce like, it causes us no surprise to see a repetition in such circles of attempts often made before—see above, section 14—to sanctify *revenge* under the name of *justice*—as if justice were at bottom merely a further development of the feeling of being aggrieved—and to rehabilitate not only revenge but all *reactive* affects in general. To the latter as such I would be the last to raise any objection: in respect to the entire biological problem (in relation to which the value of these affects has hitherto been underrrated) it even seems to me to constitute a *service*. All I draw attention to is the circumstance that it is the spirit of *ressentiment* itself out of which this new nuance of scientific fairness (for the benefit of hatred, envy, jealousy, mistrust, rancor, and revenge) proceeds. For this "scientific fairness" immediately ceases and gives way to accents of deadly enmity and prejudice once it is a question of dealing with another group of affects, affects that, it seems to me, are of even greater biological value than those reactive affects and consequently deserve even more to be *scientifically* evaluated and esteemed: namely, the truly *active* affects, such as lust for power, avarice, and the like.

So much against this tendency in general: as for Dühring's specific proposition that the home of justice is to be sought in the sphere of the reactive feelings, one is obliged for truth's sake to counter it with a blunt antithesis: the *last* sphere to be conquered by the spirit of justice is the sphere of the reactive feelings! When it really happens that the just man remains just even toward those who have harmed him (and not merely cold, temperate, remote, indifferent: being just is always a *positive* attitude), when the exalted, clear objectivity, as penetrating as it is mild, of the eye of justice and *judging* is not dimmed even under the assault of personal injury, derision, and calumny, this is a piece of perfection and supreme mastery on earth—something it would be prudent not to expect or to *believe* in too readily. On the average, a small dose of aggression, malice, or insinuation certainly suffices to drive the blood into the eyes—and fairness out of the eyes—of even the most upright people. The active, aggressive, arrogant man is still a hundred steps closer to justice than the reactive man; for he has absolutely no need to take a false and prejudiced view of the object before him in the way the reactive man does and is bound to do. For that reason the aggressive man, as the stronger, nobler, more courageous, has in fact also had at all times a *freer* eye, a *better* conscience on his side: conversely, one can see who has the invention of the "bad conscience" on his conscience—the man of *ressentiment*!

Finally, one only has to look at history: in which sphere has the entire administration of law hitherto been at home—also the need for law? In the sphere of reactive men, perhaps? By no means: rather in that of the active, strong, spontaneous, aggressive. From a historical point of view, law represents on earth—let it be said to the dismay of the above-named agitator (who himself once confessed: "the doctrine of revenge is the red thread of justice that runs through all my work and efforts")— the struggle *against* the reactive feelings, the war conducted against them on the part of the active and aggressive powers who employed some of their strength to impose measure and bounds upon the excesses of the reactive pathos and to compel it to come to terms. Wherever justice is practiced and maintained one sees a stronger power seeking a means of putting an end to the senseless raging of *ressentiment* among the weaker powers that stand under it (whether they be groups or individuals)—partly by taking the object of *ressentiment* out of the hands of revenge, partly by substituting for revenge the struggle against the enemies of peace and order, partly by devising and in some cases imposing settlements, partly by elevating certain equivalents for injuries into norms to which from then on *ressentiment* is once and for all directed. The most decisive act, however, that the supreme power performs and accomplishes against the predominance of grudges and rancor—it always takes this action as soon as it is in any way strong enough to do so—is the institution of *law*, the imperative declaration of what in general counts as permitted, as just, in its eyes, and what counts as forbidden, as unjust: once it has instituted the law, it treats violence and capricious acts on the part of individuals or entire groups as offenses against the law, as rebellion against the supreme power itself, and thus leads the feelings of its subjects away from the direct injury caused by such offenses; and in the long run it thus attains the reverse of that which is desired by all revenge that is fastened exclusively the viewpoint of the person injured: from now on the eye is trained to an ever more *impersonal* evaluation of the deed, and this applies even to the eye of the injured person himself (although last of all, as remarked above).

"Just" and "unjust" exist, accordingly, only after the institution of the law (and *not*, as Dühring would have it, after the perpetration of the injury). To speak of just or unjust *in itself* is quite senseless; *in itself*, of course, no injury, assault, exploitation, destruction can be "unjust," since life operates *essentially*, that is in its basic functions, through injury, assault, exploitation, destruction and simply cannot be thought of at all without this character. One must indeed grant something even more unpalatable: that, from the highest biological standpoint, legal conditions can never be other than *exceptional conditions*, since they constitute a partial restriction of the will of life, which is bent upon power, and are subordinate to its total goal as a single means: namely, as a means of creating *greater* units of power. A legal order thought of as sovereign and universal, not as a means in the struggle between power-complexes but as a means of *preventing* all struggle in general—perhaps after the communistic cliché of Dühring, that every will must consider every other will its equal—would be a principle *hostile to life*, an agent of the dissolution and destruction of man, an attempt to assassinate the future of man, a sign of weariness, a secret path to nothingness.—

> . . . Let us eliminate the concept of *sin* from the world—and let us soon dispatch the concept of *punishment* after it! May these exiled monsters live somewhere else henceforth and not among men—if they insist on living and will not perish of disgust with themselves! . . .
>
> *Dawn*, section 202

> *That man be delivered from revenge*, that is for me the bridge to the highest hope . . .
>
> *Zarathustra* II, "On the Tarantulas"

The U.S. Supreme Court,
Gregg v. *Georgia* (1976)

Gregg v. *Georgia* was something of a "landmark" back in 1976, when the U.S. Supreme Court reaffirmed the constitutional validity of the death penalty. The challenge had been that the death penalty itself was "cruel and unusual punishment" and thus explicitly prohibited by the Bill of Rights. The arguments behind this decision come from a variety of sources—from the long history of punishment in the world as well as in the United States, from the original intentions of the framers of the Constitution, and from current endorsements of the penalty by state legislators and juries. The Court emphasizes the importance of sensitivity to changing mores and the fact that what is considered humane and legitimate to one generation may not seem so to another a century or more later. Nevertheless, the decision was that the death penalty is not as such in violation of the Constitution. The Court argued that it served *both* the social purposes of deterrence and retribution, and it quotes an importance passage from the earlier case of *Furman* v. *Georgia* (1972) insisting on "the instinct for retribution." Then justices admit, however, that statistical studies concerning the effects of the death penalty on the deterrence of future crimes have been "inconclusive."

The affirmation of the Court was not, however, unanimous, and several justices disagreed with the majority opinion. (The vote was 7–2.) Writing for the dissenting minority, Justice Thurgood Marshall opposes the death penalty, as he had done earlier and as he has continued to do to this day. Marshall argues that the death penalty is, by any measure, excessive, and he disagrees with the majority that the American people actually support the principle. Their seeming support, he argues, is the result of their lack of information. He points out that studies show that there is "no correlation between the existence of capital punishment and lower rates of capital crime," thus undermining "deterrence" as the justification of the death penalty, and he argues that it "defies belief" to argue that the death penalty is necessary to prevent people from taking the law into their own hands. The extreme penalty of life imprisonment, Marshall argues, is just as strong an expression of the moral outrage of the community as the penalty of death. Finally, he reiterates an argument as old as Socrates, that it is wrong to return evil for evil.

Majority Opinion

The issue in this case is whether the imposition of the sentence of death for the crime of murder under the law of Georgia violates the Eighth and Fourteenth Amendments.

I

The petitioner, Troy Gregg, was charged with committing armed robbery and murder. In accordance with Georgia procedure in capital cases, the trial was in two stages, a guilt stage and a sentencing stage. . . .

. . . The jury found the petitioner guilty of two counts of armed robbery and two counts of murder.

At the penalty stage, which took place before the same jury, . . . the trial judge instructed the jury that it could recommend either a death sentence or a life prison sentence on each count. . . . The jury returned verdicts of death on each count.

II

. . . The Georgia statute, as amended after our decision in *Furman* v. *Georgia* (1972), retains the death penalty for six categories of crime: murder, kidnaping for ransom or where the victim is harmed, armed robbery, rape, treason, and aircraft hijacking. . . .

III

We address initially the basic contention that the punishment of death for the crime of murder is, under all circumstances, "cruel and unusual" in violation of the Eighth and Fourteenth Amendments of the Constitution.

The Court on a number of occasions has both assumed and asserted the constitutionality of capital punishment. In several cases that assumption provided a necessary foundation for the decision, as the Court was asked to decide whether a particular method of carrying out a capital sentence would be allowed to stand under the Eighth Amendment. But until *Furman* v. *Georgia* (1972), the Court never confronted squarely the fundamental claim that the punishment of death always, regardless of the enormity of the offense or the procedure followed in imposing the sentence, is cruel and unusual punishment in violation of the Constitution. Although this issue was presented and addressed in *Furman*, it was not resolved by the Court. Four Justices would have held that capital punishment is not unconstitutional *per se*; two Justices would have reached the opposite conclusion; and three Justices, while agreeing that the statutes then before the Court were invalid as applied, left open the question whether such punishment may ever be imposed. We now hold that the punishment of death does not invariably violate the Constitution.

A

The history of the prohibition of "cruel and unusual" punishment already has been reviewed at length. The phrase first appeared in the English Bill of Rights of 1689, which was drafted by Parliament at the accession of William and Mary. The English version appears to have been directed against punishments unauthorized by statute and beyond the jurisdiction of the sentencing court, as well as those disproportionate to the offense involved. The American draftsmen, who adopted the English phrasing in drafting the Eighth Amendment, were primarily concerned, however, with proscribing "tortures" and other "barbarous" methods of punishment.

In the earliest cases raising Eighth Amendment claims, the Court focused on particular methods of execution to determine whether they were too cruel to pass constitutional muster. The constitutionality of the sentence of death itself was not at issue, and the criterion used to evaluate the mode of execution was its similarity to "torture" and other "barbarous" methods. . . .

But the Court has not confined the prohibition embodied inthe Eighth Amendment to "barbarous" methods that were generally outlawed in the 18th century. Instead, the Amendment has been interpreted in a flexible and dynamic manner. The Court early recognized that "a principle to be vital must be capable of wider application than the mischief which gave it birth." Thus the Clause forbidding "cruel and unusual" punishments "is not fastened to the obsolete but may acquire meaning as public opinion becomes enlightened by a humane justice."

But our cases also make clear that public perceptions of standards of decency with respect to criminal sanctions are not conclusive. A penalty also must accord with "the dignity of man," which is the "basic concept underlying the Eighth Amendment." This means, at least, that the punishment not be "excessive." When a form of punishment in the abstract (in this case, whether capital punishment may ever be imposed as a sanction for murder) rather than in the particular (the propriety of death as a penalty to be applied to a specific defendant for a specific crime) is under consideration, the inquiry into "excessiveness" has two aspects. First, the punishment must not involve the unnecessary and wanton infliction of pain. Second, the punishment must not be grossly out of proportion to the severity of the crime.

* * *

The imposition of the death penalty for the crime of murder has a long history of acceptance both in the United States and in England. . . .

It is apparent from the text of the Constitution itself that the existence of capital punishment was accepted by the Framers. At the time the Eighth Amendment was ratified, capital punishment was a common sanction in every State. Indeed, the First Congress of the United States enacted legislation providing death as the penalty for specified crimes. . . .

For nearly two centuries, this Court, repeatedly and often expressly, has recognized that capital punishment is not invalid *per se*. . . .

Four years ago, the petitioners in *Furman* and its companion cases predicated their argument primarily upon the asserted proposition that standards of decency had evolved to the point where capital punishment no longer could be tolerated. The petitioners in those cases said, in effect, that the evolutionary process had come to an end, and that standards of decency required that the Eighth Amendment be construed finally as prohibiting capital punishment for any crime regardless of its depravity and impact on society. This view was accepted by two Justices. Three other Justices were unwilling to go so far; focusing on the procedures by which convicted defendants were selected for the death penalty rather than on the actual punishment inflicted, they joined in the conclusion that the statutes before the Court were constitutionally invalid.

The petitioners in the capital cases before the Court today renew the "standards of decency" argument, but developments during the four years since *Furman* have undercut substantially the assumptions upon which their argument rested. Despite the continuing debate, dating back to the nineteenth century, over the morality and utility of capital punishment, it is now evident that a large proportion of American society continues to regard it as an appropriate and necessary criminal sanction.

The most marked indication of society's endorsement of the death penalty for murder is the legislative response to *Furman*. The legislatures of at least thirty-five States have enacted new statutes that provide for the death penalty for at least some crimes that result in the death of another person. And the Congress of the United States, in 1974, enacted a statute providing the death penalty for aircraft piracy that results in death. These recently adopted statutes have attempted to address the concerns expressed by the Court in *Furman* primarily (i) by specifying the factors to be weighed and the procedures to be followed in deciding when to impose a capital sentence, or (ii) by making the death penalty mandatory for specified crimes. But all of the post-*Furman* statutes make clear that capital punishment itself has not been rejected by the elected representatives of the people. . . .

The jury also is a significant and reliable objective index of contemporary values because it is so directly involved. The Court has said that "one of the most important functions any jury can perform in making . . . a selection [between life imprisonment and death for a defendant convicted in a capital case] is to maintain a link between contemporary community values and the penal system." It may be true that evolving standards have influenced juries in recent decades to be more discriminating in imposing the sentence of death. But the relative infrequency of jury verdicts imposing the death sentence does not indicate rejection of capital punishment *per se*. Rather, the reluctance of juries in many cases to impose the sentence may well reflect the humane feeling that this most irrevocable of sanctions should be reserved for a small number of extreme cases. Indeed, the actions of juries in many States since *Furman* are fully compatible with the legislative judgments, reflected in the new statutes, as to the continued utility and necessity of capital punishment in appropriate cases. At the close of 1974 at least 254 persons had been sentenced to death since *Furman*, and by the end of March 1976, more than 460 persons were subject to death sentences.

As we have seen, however, the Eighth Amendment demands more than that a challenged punishment be acceptable to contemporary society. The Court also must ask whether it comports with the basic concept of human dignity at the core of the Amendment. Although we cannot "invalidate a category of penalties because we deem less severe penalties, adequate to serve the ends of penology," the sanction imposed cannot be so totally without penological justification that it results in the gratuitous infliction of suffering.

The death penalty is said to serve two principal social purposes: retribution and deterrence of capital crimes by prospective offenders.

In part, capital punishment is an expression of society's moral outrage at particularly offensive conduct. This function may be unappealing to many, but it is essential in an ordered society that asks its citizens to rely on legal processes rather than self-help to vindicate their wrongs.

> The instinct for retribution is part of the nature of man, and channeling that instinct in the administration of criminal justice serves an important purpose in promoting the stability of a society governed by law. When people begin to believe that organized society is unwilling or unable to impose upon criminal

offenders the punishment they "deserve," then there are sown the seeds of anarchy—if self-help, vigilante justice, and lynch law. *Furman* v. *Georgia* (Stewart, J., concurring).

"Retribution is no longer the dominant objective of the criminal law," but neither is it a forbidden objective nor one inconsistent with our respect for the dignity of men. Indeed, the decision that capital punishment may be the appropriate sanction in extreme cases is an expression of the community's belief that certain crimes are themselves so grievous an affront to humanity that the only adequate response may be the penalty of death.

Statistical attempts to evaluate the worth of the death penalty as a deterrent to crimes by potential offenders have occasioned a great deal of debate. The results simply have been inconclusive. . . .

Although some of the studies suggest that the death penalty may not function as a significantly greater deterrent than lesser penalties, there is no convincing empirical evidence either supporting or refuting this view. We may nevertheless assume safely that there are murderers, such as those who act in passion, for whom the threat of death has little or no deterrent effect. But for many others, the death penalty undoubtedly is a significant deterrent. There are carefully contemplated murders, such as murder for hire, where the possible penalty of death may well enter into the cold calculus that precedes the decision to act. And there are some categories of murder, such as murder by a life prisoner, where other sanctions may not be adequate.

The value of capital punishment as a deterrent of crime is a complex factual issue the resolution of which properly rests with the legislatures, which can evaluate the results of statistical studies in terms of their own local conditions and with a flexibility of approach that is not available to the courts. Indeed, many of the post-*Furman* statutes reflect just such a responsible effort to define those crimes and those criminals for which capital punishment is most probably an effective deterrent.

In sum, we cannot say that the judgment of the Georgia Legislature that capital punishment may be necessary in some cases is clearly wrong. Considerations of federalism, as well as respect for the ability of a legislature to evaluate, in terms of its particular State, the moral consensus concerning the death penalty and its social utility as a sanction, require us to conclude, in the absence of more convincing evidence, that the infliction of death as a punishment for murder is not without justification and thus is not unconstitutionally severe.

Finally, we must consider whether the punishment of death is disproportionate in relation to the crime for which it is imposed. There is no question that death as a punishment is unique in its severity and irrevocability. When a defendant's life is at stake, the Court has been particularly sensitive to insure that every safeguard is observed. But we are concerned here only with the imposition of capital punishment for the crime of murder, and when a life has been taken deliberately by the offender, we cannot say that the punishment is invariably disproportionate to the crime. It is an extreme sanction, suitable to the most extreme of crimes.

We hold that the death penalty is not a form of punishment that may never be imposed, regardless of the circumstances of the offense, regardless of the character of the offender, and regardless of the procedure followed in reaching the decision to impose it.

Minority Opinion

In *Furman* v. *Georgia* (1972) (concurring opinion), I set forth at some length my views on the basic issue presented to the Court in [this case]. The death penalty, I concluded, is a cruel and unusual punishment prohibited by the Eighth and Fourteenth Amendments. That continues to be my view.

I have no intention of retracing the "long and tedious journey" that led to my conclusion in *Furman*. My sole purposes here are to consider the suggestion that my conclusion in *Furman* has been undercut by developments since then, and briefly to evaluate the basis for my Brethren's holding that the extinction of life is a permissible form of punishment under the Cruel and Unusual Punishments Clause.

In *Furman* I concluded that the death penalty is constitutionally invalid for two reasons. First, the death penalty is excessive. And second, the American people, fully informed as to the purposes of the death penalty and its liabilities, would in my view reject it as morally unacceptable.

Since the decision in *Furman*, the legislatures of thirty-five States have enacted new statutes authorizing the imposition of the death sentence for certain crimes, and Congress has enacted a law providing the death penalty for air piracy resulting in death. I would be less than candid if I did not acknowledge that these developments have a significant bearing on a realistic assessment of the moral acceptability of the death penalty to the American people. But if the constitutionality of the death penalty turns, as I have urged, on the opinion of an *informed* citizenry, then even the enactment of new death statutes cannot be viewed as conclusive. In *Furman*, I observed that the American people are largely unaware of the information critical to a judgment on the morality of the death penalty, and concluded that if they were better informed they would consider it shocking, unjust, and unacceptable. A recent study, conducted after the enactment of the post-*Furman* statutes, has confirmed that the American people know little about the death penalty, and that the opinions of an informed public would differ significantly from those of a public unaware of the consequences and effects of the death penalty.

Even assuming, however, that the post-*Furman* enactment of statutes authorizing the death penalty renders the prediction of the views of an informed citizenry an uncertain basis for a constitutional decision, the enactment of those statutes has no bearing whatsoever on the conclusion that the death penalty is unconstitutional because it is excessive. An excessive penalty is invalid under the Cruel and Unusual Punishments Clause "even though popular sentiment may favor" it. The inquiry here, then, is simply whether the death penalty is necessary to accomplish the legitimate legislative purposes in punishment, or whether a less severe penalty—life imprisonment—would do as well.

The two purposes that sustain the death penalty as nonexcessive in the Court's view are general deterrence and retribution. In *Furman*, I canvassed the relevant data on the deterrent effect of capital punishment. The state of knowledge at that point, after literally centuries of debate, was summarized as follows by a United Nations Committee:

> It is generally agreed between the retentionists and abolitionists, whatever their opinions about the validity of comparative studies of deterrence, that the data

which now exist show no correlation between the existence of capital punishment and lower rates of capital crime.

The available evidence, I concluded in *Furman*, was convincing that "capital punishment is not necessary as a deterrent to crime in our society." . . .

The evidence I reviewed in *Furman* remains convincing, in my view, that "capital punishment is not necessary as a deterrent to crime in our society." The justification for the death penalty must be found elsewhere.

The other principal purpose said to be served by the death penalty is retribution. The notion that retribution can serve as a moral justification for the sanction of death finds credence in the opinion of my Brothers Stewart, Powell, and Stevens. . . . It is this notion that I find to be the most disturbing aspect of today's unfortunate [decision].

The concept of retribution is a multifaceted one, and any discussion of its role in the criminal law must be undertaken with caution. On one level, it can be said that the notion of retribution or reprobation is the basis of our insistence that only those who have broken the law be punished, and in this sense the notion is quite obviously central to a just system of criminal sanctions. But our recognition that retribution plays a crucial role in determining who may be punished by no means requires approval of retribution as a general justification for punishment. It is the question whether retribution can provide a moral justification for punishment—in particular, capital punishment—that we must consider.

My Brothers Stewart, Powell, and Stevens offer the following explanation of the retributive justification for capital punishment:

> The instinct for retribution is part of the nature of man, and channeling that instinct in the administration of criminal justice serves an important purpose in promoting the stability of a society governed by law. When people begin to believe that organized society is unwilling or unable to impose upon criminal offenders the punishment they "deserve," then there are sown the seeds of anarchy—of self-help, vigilante justice, and lynch law.

This statement is wholly inadequate to justify the death penalty. As my Brother Brennan stated in *Furman*, "[t]here is no evidence whatever that utilization of imprisonment rather than death encourages private blood feuds and other disorders." It simply defies belief to suggest that the death penalty is necessary to prevent the American people from taking the law into their own hands.

In a related vein, it may be suggested that the expression of moral outrage through the imposition of the death penalty serves to reinforce basic moral values— that it marks some crimes as particularly offensive and therefore to be avoided. The argument is akin to a deterrence argument, but differs in that it contemplates the individual's shrinking from antisocial conduct, not because he fears punishment, but because he has been told in the strongest possible way that the conduct is wrong. This contention, like the previous one, provides no support for the death penalty. It is inconceivable that any individual concerned about conforming his conduct to what society says is "right" would fail to realize that murder is "wrong" if the penalty were simply life imprisonment.

The foregoing contentions—that society's expression of moral outrage through the imposition of the death penalty preempts the citizenry from taking the law into its own hands and reinforces moral values—are not retributive in the purest sense. They are essentially utilitarian in that they portray the death penalty as valuable because of its beneficial results. These justifications for the death penalty are inadequate because the penalty is, quite clearly I think, not necessary to the accomplishment of those results.

There remains for consideration, however, what might be termed the purely retributive justification for the death penalty—that the death penalty is appropriate, not because of its beneficial effect on society, but because the taking of the murderer's life is itself morally good. Some of the language of the opinion of my Brothers Stewart, Powell, and Stevens . . . appears positively to embrace this notion of retribution for its own sake as a justification for capital punishment. They state:

> [T]he decision that capital punishment may be the appropriate sanction in extreme cases is an expression of the community's belief that certain crimes are themselves so grievous an affront to humanity that the only adequate response may be the penalty of death.

They then quote with approval from Lord Justice Denning's remarks before the British Royal Commission on Capital Punishment:

> The truth is that some crimes are so outrageous that society insists on adequate punishment, because the wrong-doer deserves it, irrespective of whether it is a deterrent or not.

Of course, it may be that these statements are intended as no more than observations as to the popular demands that it is thought must be responded to in order to prevent anarchy. But the implication of the statements appears to me to be quite different—namely, that society's judgment that the murderer "deserves" death must be respected not simply because the preservation of order requires it, but because it is appropriate that society make the judgment and carry it out. It is this latter notion, in particular, that I consider to be fundamentally at odds with the Eighth Amendment. The mere fact that the community demands the murderer's life in return for the evil he has done cannot sustain the death penalty, for as Justices Stewart, Powell, and Stevens remind us, "the Eighth Amendment demands more than that a challenged punishment be acceptable to contemporary society." To be sustained under the Eighth Amendment, the death penalty must "compor[t] with the basic concept of human dignity at the core of the Amendment"; the objective in imposing it must be "[consistent] with our respect for the dignity of [other] men." Under these standards, the taking of life "because the wrongdoer deserves it" surely must fail, for such a punishment has as its very basis the total denial of the wrongdoer's dignity and worth.

The death penalty, unnecessary to promote the goal of deterrence or to further any legitimate notion of retribution, is an excessive penalty forbidden by the Eighth and Fourteenth Amendments. I respectfully dissent from the Court's judgment upholding the [sentence] of death imposed upon the [petitioner in this case].

Hugo Bedau, from "Capital Punishment and Retributive Justice"

Hugo Bedau is one of the leading opponents of the death penalty, and he argues that its use is supported neither by considerations of deterrence nor by the supposed demand for retribution. He begins by agreeing, to get the argument going, that crime should be punished and punishment should in some sense "fit" the crime. But neither of these principles is sufficient to justify the death penalty, which, he concludes, "only tends to add new injuries of its own to the catalogue of our inhumanity to each other." He points out the fact that the death penalty has rarely been restricted to "a life for a life" and has often been employed to punish other crimes than murder, and murder itself is not always or even usually punished by death. The idea of retribution is rarely served by the death penalty (of some particularly heinous murderers, it is commonly agreed that "death is too good for him") and, perhaps most persuasive of all, there are the recently much-documented inequities in the application of the death penalty. Few, if any, white wealthy murderers have been executed, and a very large proportion of those executed are black and poor. Many of them had poor legal representation or faced an unusually ambitious prosecuting attorney, and such misfortunes as these, rather than the severity of the crimes or the dangerousness of the criminals, sealed their fates. Capital punishment, Bedau concludes, cheapens and degrades human life rather than enhances respect for it.

. . . There are two leading principles of retributive justice relevant to the capital-punishment controversy. One is the principle that crimes should be punished. The other is the principle that the severity of a punishment should be proportional to the gravity of the offense. (A corollary to the latter principle is the judgment that nothing so fits the crime of murder as the punishment of death.) Although these principles do not seem to stem from any concern over the worth, value, dignity, or rights of persons, they are moral principles of recognized weight and no discussion of the morality of capital punishment would be complete without them. Leaving aside all questions of social defense, how strong a case for capital punishment can be made on the basis if these principles? How reliable and persuasive are these principles themselves?

From Hugo Bedau, "Capital Punishment Redistributive Justice," as reprinted in *Matters of Life and Death*, ed. Tom Regan, published by McGraw-Hill Publishing Co. Copyright © 1986 by McGraw-Hill Publishing Co. Reprinted by permission of the publisher.

Crime Must Be Punished

Given [a general rationale for punishment], there cannot be any dispute over this principle. In embracing it, of course, we are not automatically making a fetish of "law and order," in the sense that we would be if we thought that the most important single thing society can do with its resources is to punish crimes. In addition, this principle is not likely to be in dispute between proponents and opponents of the death penalty. Only those who completely oppose punishment for murder and other erstwhile capital crimes would appear to disregard this principle. Even defenders of the death penalty must admit that putting a convicted murderer in prison for years is a punishment of that criminal. The principle that crime must be punished is neutral to our controversy, because both sides acknowledge it and comply with it.

It is the other principle of retributive justice that seems to be a decisive one. Under the principle of retaliation, *lex talionis*, it must always have seemed that murderers ought to be put to death. Proponents of the death penalty, with rare exceptions, have insisted on this point, and it seems that even opponents of the death penalty must give it grudging assent. The strategy for opponents of the death penalty is to show either (a) that this principle is not really a principle of justice after all, or (b) that although it is, other principles outweigh or cancel its dictates. As we shall see, both these objections have merit.

Is Murder Alone To Be Punished by Death?

Let us recall, first, that not even the Biblical world limited the death penalty to the punishment of murder. Many other nonhomicidal crimes also carried this penalty (e.g., kidnapping, witchcraft, cursing one's parents). In our own recent history, persons have been executed for aggravated assault, rape, kidnapping, armed robbery, sabotage, and espionage. It is not possible to defend any of these executions (not to mention some of the more bizarre capital statutes, like the one in Georgia that used to provide an optional death penalty for desecration of a grave) on grounds of just retribution. This entails that either such executions are not justified or that they are justified on some ground other than retribution. In actual practice, few if any defenders of the death penalty have ever been willing to rest their case entirely on the moral principle of just retribution as formulated in terms of "a life for a life." Kant seems to have been a conspicuous exception. Most defenders of the death penalty have implied by their willingness to use executions to defend limb and property, as well as life, that they did not place much value on the lives of criminals when compared to the value of both lives and things belonging to innocent citizens.

Are All Murders To Be Punished by Death?

Our society for several centuries has endeavored to confine the death penalty to some criminal homicides. Even Kant took a casual attitude toward a mother's killing of her illegitimate child. ("A child born into the world outside marriage is outside the law . . . , and consequently it is also outside the protection of the law.") (Immanuel Kant, *The Metaphysical Elements of Justice* [1797]) In our society, the development nearly 200 years ago of the distinction between first- and second-degree murder was an attempt to narrow the class of criminal homicides deserving of the death penalty. Yet those dead owing to manslaughter, or to any kind of unintentional, accidental, unpremeditated, unavoidable, unmalicious killing are just as dead as the victims of the most ghastly murder. Both the law in practice and moral reflection show how difficult it is to identify all and only the criminal homicides that are appropriately punished by death (assuming that any are). Individual judges and juries differ in the conclusions they reach. The history of capital punishment for homicides reveals continual efforts, uniformly unsuccessful, to identify before the fact those homicides for which the slayer should die. Benjamin Cardozo, a justice of the United States Supreme Court fifty years ago, said of the distinction between degrees of murder that it was

> . . . so obscure that no jury hearing it for the first time can fairly be expected to assimilate and understand it. I am not at all sure that I understand it myself after trying to apply it for many years and after diligent study of what has been written in the books. Upon the basis of this fine distinction with its obscure and mystifying psychology, scores of men have gone to their death. (Benjamin Cardozo, "What Medicine Can Do for Law" [1928])

Similar skepticism has been registered on the reliability and rationality of death-penalty statutes that give the trial court the discretion to sentence to prison or to death. As Justice John Marshall Harlan of the Supreme Court observed a decade ago,

> Those who have come to grips with the hard task of actually attempting to draft means of channeling capital sentencing discretion have confirmed the lesson taught by history. . . . To identify before the fact those characteristics of criminal homicide and their perpetrators which call for the death penalty, and to express these characteristics in language which can be fairly understood and applied by the sentencing authority, appear to be tasks which are beyond present human ability. (*McGautha* v. *California*, 402 U.S. 183 [1971])

The abstract principle that the punishment of death best fits the crime of murder turns out to be extremely difficult to interpret and apply.

If we look at the matter from the standpoint of the actual practice of criminal justice, we can only conclude that "a life for a life" plays little or no role whatever. Plea bargaining (by means of which one of the persons involved in a crime agrees to accept a lesser sentence in exchange for testifying against the others to enable the prosecutor to get them all convicted), even where murder is concerned, is widespread. Studies of criminal justice reveal that what the courts (trial or appellate) decide on a given day is first-degree murder suitably punished by death in a given jurisdiction could just as well be decided in a neighboring jurisdiction on another day either as second-degree murder or as first-degree murder but without the death penalty. The factors that influence prosecutors in determining the charge under which they will prosecute go far beyond the simple principle of "a life for a life." Nor can it be objected that these facts show that our society does not care about justice. To put it succinctly, either justice in punishment does not consist of retribution, because there are other principles of justice; or there are other moral considerations besides justice that must be honored; or retributive justice is not adequately expressed in the idea of "a life for a life."

Is Death Sufficiently Retributive?

Given the reality of horrible and vicious crimes, one must consider whether there is not a quality of unthinking arbitrariness in advocating capital punishment for murder as the retributively just punishment. Why does death in the electric chair or the gas chamber of before a firing squad or on a gallows meet the requirements of retributive justice? When one thinks of the savage, brutal, wanton character of so many murders, how can retributive justice be served by anything less than equally savage methods of execution for the murderer? From a retributive point of view, the oft-heard exclamation, "Death is too good for him!" has a certain truth. Yet few defenders of the death penalty are willing to embrace this consequence of their own doctrine.

The reason they do not and should not is that, if they did, they would be stooping to the methods and thus to the squalor of the murderer. Where criminals set the limits of just methods of punishment, as they will do if we attempt to give exact and literal implementation to *lex talionis*, society will find itself descending to the cruelties and savagery that criminals employ. But society would be deliberately authorizing such acts, in the cool light of reason, and not (as is often true of vicious criminals) impulsively or in hatred and anger or with an insane or unbalanced mind. Moral restraints, in short, prohibit us from trying to make executions perfectly retributive. Once we grant the role of these restraints, the principle of "a life for a life" itself has been qualified and no longer suffices to justify the execution of murderers.

Other considerations take us in a different direction. Few murders, outside television and movie scripts, involve anything like an execution. An execution, after all, begins with a solemn pronouncement of the death sentence from a judge, is followed by long detention in maximum security awaiting the date of execution, various appeals, perhaps a final sanity hearing, and then "the last mile" to the execution chamber itself. As the French writer Albert Camus remarked,

> For there to be an equivalence, the death penalty would have to punish a criminal who had warned his victim of the date at which he would inflict a horrible death on him and who, from that moment onward, had confined him at his mercy for months. Such a monster is not encountered in private life. (Albert Camus, *Resistance, Rebellion, and Death* [1961])

Differential Severity Does Not Require Executions

What, then, emerges from our examination of retributive justice and the death penalty? If retributive justice is thought to consist in *lex talionis*, all one can say is that this principle has never exercised more than a crude and indirect effect on the actual punishments meted out. Other principles interfere with a literal and single-minded application of this one. Some murders seem improperly punished by death at all; other murders would require methods of execution too horrible to inflict; in still other cases any possible execution is too deliberate and monstrous given the nature of the motivation culminating in the murder. Proponents of the death penalty rarely confine themselves to reliance on this principle of just retribution and nothing else, since they rarely confine themselves to supporting the death penalty only for all murders.

But retributive justice need not be thought to consist of *lex talionis*. One may reject that principle as too crude and still embrace the retributive principle that the severity of punishments should be graded according to the gravity of the offense. Even though one need not claim that life imprisonment (or any kind of punishment other than death) "fits" the crime of murder, one can claim that this punishment is the proper one for murder. To do this, the schedule of punishments accepted by society must be arranged so that this mode of imprisonment is the most severe penalty used. Opponents of the death penalty need not reject this principle of retributive justice, even though they must reject a literal *lex talionis*.

Equal Justice and Capital Punishment

During the past generation, the strongest practical objection to the death penalty has been the inequities with which it has been applied. As Supreme Court Justice William O. Douglas once observed, "One searches our chronicles in vain for the execution of any member of the affluent strata of this society." (*Furman* v. *Georgia*, 408 U.S. 238 [1972]) One does not search our chronicles in vain for the crime of murder committed by the affluent. Every study of the death penalty for rape has confirmed that black male rapists (especially where the victim is a white female) are far more likely to be sentenced to death (and executed) than white male rapists. Half of all those under death sentence during 1976 and 1977 were black, and nearly half of all those executed since 1930 were black. All the sociologial evidence points to the conclusion that the death penalty is the poor man's justice; as the current street saying has it, "Those without the capital get the punishment."

Let us suppose that the factual basis for such a criticism is sound. What follows for the morality of capital punishment? Many defenders of the death penalty have been quick to point out that since there is nothing intrinsic about the crime of murder or rape that dictates that only the poor or racial-minority males will commit it, and since there is nothing overtly racist about the statutes that authorize the death penalty for murder or rape, it is hardly a fault in the idea of capital punishment if in practice it falls with unfair impact on the poor and the black. There is, in short, nothing in the death penalty that requires it to be applied unfairly and with arbitrary or discriminatory results. It is at worst a fault in the system of administering criminal justice (and some, who dispute the facts cited above, would deny even this).

Presumably, both proponents and opponents of capital punishment would concede that it is a fundamental dictate of justice that a punishment should not be unfairly—inequitably or unevenly—enforced and applied. They should also be able to agree that when the punishment in question is the extremely severe one of death, then the requirement to be fair in using such a punishment becomes even more stringent. Thus, there should be no dispute in the death penalty controversy over these principles of justice. The dispute begins as soon as one attempts to connect these principles with the actual use of this punishment.

In this country, many critics of the death penalty have argued, we would long ago have got rid of it entirely if it had been a condition of its use that it be applied equally and fairly. In the words of the attorneys who argued against the death penalty in the Supreme Court during 1972, "It is a freakish aberration, a random extreme act of violence, visibly arbitrary and discriminatory—a penalty reserved for unusual application because, if it were usually used, it would affront universally shared standards of public decency." It is difficult to dispute this judgment, when one considers that there have been in the United States during the past fifty years about half a million criminal homicides but only about 4,000 executions (all but 50 of which were of men).

We can look at these statistics in another way to illustrate the same point. If we could be assured that the 4,000 persons executed were the worst of the worst, repeated offenders without exception, the most dangerous murderers in captivity— the ones who had killed more than once and were likely to kill again, and the least likely to be confined in prison without imminent danger to other inmates and the staff—then one might accept half a million murders and a few thousand executions with a sense that rough justice had been done. But the truth is otherwise. Persons are sentenced to death and executed not because they have been found to be uncontrol- lably violent, hopelessly poor parole and release risks, or for other reasons. Instead, they are executed for entirely different reasons. They have a poor defense at trial; they have no funds to bring sympathetic witnesses to court; they are immigrants or strangers in the community where they were tried; the prosecuting attorney wants the publicity that goes with "sending a killer to the chair"; they have inexperienced or overworked counsel at trial; there are no funds for an appeal or for a transcript of the trial record; they are members of a despised racial minority. In short, the actual study of why particular persons have been sentenced to death and executed does not show any careful winnowing of the worst from the bad. It shows that the executed were usually the unlucky victims of prejudice and discrimination, the losers in an arbitrary lottery that could just as well have spared them as killed them, the victims of the disadvantages that almost always go with poverty. A system like this does not enhance respect for human life; it cheapens and degrades it. However heinous murder and other crimes are, the system of capital punishment does not compensate for or erase those crimes. It only tends to add new injuries of its own to the catalogue of our inhumanity to each other.

Ernst van der Haag,
from "Deterrence and Uncertainty"

Ernst van der Haag has long been a proponent of the death penalty, and in the following excerpt he defends the use of execution in the most serious criminal cases, despite the inconclusive statistics that show that the use of the death penalty may in fact not deter crime. Van der Haag points out that this uncertainty is to be found in any method of punishment, and there are no clear statistics either to prove that the threat of a 6-year prison term deters crime any more than the threat of a 3-year term. In every case, there is uncertainty, and though in the case of the death penalty more proof is demanded than for other penalties (because of the irrevocability of the punishment), the argument would still be that the calculated risks favor the death penalty. It may be that there is no deterrent effect, but, then again, there may be such an effect, and surely the life (or lives) of the murderer's future victims (or other murderer's vicims) are worth more than the life of the murderer him or herself.

. . . If we do not know whether the death penalty will deter others [in a uniquely effective way], we are confronted with two uncertainties. If we impose the death penalty, and achieve no deterrent effect thereby, the life of a convicted murderer has been expended in vain (from a deterrent viewpoint). There is a net loss. If we impose the death sentence and thereby deter some future murderers, we spared the lives of some future victims (the prospective murderers gain too; they are spared punishment because they were deterred). In this case, the death penalty has led to a net gain, unless the life of a convicted murderer is valued more highly than that of the unknown victim, or victims (and the non-imprisonment of the deterred non-murderer).

The calculation can be turned around, of course. The absence of the death penalty may harm no one and therefore produce a gain—the life of the convicted murderer. Or it may kill future victims of murderers who could have been deterred, and thus produce a loss—their life.

To be sure, we must risk something certain—the death (or life) of the convicted man, for something uncertain—the death (or life) of the victims of murderers who may be deterred. This is in the nature of uncertainty—when we invest, or gamble, we risk the money we have for an uncertain gain. Many human actions, most commit-

Reprinted by special permission of Northwestern University, School of Law, 60, *Journal of Criminal Law, Criminology and Police Science*, 141, (1969)—selected passages.

ments—including marriage and crime—share this characteristic with the deterrent purpose of any penalization, and with its rehabilitative purpose (and even with the protective).

More proof is demanded for the deterrent effect of the death penalty than is demanded for the deterrent effect of other penalties. This is not justified by the absence of other utilitarian purposes such as protection and rehabilitation; they involve no less uncertainty than deterrence. (Rehabilitation or protection are of minor importance in our penal system (though not in our theory). We confine many people who do not need rehabilitation and against whom we do not need protection (e.g., the exasperated husband who killed his wife); we release many unrehabilitated offenders against whom protection is needed. Certainly rehabilitation and protection are not, and deterrence is, the main actual function of legal punishment, if we disregard nonutilitarian purposes.)

Irrevocability may support a demand for some reason to expect more deterrence than revocable penalties might produce, but not a demand for more proof of deterrence, as has been pointed out above. The reason for expecting more deterrence lies in the greater severity, the terrifying effect inherent in finality. Since it seems more important to spare victims than to spare murderers, the burden of proving that the greater severity inherent in irrevocability adds nothing to deterrence lies on those who oppose capital punishment. Proponents of the death penalty need show only that there is no more uncertainty about it than about greater severity in general.

The demand that the death penalty be proved more deterrent than alternatives cannot be satisfied any more than the demand that six years in prison be proved to be more deterrent than three. But the uncertainty which confronts us favors the death penalty as long as by imposing it we might save future victims of murder. This effect is as plausible as the general idea that penalties have deterrent effects which increase with their severity. Though we have no proof of the positive deterrence of the penalty, we also have no proof of zero, or negative effectiveness. I believe we have no right to risk additional future victims of murder for the sake of sparing convicted murderers; on the contrary, our moral obligation is to risk the possible ineffectiveness of executions. However rationalized, the opposite view appears to be motivated by the simple fact that executions are more subjected to social control than murder. However, this applies to all penalties and does not argue for the abolition of any.

Robert Nozick, "Retribution and Revenge," from *Philosophical Explanations*

While almost every author who defends a "retributive" theory of punishment takes considerable pains to distinguish between retribution and revenge, too few try to spell out what the difference is. Both involve punishment inflicted for some past offense, but retribution is supposed to have rational support and be essential to justice, while revenge or vengeance, we are told, is just an emotion, an instinct, and wholly unworthy of the honorable name "justice." In the following section from *Philosophical Explanations*, Robert Nozick marks a variety of distinctions, including not only the emotionality of revenge in contrast to the possible impersonality and coolness of retribution, but also the insistence that retribution is always done for a wrong and not merely for a personal harm or offense, and that retribution requires strict limits to punishment while revenge "by its nature need set no limits, although the revenger may limit what he inflicts for external reasons." Nozick allows that punishment is often done for "mixed motives," but he ultimately defends only retribution, on the grounds that it "sends a message" to the criminal (and to others, too), thus slightly blurring the usual distinction between retribution and deterrence (though Nozick later criticizes deterrence theory as such) and ends by asking the question why, for this purpose, punishment is ultimately necessary at all.

Retribution and Revenge

The view that people deserve punishment for their wrongful acts, independently of the deterrent effect of such punishment, strikes some people as a primitive view, expressive only of the thirst for revenge. Before pursuing the underlying rationale of retribution, punishment inflicted as deserved for a past wrong, we should consider some ways in which retribution differs from revenge.

(1) Retribution is done for a wrong, while revenge may be done for an injury or harm or slight and need not be for a wrong.

(2) Retribution sets an internal limit to the amount of the punishment, according to the seriousness of the wrong, whereas revenge internally need set no limit to what is inflicted. Revenge by its nature need set no limits, although the revenger may limit what he inflicts for external reasons.

(3) Revenge is personal: "this is because of what you did to my ——" (self, father, group, and so on). Whereas the agent of retribution need have no special or personal tie to the victim of the wrong for which he exacts retribution.

Do not say he exacts the penalty because of the injury done to his own moral code; that overextends the notion of personal tie. Steps sometimes are taken to exclude the personal tie from intruding in a process of retribution and clouding the nature of what is happening by blurring the distinctness of retribution from revenge. Thus, under a system of capital punishment, if the sister of the official executioner is murdered and the killer is apprehended, someone else will be substituted to perform that execution.

This third point has two aspects: revenge can be desired only by someone with a personal tie (others can desire that some such person inflict revenge, but their desire is not a desire for revenge), and it can be inflicted only by (the agent of) someone with a personal tie. (Revenge may involve differing notions of linkage: (a) because of what you did to my ——; (b) because of what you did to me. If someone kills your father, under linkage *a* you kill him while under *b* you kill his father.) Retribution, on the other hand, may be desired or inflicted by people without such a tie. This personal factor also enters into the revenger's desire, noted below, that his connection to the victim for whom revenge is being exacted be known to the recipient of revenge.

(4) Revenge involves a particular emotional tone, pleasure in the suffering of another, while retribution either need involve no emotional tone, or involves another one, namely, pleasure at justice being done. Therefore, the thirster after revenge often will want to experience (see, be present at) the situation in which the revengee is suffering, whereas with retribution there is no special point in witnessing its infliction.

This connects with the previous point about the personal tie; one purpose of revenge may be to produce a psychological effect in the person who seeks revenge (that particular emotional tone, for example), while retribution has no such personal purpose.

(5) There need be no generality in revenge. Not only is the revenger not committed to revenging any similar act done to anyone; he is not committed to avenging all done to himself. Whether he seeks vengeance, or thinks it appropriate to do so, will depend upon how he feels at the time about the act of injury. Whereas the imposer of retribution, inflicting deserved punishment for a wrong, is committed to (the existence of some) general principles (prima facie) mandating punishment in other similar circumstances. Furthermore, if possible these general standards will be made known and clear in the process of retribution; even those who act in retribution against the guilty agents of a torturing dictatorship, keeping their own identities secret, will make the principles known.

In drawing these contrasts between retribution and revenge, I do not deny that there can be mixed cases, or that people can be moved by mixed motives, partially a desire for retribution, partially a desire for revenge, or that a stated desire can mask another one that is operative. Usually, it is charged that those favoring retribution really crave revenge; but this will be especially implausible in the absence of a special tie to the victim. (The charge never is made in the other direction, that some who call for revenge really are seeking retribution but are embarrassed at appearing moralistic.) The charge itself, though, recognizes the distinction, even as it seeks to blur it. That retribution can be distinguished from revenge and is, on its surface at least, less primitive neither shows that, nor explains why, retribution is justified. Nor does it explain why retribution and revenge so often have been confused.

Retribution and revenge share a common structure: a penalty is inflicted for a reason (a wrong or injury) with the desire that the other person know why this is occurring and know that he was intended to know. (In the comic books of my youth, the villain seeking revenge always was thwarted by his desire that the hero not merely die but realize why he was dying and at whose hand, in prolonged agony—this gave the hero extended opportunity to escape.) I shall spell out that common structure as it is exemplified by retribution; this must be modified in accordance with the contrasts we have listed to obtain an account of revenge.

* * *

"Poetic justice" involves the wrongdoer's undergoing a consequence that approximately could be visited upon him in retribution but which was not produced in that way, usually owing to the failure of one of the first two conditions of retribution. A system of karma, whereby the moral quality of acts produces effects automatically in (this or) another lifetime, is not a system of poetic justice. It is crucial to poetic justice that the (penalty) effect is not a result of the moral quality of the act, even though it appropriately would fit that moral quality. Thus, although very many poetically just things could occur, there could not be a system of poetic justice. The generality a system involves (supporting subjunctives about what would occur) could stem only from the (appropriate) effects being due to the moral quality of the acts, qua moral quality, and so the justice done would not be merely "poetic."

The conditions demarcating retribution explain what otherwise appears to be a ludicrous phenomenon. If someone sentenced to death falls perilously ill or is accidentally injured or attempts suicide the day before the scheduled execution, then the execution is postponed and measures are taken to bring the condemned person back to health so that he then can be executed. Although due-process reasons might be conjured up for this, I believe the reason is that his punishment is to involve something's being visited upon him by others because of the wrongness of his act. His death by natural causes or by his own hand would avoid this, so measures are taken to restore him for punishment.

* * *

Retributive punishment is an act of communicative behavior. Revenge also fits this communicative structure, though with a somewhat different message; this provides an explanation of why the two are so often confused.

What is the message of retributive punishment, and why is it communicated in that especially forceful and unwelcome way? The (Gricean) message is: this is how wrong what you did was.

* * *

In the case of retributive matching punishment where, to the extent feasible, the penalty inflicted on the wrongdoer is the same as the wrong or harm he did, perhaps the message then is: this is (precisely) the wrong you did. But if our intention is to mean his act was that (magnitude of) wrong, why don't we just say so and spare him the penalty? (Don't say we first must get his attention.) What justifies us in inflicting upon him so unwelcome a mode of communication?

We may view different "theories" of punishment as focusing upon different aspects of communication: the sender of a message, the recipient of this message, the transmission itself. Some have pointed out that punishment has an expressive function, wherein the sender condemns the crime. More frequently, the literature focuses upon the recipient. Under this rubric, we might see punishment as an attempt to demonstrate to the wrongdoer that his act was wrong, not only to mean the act is wrong but to *show* him its wrongness. Some retributive theorists see the showing as having a further goal: the moral improvement of the offender. Punishment is supposed to achieve this goal by bringing home to the offender the nature of what he has done, from which he is to realize its wrongness. Since these theorists see the central purpose of punishment in its further consequences, they have been termed teleological retributivists.

* * *

The (Gricean) message of teleological retributive punishment is delivered in a way so that the delivery is evidence that or shows that it is true. (Compare a telegram that says "you have just received a telegram.") Receiving the message (sent that way), "this is how wrong what you did was", is supposed to convince one that it is true; the message, via its sending, is to be self-supporting.

Albert Camus, from
"Reflections on the Guillotine"

Albert Camus was one of the great humanist writers of the mid-century. Caught in the middle of the French-Algerian war (he was a French Algerian or "*pied noir*" by birth) and having lived in France during the Nazi occupation, he became the conscience and spokesman for those who found the ideologies of both left and right intolerably extreme. "Neither victims nor executioners," the well-chosen title of one of his essays, expresses a humanist sentiment shared by millions of his fellow French and many more millions besides. He has long been identified with the "existentialist" movement in France, but he denied the affiliation and openly broke off his friendship with existentialist Godfather Jean-Paul Sartre precisely on such political issues. Sartre was an ideologue of the Left—albeit an unorthodox one; Camus refused to side with the left because of its inexcusable violence. It was in this same frame of mind, "neither victims nor executioners," that he wrote one of his most quoted essays, a long argument against the death penalty. A brief excerpt is included here.

If there is a desire to maintain the death penalty, let us at least be spared the hypocrisy of a justification by example. Let us be frank about that penalty which can have no publicity, that intimidation which works only on respectable people, so long as they are respectable, which fascinates those who have ceased to be respectable and debases or deranges those who take part in it. It is a penalty, to be sure, a frightful torture, both physical and moral, but it provides no sure example except a demoralizing one. It punishes, but it forestalls nothing; indeed, it may even arouse the impulse to murder. It hardly seems to exist, except for the man who suffers it—in his soul for months and years, in his body during the desperate and violent hour when he is cut in two without suppressing his life. Let us call it by the name which, for lack of any other nobility, will at least give the nobility of truth, and let us recognize it for what it is essentially: a revenge.

A punishment that penalizes without forestalling is indeed called revenge. It is a quasi-arithmetical reply made by society to whoever breaks its primordial law. That reply is as old as man; it is called the law of retaliation. Whoever has done me harm must suffer harm; whoever has put out my eye must lose an eye; and whoever has killed must die. This is an emotion, and a particularly violent one, not a principle.

Retaliation is related to nature and instinct, not to law. Law, by definition, cannot obey the same rules as nature. If murder is in the nature of man, the law is not intended to imitate or reproduce that nature. It is intended to correct it. Now, retaliation does no more than ratify and confer the status of a law on a pure impulse of nature. We have all known that impulse, often to our shame, and we know its power, for it comes down to us from the primitive forests. In this regard, we French, who are properly indignant upon seeing the oil king in Saudi Arabia preach international democracy and call in a butcher to cut off a thief's hand with a cleaver, live also in a sort of Middle Ages without even the consolations of faith. We still define justice according to the rules of a crude arithmetic. Can it be said at least that that arithmetic is exact

> A few years ago I asked for the reprieve of six Tunisians who had been condemned to death for the murder, in a riot, of three French policemen. The circumstances in which the murder had taken place made difficult any division of responsibilities. A note from the executive office of the President of the Republic informed me that my appeal was being considered by the appropriate organization. Unfortunately, when that note was addressed to me I had already read two weeks earlier that the sentence had been carried out. Three of the condemned men had been put to death and the three others reprieved. The reasons for reprieving some rather than the others were not convincing. But probably it was essential to carry out three executions where there had been three victims.

and that justice, even when elementary, even when limited to legal revenge, is safeguarded by the death penalty? The answer must be no.

Let us leave aside the fact that the law of retaliation is inapplicable and that it would seem just as excessive to punish the incendiary by setting fire to his house as it would be insufficient to punish the thief by deducting from his bank account a sum equal to his theft. Let us admit that it is just and necessary to compensate for the murder of the victim by the death of the murderer. But beheading is not simply death. It is just as different, in essence, from the privation of life as a concentration camp is from prison. It is a murder, to be sure, and one that arithmetically pays for the murder committed. But it adds to death a rule, a public premeditation known to the future victim, an organization, in short, which is in itself a source of moral sufferings more terrible than death. Hence there is no equivalence. Many laws consider a premeditated crime more serious than a crime of pure violence. But what then is capital punishment but the most premeditated of murders, to which no criminal's deed, however calculated it may be, can be compared? For there to be equivalence, the death penalty would have to punish a criminal who had warned his victim of the date at which he would inflict a horrible death on him and who, from that moment onward, had confined him at his mercy for months. Such a monster is not encountered in private life.

Susan Jacoby, from *Wild Justice*

Susan Jacoby has studied the long history of vengeance, or "wild justice" as she calls it (after Francis Bacon), and from that long historical perspective concluded that we do ourselves damage by denying so basic an urge or instinct. She notes with disapproval the contempt which accompanies all discussions of revenge and, even more so, all appeals to and expressions of revenge. She abuses the "foolish cliché," to "forgive and forget," and she marks with considerable irony the now established distinction—even opposition—between vengeance on the one hand and justice on the other. She clearly defends retribution (though not always marking or answering the distinctions between retribution and revenge), and attacks those many legal theorists who see the law not as legalized revenge but as a deterrent to future criminal action. She then turns back to history and gives us some sense of the long historical emphasis placed on personal retribution, with all of its dangers and excesses; but she makes it clear that our current denial of the importance of such satisfaction and settlement is akin to the Victorian denial of the importance of sexual satisfaction, and the psychological penalties may turn out to be just as severe.

Forgive and forget. This admonition surely ranks as one of the most foolish clichés in any language. Remembrance is unquestionably a form of revenge, but, in one of the great paradoxes of civilized life, it is equally indispensable to the attainment of true forgiveness. The concept of "just deserts," which was not a philosophical abstraction but a fact of life throughout most of human history, evokes a deep unease in modern men and women. Who wants to be confronted by an Old Testament prophet, whether across the dinner table or on the evening news? We are more comfortable with the notion of forgiving and forgetting, however unrealistic it may be, than with the private and public reality of revenge, with its unsettling echoes of the primitive and its inescapable reminder of the fragility of human order.

Justice is a legitimate concept in the modern code of civilized behavior. Vengeance is not. We prefer to avert our eyes from those who persist in reminding us of the wrongs they have suffered—the mother whose child disappeared three years ago on a New York street and who, instead of mourning in silence, continues to appear on television and appeal for information about her missing son; the young Sicilian woman who, instead of marrying her rapist as ancient local custom dictates, scandalizes the town by bringing criminal charges; the concentration-camp survivors who, instead of putting the past behind them, persist in pointing their fingers at ex-Nazis living comfortable lives on quiet streets. Such people are disturbers of the peace; we wish they would take their memories away to a church, a cemetery, a psychotherapist's office and allow us to return justice and vengeance to the separate compartments they supposedly occupy in twentieth-century life.

* * *

The very word "revenge" has pejorative connotations. Advocates of draconian punishment for crime invariably prefer "retribution"—a word that affords the comfort of euphemism although it is virtually synonymous with "revenge." In this century, which has produced a new dictionary of terms like "resettlement," "special treatment," "protective reaction strike," and "re-education centers," euphemistic language has come to be regarded primarily as a tool of governments and officials attempting to conceal or at least to blur the outlines of their evil deeds. But euphemism is as likely to be the product of confusion as of design—especially when it concerns ethical dilemmas that fall within the realm of private as well as public behavior. The relationship between "retribution" and "revenge" is analogous to the only recently obsolete substitution of "protection" for "birth control"; it has less to do with good and evil than with ambivalence about violations of social piety and propriety grown so widespread that they have become the rule rather than the exception.

The proper relationship between justice and revenge has been a major preoccupation of literature, religion, and law throughout the recorded history of the West. Establishment of a balance between the restraint that enables people to live with one another and the ineradicable impulse to retaliate when harm is inflicted has always been one of the essential tasks of civilization. The attainment of such a balance depends in large measure on the confidence of the victimized that someone else will act on their behalf against the victimizers. Laws are designed not to weed out the impulse toward revenge but to contain it in a manner consistent with the maintenance of an orderly and humane society. Even the classic Christian statement on the subject—"Dearly beloved, avenge not yourselves, but rather give place unto wrath: for it is written, Vengeance is mine; I will repay, saith the Lord"—usually cited as an injunction against revenge, is as much a pledge of divine action as a prohibition of human retribution. In the absence of a divinity who dispenses floods, fires, and plagues to punish offenders, civil authorities have been regarded as legitimate stand-ins. How that legitimacy may be established is a separate question, whether it is framed in secular or religious terms.

One measure of a civilization's complexity is the distance between aggrieved individuals and the administration of revenge. But distance alone is not sufficient to establish a society's control over the vengeful impulses of its members, and it is certainly no measure of a civilization's humaneness. In determining the role of retribution in society, the ultimate aim of punishment is no less important than the procedure by which it is imposed. The fact that a judge rather than a mob designates drawing-and-quartering as a proper mode of execution is, in strict legal terms, an advance in the social control of revenge, but it also means that the values of those who control the social order are scarcely more advanced than those of the mob.

The United States Supreme Court evaded this issue when, in a 1976 decision upholding capital punishment, it described executions of murderers as an expression of a deep emotional need for retribution that is "neither a forbidden objective nor one inconsistent with our respect for the dignity of men." One might easily construct an argument in which society's deep emotional need for retribution can only be satisfied by subjecting murderers to precisely the same torments they inflict on their victims. Until the eighteenth century, judicial bodies in England and continental Europe did exactly that. The question the Court ought to have addressed is not whether retribution *per se* is a "forbidden" objective of criminal justice but which forms of revenge are consistent with the aims of a just society.

* * *

Legalized revenge. The expression invariably resounds as an indictment, embodying either the general conviction that vengeance has no place in modern jurisprudence or a more limited argument against specific legal penalties. The specific argument, most frequently directed against the death penalty or other cruel (though not necessarily unusual) punishments, rests squarely within the mainstream of progressive legal tradition—while the general argument is founded on excessive optimism about the prospects for modifying human behavior. The specific approach, which focuses on the attainable, has made the most enduring contributions to legal control of both private and public vengeance. In the struggle against the vicious punishments human beings repeatedly inflict upon one another, the most effective reformers rarely waste their words in pious pronouncements about the immorality of legalized revenge. Instead, they concentrate on defining the forms of retribution permitted by law and on limiting the powers of those who apprehend criminals, pass judgment, and pronounce sentence.

In societies—still few in number—with a highly developed concept of individual worth, law extends itself to uphold a wide variety of personal rights and liberties. This aspect of the legal system, which falls within the broad classification of "human rights," is a sophisticated, relatively recent addition to the three irreducible functions of law—punishment of the guilty, exoneration of the innocent, and deterrence of those who might, in the absence of sufficiently reliable and unpleasant penalties, sustain themselves by preying on their fellow citizens.

Insofar as humanly possible (and the task has often seemed humanly impossible), law attempts to remove personal animus from the process of apportioning blame and exacting retribution. It is the removal of private animus, not the absence of vengeance, that distinguishes the rule of law from the rule of passion. Laws may, of course, be unjust even though they are not founded on personal enmity; the old Jim Crow laws of the American South and the current apartheid laws of South Africa are cases in point. Inequitable retribution is one of the chief features of such laws; punishment is determined not only on the basis of what a person does but of what he or she *is*—or is perceived to be under the ruling biases of the land. Equitable retribution requires a fusion of libertarian and punitive concerns—a rare blend in a world that tends to regard freedom and order as natural antagonists. Retribution per se is an integral component of just as well as unjust legal systems. Legalized revenge, however repellent it sounds, is not an accusation but a fact.

* * *

The conceptual shift upon which all legal restraint of revenge is based is the movement from diffuse to specific responsibility, from hereditary guilt and punishment to individual accountability. The concept of the "born" criminal—and particularly of the born political criminal—has proved to be one of the most tenacious in human history. During the age of Pericles, collective and hereditary vengeance had long been abolished for murder but a form of it still prevailed for the more serious (to the Greeks) crime of treason. No descendant of a convicted traitor could ever be permitted to live in or hold property in Athens. In thirteenth-century Florence, an aristocrat named Giano della Bella declared himself a commoner and led the political fight to widen the base of the electorate. He then promulgated a new code of laws making every citizen liable for the crimes of his or her relatives. Special boxes were set up to receive secret denunciations; dozens of prominent families were deprived of their civic rights; old aristocrats took plebeian names to disguise their identity, in much the same manner as the Jewish *conversos* did during the Spanish and Portuguese inquisitions.

The practice of imprisoning or executing relatives of alleged traitors is, of course, equally familiar in our own century. The Nazi procedure of *Sippenhaft* (which literally translates as "kin-arrest or apprehension"); Stalin's arrest of wives and children of real or presumed opponents as "class enemies"; the Red Guard attacks during the Chinese cultural revolution on relatives of scholars and former landowners or government officials—these instances proclaim the durability of the practice of government-imposed or institutionalized familial punishment.

* * *

Belief in hereditary responsibility has always impeded the historical effort to place revenge within a legal rather than a mystical, quasi-religious framework in which "blood cries out for blood." Hereditary guilt is inextricably linked with hereditary punishment; the generalization can be applied to the most serious and the most trivial crimes. The decline of belief in hereditary responsibility for sin, or crime, was a pre-condition for what was probably the single most important advance in the legal control of revenge: the abolition of private settlement for murder.

The historical meaning of "private settlement" is an agreement between families, or tribal clans, on appropriate retribution and/or restitution for the death of a member of one of the groups. A group, not an individual, therefore becomes the injured party or the offender. Private settlement was itself an advance over unrestrained vendetta; without such settlement, human beings would undoubtedly have rendered themselves extinct. The obvious deficiency of private settlement is that it does not always work; instead of a murder's being settled by restitution and a peace agreement, it is settled by yet another murder and the always unsettling resumption of collective vendetta. One need not look to ancient history for examples; in cultures where *machismo* is an unwritten law unto itself, vendetta is alive and well in the 1980s. In 1981, the Brazilian military police were forced to assume control of a town named Exu after nine people were killed in the resumption of a thirty-two-year-old blood vendetta. A private settlement—a "moral pact of nonaggression" negotiated under the auspices of the Roman Catholic primate of Brazil—had broken down. The legal basis for the military takeover, the *New York Times* soberly reported, was the fact that the local city council could no longer produce a quorum. Two members had been murdered in a four-month period and four resigned under death threats. One gave notice of his resignation by mail after he had fled to Rio de Janeiro, two thousand miles from Exu.

In the ancient world, as in the modern, the abolition of private settlement was obviously intended to prevent members of the populace from killing one another. However, private settlement was not always ineffective, and the assumption of retributive powers by the state was frequently accompanied by more rather than less stringent penalties for crimes of bloodshed. Exile, not death, had been a standard penalty for murder in many tribal cultures. In Greece, the rise of the pollution doctrine and the prohibition of private settlement gave sanction to state executions for murder. In tribal cultures such as those depicted in the *Iliad* and the *Odyssey*, there was no provision for corporal punishment or restitution for manslaughter committed within the clan. How could there be, when the only lawful recipients of restitution were kin to both victim and killer?

In Greece, the abolition of private settlement even injected the state into disputes between spouses—something that would have been unthinkable in tribal culture. When a husband or wife killed a spouse, either by accident or in a sudden rage, permanent exile from the family and temporary exile from the state were the legal penalties. According to Plato's *Laws*, composed in the fourth century B.C., a husband or wife who had slain a spouse could never return home to dwell with the family. (Greek law was more flexible on the issue of return to one's former abode if the victim and killer were not married.)

The abolition of private settlement made it difficult in all cultures, and impossible in some, to atone for the taking of a life by a payment of money or property. Both Hebrew and Roman law ruled out financial restitution as a substitute for corporal punishment of the slayer (although they did not rule out restitution in addition to legal retribution, just as modern law does not prohibit the bringing of a civil action for damages following a criminal prosecution). The Covenant Code of the Israelites, the early legal structure established between the exodus from Egypt and the establishment of a monarchy in Jerusalem, explicitly forbade absolution (there was no difference at that point between the legal and religious meanings of the term) for murder through payment or through substituting another man's life for the life of the murderer. This represented an enormous advance over contemporary law codes in the Fertile Crescent, including those of the Assyrians and the Hittites, which allowed a powerful murderer to arrange the execution of someone else in his place. This prohibition was reinforced in later rabbinical law, which outlawed the settlement of any murder by financial restitution and which also emphasized the legal irrelevance of private forgiveness by the victim's relatives. "The court is warned against accepting ransom from a murderer," Maimonides emphasized, "even if he offers all the money in the world and *even if the avenger of blood agrees to let him go free*" (italics mine).

The abolition of private settlement for murder was the first, crucial step in the long process by which the domain of the state was gradually extended to all crimes. For murder was not, of course, the only offense that could provoke revenge and civic violence—and the extension of state jurisdiction to lesser crimes has not been fully accomplished even now.

Robert C. Solomon, "Justice and the Passion for Vengeance"

In the following excerpt, the author argues for the importance of vengeance, acknowledging the legitimacy of the distinction between retribution and "mere" revenge, but arguing that much of the supposed irrationality and excess of revenge has been overstated, and the denial of its role in justice leads to a false dichotomy between impersonal justice and merely personal revenge.

> "I think that the deterrent argument is simply a rationalization. The motive for punishment is revenge—not deterrence. . . . Punishment is hate."
>
> A. S. NEIL

Vengeance is the original meaning of justice. The word "justice" in the Old Testament and in Homer too virtually always refers to revenge. Throughout most of history the concept of justice has been far more concerned with the punishment of crimes and the balancing of wrongs than it has been with the fair distribution of goods and services. "Getting even" is and has always been one of the most basic metaphors of our moral vocabulary, and the frightening emotion of righteous, wrathful anger is an essential part of the emotional basis for our sense of justice, just as much as benign compassion and sympathy. Our resentment of injustice is a necessary precondition of our passion for justice, and the urge to retribution its essential consequence. "Don't get mad, get even"—whether or not it is prudent advice—is conceptually confused. Getting even is just an effective way of being angry, and getting angry, as Aristotle argued long ago, already includes the desire for vengeance.

Like it or not, I think that we have to agree with Arthur Lelyveld when he writes, "there is no denying the aesthetic satisfaction, the sense of poetic justice, that pleasures us when evildoers get the comeuppance they deserve. The impulse to punish is primarily an impulse to even the score . . . That satisfaction is heightened when it becomes possible to measure out punishment in exact proportion to the size and shape of the wrong that has been done." The immense pleasure, the aesthetic satisfaction points to the depth of the passion, and the need for "proportion" already indicates the intelligence involved in this supposedly most irrational and uncontrollable emotion. This is not to say, of course, that the motive of revenge is therefore legitimate or the action of revenge always justified. Sometimes vengeance is wholly

Robert C. Solomon, *A Passion for Justice*, © 1989, Addison Wesley Publishing Co., Inc., Reading, Massachusetts. Reprinted with permission of the publisher.

called for, even obligatory, and revenge is both legitimate and justified. Sometimes it is not, notably when one is mistaken about the offender or the offense. But to seek vengeance for a grievous wrong, to revenge oneself against evil—that seems to lie at the very foundation of our sense of justice, indeed, of our very sense of ourselves, our dignity and our sense of right and wrong. Even Adam Smith writes, in his *Theory of the Moral Sentiments* "The violation of justice is injury . . . it is, therefore, the proper object of resentment, and of punishment, which is the natural consequence of resentment." We are not mere observers of moral life, and the desire for vengeance seems to be an integral aspect of our recognition of evil. But it also contains—or can be cultivated to contain—the elements of its own control, a sense of its limits, a sense of balance. Thus the Old Testament instruction that revenge should be *limited to* "an eye for an eye, a tooth for a tooth, hand for hand, foot for foot, burning for burning, wound for wound, stripe for stripe" ("*Lex Talionis*"), *Exodus* 21:24–5). The New Testament demands even more restraint, the abstention from revenge oneself and the patience to entrust it to God. Both the Old and New Testaments (more the latter than the former) also encourage "forgiveness," but there can be no forgiveness if there is not first the desire (and the warrant) for revenge.

Vengeance is not just punishment, no matter how harsh. It is a matter of emotion, and like punishment, it is always *for* some offense, not just hurting for its own sake (even if, in some other sense, it is deserved). Vengeance, then, always has its reasons (though, to be sure, these can be mistaken, irrelevant, out of proportion or otherwise bad reasons). Vengeance is no longer a matter of obligation and it certainly can't claim to be rational as such but neither is it opposed to a sense of obligation (e.g. in matters of family honor) or rationality (insofar as rationality is to be found in every emotion, even this one). Vengeance is the emotion of "getting even," putting the world back in balance, and this simple phrase already embodies a whole philosophy of justice, even if (as yet) unarticulated and unjustified. Philosophers have been much too quick to attribute this sense of "balance" or "retribution" to reason, but I would want to argue that it is rather a function of emotion. Kant, of course, immediately opts for the former, dismissing the latter suggestion virtually altogether. Vengeance, he suggests, is purely subjective, wholly irrational, undependable and unjustifiable. It is wholly without measure or reason, devoid of any sense of balance or justice. But I want to suggest that vengeance just is that sense of measure or balance that Kant (and so many other philosophers) attribute to reason alone. But, of course, it is ultimately the same old dichotomy that is most at fault here, the supposed antagonism between reason on the one side and passions on the other. Where would our reasoning about punishment begin if not for our emotional sense of the need for retaliation and retribution? (We should stress here that retaliation and retribution should not be confused with reparation and mere compensation, which may in some cases "undo" the damage but in no case by themselves count as punishment.) And what would our emotion be if it were not already informed and cultivated by a keen sense of its object and its target, as well as the mores and morals of the community in which the offense in question is deserving of revenge?

Vengeance, unlike justice, is said to be "blind" (though it is worth reminding ourselves which of the two is depicted in established mythology as blind-folded). Vengeance, it is said, knows no end. It is not just that it gets out of hand; it cannot be held "in hand" in the first place. And, of course, we agree that there is danger in vengeance: It is by its very nature violent, disrupting the present order of things in an often impossible attempt to get back to a prior order which has itself been violently disrupted. Such an impossibility breeds frustration, and violence—even justified as vengeance (if, indeed, revenge is possible), and this typically leads to more violence. Too often, an act of revenge (even if legitimate) results in a new offense to be righted. And when the act is perpetrated not against the same person who did the offense but against another who is part of the same family, tribe or social group (the logic of "vendetta"), the possibilities for escalation are endless. (Ironically, much of the traditional danger of escalating vengeance has been reduced or eliminated by our very strong contemporary notion of individual (rather than collective) responsibility at the same time that the emphasis on individual responsibility has increasingly denied its attachments to vengeance.) It is because of the likelihood of escalation as well as the possibilities of mistakes on a purely personal level (what John Locke called "the inconveniences" of rights-enforcement in the state of nature) that the limitation of revenge through institutionalization becomes necessary. But it does not follow that vengeance itself is illegitimate or without measure or of no importance in considerations of punishment. To the dangers of vengeance unlimited it must be countered that if punishment no longer satisfies vengeance, if it ignores not only the rights but the emotional needs of the victims of crime, then punishment no longer serves its primary purpose, even if it were to succeed in rehabilitating the criminal and deterring other crime (which it evidently, in general, does not). The restriction of vengeance by law is entirely understandable, but, again, the wholesale denial of vengeance as a legitimate motive may be a psychological disaster.

These preliminary comments are intended to unearth a number of bad arguments against vengeance (which are often generalized into even worse bad arguments concerning 'negative' and violent emotions as a unified class):

1. *Vengeance is (as such) irrational, and, consequently, it is never legitimate.* Only a moment's reflection is necessary to realize that we all recognize (whether or not we recognize at the time) the difference between justified and unjustified revenge. Vengeance is not just the desire to harm but the desire to harm *for a reason*, and a reason of a very particular sort. To flunk a student because he has an orange punk hairdo or because he disagreed with one's pet theory in class is not justified, but to expel him for burning down the department library is another matter. But what about the fact that sometimes, while in the "grip" of revenge, we fail to recognize or properly exercise the reason and warrant for our vengeance? But the point is the word "sometimes," for there is nothing essential to vengeance that requires such failure. In indisputably rational contexts a decision-maker mistakes a means for an end, or becomes so distracted in his pursuit of the end that he neglects or simply misses the most appropriate means. In vengeance one can also get caught up in the means or obsessed and distracted by the end, but the logic of "reasons" and appropriateness is nevertheless present as a standard. Accordingly, the question is not whether vengeance is ever legitimate but rather *when* it is legitimate, when those standards and reasons are in fact appropriate and warranted.

2. *There is no "natural" end to it.* But, of course, there is. The idea that vengeance leads to a total loss of inhibition and control ignores the built-in and cultivated satisfactions of revenge, and seems to confuse the fact that mutually vengeful acts tend to escalate with the fact that a single act of vengeance typically has its very specific goals and, consequently, its own built-in standard of satisfaction.

Vengeance is *rational*. I think that we are misled here by the conflation of vengeance, which is always particular and aimed at some particular offense (or series of offenses) and the familiar *feud*, which is an on-going form of personal, family or tribal hostility whose origins may well be forgotten. In other words, we conflate hatred (which may or may not be part of vengeance) with vengeance itself. We might note here too that one reason for the escalation of violence in vengeance, even when it seems that both sides have good reason to be vengeful, is that they may have, as Elizabeth Wolgast suggests, "different arithmetics." That is, what seems to count as "getting even" to one side is evidently an overpayment according to the other, or else the frameworks in which the "balance" is calculated are themselves very different. One of Wolgast's examples is the debt of Agamemnon to Apollo, which he "pays off" by sacrificing his daughter Iphegenia. But Iphegenia's mother, Clytemnestra, does not recognize the legitimacy of that debt; she sees only the murder of her daughter. So she murders Agamemnon, which her other children Orestes and Electra then avenge by murdering her. I do not deny that such horrible sequences are possible, but there is a reason why this particular sequence should have come down to us as a *tragedy*. It is not the ordinary course or complications of revenge, and while it is easy enough to identify conflicts of "arithmetics" in our own day (e.g. in the Middle East) I believe that it is an enormous mistake to see these as the paradigm or conceptual prototype of vengeance as such.

3. *Vengeance is always violent.* The blood-thirsty acts of the Clint Eastwood character or the Ninja assassin may hold dramatic sway over our fantasies, but the more usual act of revenge is a negative vote in the next department meeting, a forty-five minute delay in arriving to dinner or a hurtful comment or letter, delivered with a vicious twist of phrase, perhaps, but rarely the twist of a blade, except, of course, metaphorically. One might, given the current tendency to inflate the meaning of words and numb our sensitivities to moral differences, argue that such acts do indeed constitute "violence", but this certainly drains the substance from this standard objection against vengeance. Of course, there is an important if problematic distinction that is at stake here, between actually doing harm to an offender and depriving him or her of some essential human good, like liberty, for example. It is probably true that vengeance generally aims at harm whereas the cooler claims of the law tend to prefer punishments by way of deprivation—the payment of fines, for example, or depriving the criminal of certain privileges or freedoms (to drive a car, for example, or to use the university library), even depriving him of freedom as such—though prisons today as probably always seem to be just as much infliction of harm as deprivation. (The continuing rucus over the death penalty is, in part, due to the fact that it is one of the few punishments today that inflicts harm—indeed the ultimate harm—rather than imposing some deprivation.) But, then, it is obvious that this important distinction will not stand too much scrutiny, and it would be hard to insist for very long that sending a man to prison is not in fact actually *harming* him. And by the same token, there is much that counts as punishment and consequently satisfies the desire for revenge which need not be violence.

4. *It takes the law "into our own hands."* (The use of "in hand" metaphors seem to abound in such discussions.) It is worth noting that historically, punishing the perpetrator for almost any offense against an individual, from an obscene gesture to rape and murder, rested with the family, and it was considered not only inappropriate but unjustifiable intrusion into private matters for the state to step in. It is a

relatively recent development—and an obviously very political move—that punishment of such crimes should have become the *exclusive* province of the state. Moral objctions against vengeance and the desire for public order seem to me to have far less to do with this than the usual arrogance of the state in abrogating individual rights and its desire for control. Indeed, it is a point worth pondering that major crimes against the person are in law crimes against the state. When current criminal law reduces the victim of such crimes to a mere bystander (if, that is, the victim has survived the crime), the problem is not that in vengeance we take the law "into our own hands" but rather that without vengeance justice seems not only to be taken out of our hands but eliminated as a consideration altogether. Current concerns with punishment, even those that claim to take "retribution" seriously, seem to serve the law and sanction respect for the law (or reason) rather than the need for justice. Not that the law and respect for the law are unimportant, of course, but one should not glibly identify these with justice and dismiss the passion for vengeance as something quite different and wholly illegitimate.

"Retribution" is the pivotal term in all arguments about punishment. On the one hand, "backward-looking" retribution is juxtaposed against the "forward-looking" utilitarian concerns of deterrence and rehabilitation. That is, retribution is characterized in terms of its "undoing" a past offense; deterrence and criminal reform are concerned with preventing future offenses. Much of this dispute, I would argue, is purely academic; it is almost impossible to imagine an instance in which both responsibility for the offense and concern for the future not both are at stake. (It is to get around this practical point that Kant shocks us by insisting: "even if a civil society were to dissolve itself by common agreement of all its members, the last murderer remaining in prison must first be executed.") What gets left out of the argument, except marginally, is the all-important question of *character*—which obviously refers both to one's past behavior and one's disposition to future behavior. Punishment theorists, too, like so many theorists of justice, are too caught up in the abstractions—"punishment," "deterrence"—and too dismissive of the personal motives and character of both the criminal and those who do the punishing. Judges, of course, virtually always use "discretion"—which means, essentially, that they weight many factors, past and future, in determining punishments. Retribution, so considered, is not just "getting even" and "blind" to the future but roundly concerned with the whole question of the personalities involved in guilt and punishment.

How did our passion for retribution—our need for vengeance, come about? I think that our evolutionary speculations in Chapter 3 go a long way in answering this question. In that chapter, I was primarily concerned to account for our "natural" sympathies and our sense of fellowship with others, as opposed to the antagonistic, competitive view of the "state of nature" described by writers like Hobbes. But I hope that I was sufficiently careful not to give the impression that we are naturally "nice" in any ridiculous sense, and without retreating to the too-prevalent presumptions of self-interest I nevertheless wanted to show that there is some demonstrable advantage for groups and species—if not always for us as individuals—in the evolution of cooperation. But cooperation has two sides, the willingness to cooperate, first of all, but then the resentment and punishment of those who do not cooperate as well. (This includes the expectation that one will be punished oneself if one does not cooperate.) One cannot imagine the evolution of the cooperation without the evolution of punishment, and Robert Axelrod's now-classic "tit-for-tat" model of the former explains as well the latter. In a repetitive "prisoner's dilemma" type situation, or in any on-going situation in which one person frequently has the ability to "cheat" the other(s), an optimum strategy for discouraging such cheating is to respond, dependably, with retribution. A creature endowed only with compassion, who would "understand" the motives of the criminal in every case, would be just as much of an evolutionary failure as a creature who did nothing but watch out for his or her own advantage and cheated every time. Swift and dependable retaliation is thus in the nature of social animals as well as the lesson of game theory. Vengeance is not the antagonist to rationality but its natural manifestation. To breed a social animal with "the right to make promises," according to Nietzsche, "nature's paradoxical task," is to understand the evolution of a creature who has the natural urge to punish as well as natural sympathy and a sense of social solidarity.

Vengeance is our natural sense of retribution. Needless to say, few philosophers have acknowledged this kinship. Kant, notably, defends retribution as a rational necessity but insists that it is wholly rational and not at all emotional. Virtually all "retributivists," in fact, insist on separating retribution from vengeance. On the one hand, they insist that the notion of retribution lies right at the heart of punishment—indeed, it has often been argued that it is just another word for "punishment." On the other hand, retribution is therefore defended as a product of reason, even a matter of metaphysics—the undoing of some "corrupt" or "unnatural" state caused by crime. But even retribution has often been called "barbaric," and it remains, as R. S. Gerstein recently writes, "the most unfashionable theory in philosophical and other circles" and "frequently dismissed with contempt." So why do we punish? Why not forget about the past misdeeds—which after all cannot be undone—and move on to the future? The usual argument is this: if punishment were essentially a way of preventing future crime rather than punishment for a particular past crime, then wouldn't it be justifiable to punish a wholly innocent man just as a warning to potential law-breakers? and the obvious answer to this (usually) rhetorical question is, "no! That would be the height of injustice." And so even if retribution is "barbaric," "unfashionable" and "dismissed with contempt," it has long held the dominant position against its rival theories (which emphasize deterrence and rehabilitation respectively). Too often it is defended in a purely trivial way, as the very meaning of the word "punishment," and too often it is defended with great obscurity in metaphysical terms that are no longer even intelligible to most people. But, without assuming that what is natural is therefore desirable, I think we can argue and argue persuasively that our natural sense of vengeance—even if it requires careful correction and containment under the auspices of reason and tradition and through the machinery of the law—already gives us good grounds for punishment, even if, within the thrall of the emotion itself, we cannot articulate those grounds.

The argument that vengeance and retribution are "backward-looking" does not, I think, hold up to scrutiny. The "tit-for-tat" strategy may be a response to a past offense but it is a *strategy* just insofar as it is a way of planning a future. Thus it is not that retributivists need deny the importance or desirability of deterring crime or changing the character or at least the behavior of criminals. It is just that such activities, no matter how well-intended or conceived, do not by themselves count as punishment. But, again, why punish? I do not think that we can answer this question without reference to that supposedly primitive passion for vengeance and its implicit strategy. The idea of "evening the score" is also a way of "teaching the offender a lesson." This phrase does, of course, capture something of deterrence and rehabilitation theories as well as the idea of "getting back" at the offender, but the key point is that this is not primarily motivated by either deterrence or rehabilitation or the rationality of retribution. As Nietzsche tells us in his *Genealogy*, "the urge to punish" comes first; the reasons and attempts at justification come later.

The rationale of retribution, the "intelligence" embodied in vengeance, is the idea that "the punishment should fit the crime." This is an idea that goes back (at least) to the Biblical "eye for an eye" injunction, but it has also been long under fire as well. Since ancient times, there has been Socrates's oft-repeated objection—that punishment is the return of evil for evil and so never legitimate, no matter how horrendous the crime. So conceived, of course, punishment is just another wrong, whether carried out as a personal act of revenge or under the cool, deliberate auspices of the state legal system. (The argument that "murder is wrong, no matter who does it" has often been used as a central argument against capital punishment.) But punishment is not the return of "evil for evil." The German philosopher Hegel argued, I think convincingly, that this is one sure way to misunderstand the nature of punishment. "Harm for harm" perhaps, but it is justified and legitimate harm in return for unjustified and illegitimate harm. The hard question is, how does one get one harm to "fit" another? What does "an eye for an eye" really mean?

It seems to be false in practice that the punishment is designed to fit the crime: Punishment also has to fit the criminal and the circumstances and serve the public good. It is, accordingly, not true that the crime and nothing but the crime justifies infliction of punishment, but what this would seem to mean is that we should enlarge our sense of "fit" to include not just the offense narrowly construed but the whole context in which the crime was committed, including, first and foremost, the character of the criminal. The likelihood of a repetition of the crime and questions of deterrence are consequences of this concern rather than its motivation. But even so expanded, the idea of balance or "fit" comes under fire. How does one calculate how many years in prison it will take to "equal" the crime of armed robbery, however we take into account the history and character of the criminal and the circumstances of the crime? In what conceivable sense does time in prison "equal" the harm inflicted by the offender on his victims or on the community? My favorite objection to the "eye for an eye" formulation was levelled against retributivism by Lord Blackstone back in the eighteenth century. He queried, "what is the equivalent harm when a two-eye'd man knocks out the eye of a one-eye'd man?" Cute, to be sure, but hardly a knock-down argument against the idea that the punishment should fit the crime. Granting the gruesomeness of the concept, we do feel that familiar if embarrassing satisfaction when a vicious criminal who has intentionally blinded his innocent victim is blinded himself in an accident. To be sure, few of us would suggest that the law itself undertake to inflict such a punishment, but the notion of "fit" is evident enough to us, if only as "poetic" justice.

Kant puts the case for retribution succinctly: "only the law of retribution (*jus talionis*) can determine exactly the kind and degree of punishment," though (he carefully adds) such a determination must be made in a court of justice and not as a private judgment. It is this sense of what Kant calls "equality" that makes punishment rational, the idea of "fitting" the crime. If a man has committed a murder, Kant argues, he must die; "there is no substitute that will satisfy the requirements of legal justice. There is no sameness of kind between death and remaining alive even under the most miserable conditions." We may not agree. But we must agree that a serious crime such as murder demands a serious punishment, while a minor crime does not. Hanging a poor man for stealing a loaf of bread now strikes us as barbaric, not because we reject the idea of retribution but because we accept it (and not just because we feel compassion for the man's plight). Hanging a man for stealing food is barbaric because it doesn't "fit." It is monstrously excessive. And that means that we do indeed use the standard of "equality" in making such judgments, and we would not understand the meaning of "punishment" without it.

Much of the language with which we describe both our passion for vengeance and the concept of retribution consists largely of an interrelated group of metaphors—notably the images of "equality," "balance" and "fit." These are too often thrown together as one, and too often treated as if they were not metaphors at all but literal truths. I want to wrap up this section by distinguishing and discussing briefly four of these:

1. The "*debt*" metaphor: to punish is to "repay" a wrong. There is some dispute (for example, in Nietzsche, *Genealogy*, Essay II) whether the notion of legal obligation preceded or rather grew out of this idea of a debt, but with regard to punishment, the metaphorical character of "repayment" is quite clear. The suggestion that there is an implicit contract (whether via Hobbes and Locke or Plato's *Crito*) is just to repeat the question, how the metaphor of repayment can be rationally justified. In contrast with the other metaphors, punishment does not "balance the books" nor does it "erase" the wrong in question, as repayment of a debt surely does. The "debt" metaphor, by the way, is not restricted to capitalist societies; "debt" is not the same as "consumer debt" and applies to the New Guinea custom of giving a pig for a wrongful death as well as financial liability and "punitive damages." Debt, in most societies, is a moral measure rather than a monetary arrangement.

2. The "*fit*" metaphor, again, is the popular idea that the punishment should "fit" the crime. (W. S. Gilbert's *Mikado*: "an object all sublime/make the punishment fit the crime.") "Eye for eye, tooth for tooth," we are told since biblical times. But as many opponents and critics of retributivism have pointed out, such punishments are administered—or even make sense—only in a very limited number of crimes, e.g. intentional murder. But even then, as Albert Camus has famously pointed out,

> "For there to be equivalence, the death penalty would have to punish a criminal who had warned his victim of the date at which he would inflict a horrible death on him and who, from that moment onward, had confined him at his mercy for months. Such a monster is not encountered in private life."

Defenders of retributivism gladly weaken the demands of "fit," suggesting, for instance, that it provides only a general measure, that the crucial concept here is one of "proportion"—so that petty theft is not (as it once was) punished with the harshness of a violent crime. With this, of course, we all agree, but "proportion" too misleadingly suggests quantification where there often is none, and it too summarizes rather than solves the problem of punishment. To be sure, it is somehow "fit" to trade a life for a life (or more accurately, "a death for a death"), but is it just? And there are so many qualifications and extenuating circumstances concerning intentions, risk and appropriate caution that qualify each and every case, does the singular image of "fit" make sense? But again, we recognize the *lack* of fit: witness our horror, e.g. when Afghan tribesmen summarily execute [by beheading] the hapless tourist involved in an automobile accident, even one which [were the driver to have made an insurance claim] might have been proclaimed "faultless".

3. The *"balance"* metaphor: punishment makes things "even" again. It is through punishment that one "gets even." In epic literature, it is by punishing the villain that one "balances the forces of good and evil." One problem is that this moral balance is often simply equated, in the crudest utilitarian fashion, with a balance of pleasure and pain, as if the application of an amount of pain to the villain which is equal to the amount of pain he or she has caused (which is not the same as the amount of pleasure gotten from his crime) balances the scales. (We should remember again the standard allegorical figure of justice, this time with her scales.) Where the crime is strictly pecuniary, it might seem that balance (like the repayment of a debt) might be literally appropriate, but this, of course, isn't so. One can pay back the amount of money stolen or otherwise taken, but this does not yet take into account the offense itself. As soon as one must pay back even an extra penny, the literalness of the "balance" again comes into question. Granted the original sum has been repaid, but now what is the "cost" or the "price" of the crime? Again, my point is not that the metaphor of "balance" isn't applicable or revealing of how we think about justice, but it underscores rather than solves our problems.

4. The *"erasure"* metaphor: the idea that we can "annul the evil" through punishment. Vendetta cultures talk about "bloodmarks (also "blood debts") not as a sign of guilt but rather of unrevenged wrong. But the obvious question is, can we undo a crime—for instance, rape or murder, in any sense whatever? In financial crimes, again, one can "erase" the debt by paying it back, but not the crime itself. For example, how does one erase the terror one suffered in an armed robbery, even if the money were to be politely returned. ("Here you are, Miss; I'm a student at the local police academy and I wanted to experience what the criminal felt like.") Indeed, how does one measure the fear that one suffers in such crimes, even if there is no "harm done" in any of the usual senses? And yet, the idea of "annulment" looms large in the history of theories of punishment, from the ancient world (in which one quite literally made the offender disappear) to modern conceptions in Hegel and Bosanquet and others. The punishment doesn't literally eliminate the crime of course, but Bosanquet's suggestion that it eliminates it *as a precedent* has the virtue of showing us how this metaphor too, hardly intelligible when taken literally, can nonetheless be given a good intelligible interpretation that is not strictly retributivist but takes future behavior into account as well.

Retributive justice consists, first of all, of retribution. That isn't much of a claim, but it is remarkable how often it is denied by our best current legal theorists. But retributivism is wrongly presented as a theory or a set of principles when it ought to be considered, first of all, as the expression of revenge. Or, rather, retribution and revenge are one and the same. (One might, adopting a distinction from Robert Nozick, say that retribution is justified revenge, where revenge is strictly personal but retribution is not.) Of course, retribution can be turned into a theory (no one would accuse Kant of mere metaphor mongering) but retributivism, I want to argue, is primarily a set of concepts and judgments embodied in a feeling, not a theory. Perhaps it was overstated in the majority opinion in United States Supreme Court decision *Gregg* vs *Georgia* (1976):

> "the instinct for retribution is part of the nature of man, and channeling that instinct in the administration of criminal justice serves an important purpose in promoting the stability of a society governed by law. When people begin to believe that organized society is unwilling or unable to impose upon criminal offenders the punishment they 'deserve,' then there are sown the seeds of anarchy—of self-help, vigilante justice, and lynch law."

But at least the emotion of vengeance was taken seriously and not merely sacrificed to the dispassionate authority of the law. Retributive justice, however rationalized, is not as such a purely "rational matter"—but neither is it thereby "irrational" either. Most of the arguments that have been advanced against vengeance could, with only slight modifications, be applied to the standard notions of retributive justice as well—which is not surprising if vengeance and retributive justice are in the end identical. But in the end, it is perhaps not just a question of whether revenge is rational or not, but whether it is—at the bottom of our hearts as well as off the top of our heads—an undeniable aspect of the way we react to the world, not as an instinct but as such a basic part of our world-view and our moral sense of ourselves that it is, in that sense, unavoidable.

I have not tried here to defend vengeance as such, but my claim is that vengeance deserves its central place in any theory of justice and, whatever else we are to say about punishment, the desire for revenge must enter into our deliberations along with such emotions as compassion, caring and love. Any system of legal principles that does not take such emotions into account, which does not motivate itself on their behalf, is not—whatever else it may be—a system of justice. But vengeance as such, I do not deny, is in any case dangerous. As the Chinese used to say (and no doubt still do) "if you seek vengeance, dig two graves." But I think that the dangers and destructiveness of vengeance are much overblown and its importance for a sense of one's own self-esteem and integrity ignored. Many people believe that vengeance is the primary cause of the world's troubles today, unending feuds and vendettas that block every rational effort at resolution and peace. But in addition to my insistence that vengeance is not the same as vendetta and feuds are not the same as vengeance, I would argue that the passionate hostilities of the world that are fueled by revenge are only secondary and in many cases caused or at any rate aggravated and rendered unresolvable not by passion at all so much as by supposedly rational *ideology*, abstracting and elevating personal prejudices to the status of absolute truths and giving vengeance a set of reasons far less negotiable than any feud or mere urge to "get even." Vengeance, at least, has its measure. Ideology, however "reasonable," may not. Goldwater: "Extremism in the name of liberty is no vice."

Vengefulness, no doubt, is a vice, and because vengeance is so often dangerous and destructive it makes perfectly good sense for moralists to urge us to rise "above" the urge to vengeance. But the argument behind this need not involve forgiveness and mercy nor need it appeal to Nietzschean "master" morality and "self-overcoming." Neither need it be anything so detached and saintly as "turning the other cheek." The argument is that instead of following that often narrow path to personal retribution we would much better embrace that expansive form of social consciousness in which we become more aware of the real desperation of others than we are of our own (usually petty) complaints. To transcend revenge is to become keenly aware of the suffering of others with an urgency that eclipses the blows to our own fragile egos and gives our sense of compassion priority over the urge to vengeance. But it is not to give up vengeance as such and, as in tit-for-tat, it is the ready willingness to retaliate that provides stability to both the social system and one's personal sense of integrity and control. Despite volumes of propaganda to the contrary, experience seems to show that to see oneself as a helpless victim makes one less and not more likely to open one's heart to others. But we do not have to be or see ourselves as victims, and it is vengeance or at least fantasies of vengeance that make this possible. Our concept of injustice is inextricably tied up with the concept of blame and with the concept of punishment, and where the injustice is personal so is the felt need for retribution. In a world in which justice is getting ever more impersonal and statistical, vengeance retains the virtue of being personal. But so, of course, does compassion, which commands not the impersonal but a more expanded sense of the personal. Between vengeance and compassion there is no doubt which is the greater virtue, but vengeance is nevertheless necessary, and compassion for one's own offender (the object of one's revenge) is often foolish rather than noble. Justice is not forgiveness nor even forgetting but rather it is getting one's emotional priorities right, putting blame aside in the face of so much other human suffering and thereby giving up vengeance for the sake of larger and more noble emotions.

PART V

Today, the central debate about justice emerges from and, with remarkable fidelity, sticks to the text and strategy of John Rawls's monumental *Theory of Justice*. Virtually nothing written today about justice manages to escape a direct confrontation with Rawls, whether the author intends to agree with or merely amend Rawls's work, tries to deny its basic principles or strategy or both, insists that Rawls's vision of justice is mistaken at its very foundation, or wants to counter Rawls's influence with an entirely different approach or conception. Robert Nozick would certainly seem to fall into the last category, for both his approach to social philosophy and his conception of justice are radically at odds with those of his colleague in Emerson Hall. But from a longer perspective it turns out that Rawls and Nozick may be more similar than different, and what they jointly presuppose and exclude from their theories may be more telling than their now famous disagreements about the justifiability of redistribution. This is the claim made by Alasdair MacIntyre in his already classic study, *After Virtue*, and it finds an able voice again in the neo-Aristotelian writings of Charles Taylor. Thomas Nagel, once a student of Rawls at Harvard, takes a much more sympathetic view of *A Theory of Justice*, raising doubts about its methodology but doubtless in agreement with its essential liberal sympathies. Michael Sandel is, by contrast, not so sympathetic at all, and here he argues his "communitarian" alternative to the liberal individualism that underlies both Rawls and Nozick. (MacIntyre and Taylor also defend the importance of community and tradition to any adequate conception of justice, but neither of them defends the generic conception of community as such that we find in Sandel.) Michael Walzer also provides an alternative to Rawls's all-embracing theory, but for him the defining feature of justice turns out to be not so much community as *context*. Different "spheres of justice" have different ground rules and different concerns, and there is no reason to suppose that there is any overarching conception of justice that embraces them all. Walzer, obviously, is a radical *pluralist*, and he defends a notion of "complex equality" that is not so much an argument for universal equality as it is a refusal to try to compare concepts of justice (and injustice) in one sphere with those of another. Finally, Elizabeth Wolgast raises the question whether the question, "what is justice?" even makes sense. She explores what she calls (following Wittgenstein) the "grammar" of justice, and argues that the primary concept is not "justice" but "injustice," and the former cannot be properly derived from the latter. (An annotated bibliography follows the text.)

John Rawls, from "Justice as Fairness" and *A Theory of Justice* (1921)

In the two selections that follow, Rawls defends both his general approach to justice as a rational decision procedure and the two principles that define his theory.

"Justice as Fairness"

1. It might seem at first sight that the concepts of justice and fairness are the same, and that there is no reason to distinguish them, or to say that one is more fundamental than the other. I think that this impression is mistaken. In this paper I wish to show that the fundamental idea in the concept of justice is fairness; and I wish to offer an analysis of the concept of justice from this point of view. To bring out the force of this claim, and the analysis based upon it, I shall then argue that it is this aspect of justice for which utilitarianism, in its classical form, is unable to account, but which is expressed, even if misleadingly, by the idea of the social contract.

To start with I shall develop a particular conception of justice by stating and commenting upon two principles which specify it, and by considering the circumstances and conditions under which they may be thought to arise. The principles defining this conception, and the conception itself, are, of course, familiar. It may be possible, however, by using the notion of fairness as a framework, to assemble and to look at them in a new way. Before stating this conception, however, the following preliminary matters should be kept in mind.

Throughout I consider justice only as a virtue of social institutions, or what I shall call practices. The principles of justice are regarded as formulating restrictions as to how practices may define positions and offices, and assign thereto powers and liabilities, rights and duties. Justice as a virtue of particular actions or of persons I do not take up at all. It is important to distinguish these various subjects of justice, since the meaning of the concept varies according to whether it is applied to practices, particular actions, or persons. These meanings are, indeed, connected, but they are not identical. I shall confine my discussion to the sense of justice as applied to practices, since this sense is the basic one. Once it is understood, the other senses should go quite easily.

Justice is to be understood in its customary sense as representing but *one* of the many virtues of social institutions, for these may be antiquated, inefficient, degrading, or any number of other things, without being unjust. Justice is not to be confused with an all-inclusive vision of a good society; it is only one part of any such conception. It is important, for example, to distinguish that sense of equality which is an aspect of the concept of justice from that sense of equality which belongs to a more comprehensive social ideal. There may well be inequalities which one concedes are just, or at least not unjust, but which, nevertheless, one wishes on other grounds, to do away with. I shall focus attention, then, on the usual sense of justice in which it is essentially the elimination of arbitrary distinctions and the establishment, within the structure of a practice, of a proper balance between competing claims.

Finally, there is no need to consider the principles discussed below as *the* principles of justice. For the moment it is sufficient that they are typical of a family of principles normally associated with the concept of justice. The way in which the principles of this family resemble one another, as shown by the background against which they may be thought to arise, will be made clear by the whole of the subsequent argument.

2. The conception of justice which I want to develop may be stated in the form of two principles as follows: first, each person participating in a practice, or affected by it, has an equal right to the most extensive liberty compatible with a like liberty for all; and second, inequalities are arbitrary unless it is reasonable to expect that they will work out for everyone's advantage, and provided the positions and offices to which they attach, or from which they may be gained, are open to all. These principles express justice as a complex of three ideas: liberty, equality, and reward for services contributing to the common good.

* * *

The first principle holds, of course, only if other things are equal: that is, while there must always be a justification for departing from the initial position of equal liberty (which is defined by the pattern of rights and duties, powers and liabilities, established by a practice), and the burden of proof is placed on him who would depart from it, nevertheless, there can be, and often there is, a justification for doing so. Now, that similar particular cases, as defined by a practice, should be treated similarly as they arise, is part of the very concept of a practice; it is involved in the notion of an activity in accordance with rules. The first principle expresses an analogous conception, but as applied to the structure of practices themselves. It holds, for example, that there is a presumption against the distinctions and classifications made by legal systems and other practices to the extent that they infringe on the original and equal liberty of the persons participating in them. The second principle defines how this presumption may be rebutted.

It might be argued at this point that justice requires only an equal liberty. If, however, a greater liberty were possible for all without loss or conflict, then it would be irrational to settle on a lesser liberty. There is no reason for circumscribing rights unless their exercise would be incompatible, or would render the practice defining them less effective. Therefore no serious distortion of the concept of justice is likely to follow from including within it the concept of the greatest equal liberty.

The second principle defines what sorts of inequalities are permissible; it specifies how the presumption laid down by the first principle may be put aside. Now by inequalities it is best to understand not *any* differences between offices and positions, but differences in the benefits and burdens attached to them either directly or indirectly, such as prestige and wealth, or liability to taxation and compulsory services. Players in a game do not protest against there being different positions, such as batter, pitcher, catcher, and the like, nor to there being various privileges and powers as specified by the rules; nor do the citizens of a country object to there being the different offices of government such as president, senator, governor, judge, and so on, each with their special rights and duties. It is not differences of this kind that are normally thought of as inequalities, but differences in the resulting distribution established by a practice, or made possible by it, of the things men strive to attain or avoid. Thus they may complain about the pattern of honors and rewards set up by a practice (e.g., the privileges and salaries of government officials) or they may object to the distribution of power and wealth which results from the various ways in which men avail themselves of the opportunities allowed by it (e.g., the concentration of wealth which may develop in a free price system allowing large entrepreneurial or speculative gains).

It should be noted that the second principle holds that an inequality is allowed only if there is reason to believe that the practice with the inequality, or resulting in it, will work for the advantage of *every* party engaging in it. Here it is important to stress that *every* party must gain from the inequality. Since the principle applies to practices, it implies that the representative man in every office or position defined by a practice, when he views it as a going concern, must find it reasonable to prefer his condition and prospects with the inequality to what they would be under the practice without it. The principle excludes, therefore, the justification of inequalities on the grounds that the disadvantages of those in one position are outweighed by the greater advantages of those in another position. This rather simple restriction is the main modification I wish to make in the utilitarian principle as usually understood.

* * *

3. Given these principles one might try to derive them from a priori principles of reason, or claim that they were known by intuition. These are familiar enough steps and, at least in the case of the first principle, might be made with some success. Usually, however, such arguments, made at this point, are unconvincing. They are not likely to lead to an understanding of the basis of the principles of justice, not at least as principles of justice. I wish, therefore, to look at the principles in a different way.

Imagine a society of persons amongst whom a certain system of practices is *already* well established. Now suppose that by and large they are mutually self-interested; their allegiance to their established practices is normally founded on the prospect of self-advantage. One need not assume that, in all senses of the term "person," the persons in this society are mutually self-interested. If the characterization as mutually self-interested applies when the line of division is the family, it may still be true that members of families are bound by ties of sentiment and affection and willingly acknowledge duties in contradiction to self-interest. Mutual self-interestedness in the relations between families, nations, churches, and the like, is commonly associated with intense loyalty and devotion on the part of individual members. Therefore, one can form a more realistic conception of this society if one thinks of it as consisting of mutually self-interested families, or some other association. Further, it is not necessary to suppose that these persons are mutually self-interested under all circumstances, but only in the usual situations in which they participate in their common practices.

Now suppose also that these persons are rational: they know their own interests more or less accurately; they are capable of tracing out the likely consequences of adopting one practice rather than another; they are capable of adhering to a course of action once they have decided upon it; they can resist present temptations and the enticements of immediate gain; and the bare knowledge or perception of the difference between their condition and that of others is not, within certain limits and in itself, a source of great dissatisfaction. Only the last point adds anything to the usual definition of rationality. This definition should allow, I think, for the idea that a rational man would not be greatly downcast from knowing, or seeing, that others are in a better position than himself, unless he thought their being so was the result of injustice, or the consequence of letting chance work itself out for no useful common purpose, and so on. So if these persons strike us as unpleasantly egoistic, they are at least free in some degree from the fault of envy.

Finally, assume that these persons have roughly similar needs and interests, or needs and interests in various ways complementary, so that fruitful cooperation amongst them is possible; and suppose that they are sufficiently equal in power and ability to guarantee that in normal circumstances none is able to dominate the others. This condition (as well as the others) may seem excessively vague; but in view of the conception of justice to which the argument leads, there seems no reason for making it more exact here.

Since these persons are conceived as engaging in their common practices, which are already established, there is no question of our supposing them to come together to deliberate as to how they will set these practices up for the first time. Yet we can imagine that from time to time they discuss with one another whether any of them has a legitimate complaint against their established institutions. Such discussions are perfectly natural in any normal society. Now suppose that they have settled on doing this in the following way. They first try to arrive at the principles by which complaints, and so practices themselves, are to be judged. Their procedure for this is to let each person propose the principles upon which he wishes his complaints to be tried with the understanding that, if acknowledged, the complaints of others will be similarly tried, and that no complaints will be heard at all until everyone is roughly of one mind as to how complaints are to be judged. They each understand further that the principles proposed and acknowledged on this occasion are binding on future occasions. Thus each will be wary of proposing a principle which would give him a peculiar advantage, in his present circumstances, supposing it to be accepted. Each person knows that he will be bound by it in future circumstances the peculiarities of which cannot be known, and which might well be such that the principle is then to his disadvantage. The idea is that everyone should be required to make *in advance* a firm commitment, which others also may reasonably be expected to make, and that no one be given the opportunity to tailor the canons of a legitimate complaint to fit his own special condition, and then to discard them when they no longer suit his purpose. Hence each person will propose principles of a general kind which will, to a large degree, gain their sense from the various applications to be made of them, the particular circumstances of which being as yet unknown. These principles will express the conditions in accordance with which each is the least unwilling to have his interests limited in the design of practices, given the competing interests of the others, on the supposition that the interests of others will be limited likewise. The restrictions which would so arise might be thought of as those a person would keep in mind if he were designing a practice in which his enemy were to assign him his place.

The two main parts of this conjectural account have a definite significance. The character and respective situations of the parties reflect the typical circumstances in which questions of justice arise. The procedure whereby principles are proposed and acknowledged represents constraints, analogous to those of having a morality, whereby rational and mutually self-interested persons are brought to act reasonably. Thus the first part reflects the fact that questions of justice arise when conflicting claims are made upon the design of a practice and where it is taken for granted that each person will insist, as far as possible, on what he considers his rights. It is typical of cases of justice to involve persons who are pressing on one another their claims, between which a fair balance or equilibrium must be found. On the other hand, as expressed by the second part, having a morality must at least imply the acknowledgment of principles as impartially applying to one's own conduct as well as to another's, and moreover principles which may constitute a constraint, or limitation, upon the pursuit of one's own interests. There are, of course, other aspects of having a morality: the acknowledgment of moral principles must show itself in accepting a reference to them as reasons for limiting one's claims, in acknowledging the burden of providing a special explanation, or excuse, when one acts contrary to them, or else in showing shame and remorse and a desire to make amends, and so on. It is sufficient to remark here that having a morality is analogous to having made a firm commitment in advance; for one must acknowledge the principles of morality even when to one's disadvantage. A man whose moral judgments always coincided with his interests could be suspected of having no morality at all.

Thus the two parts of the foregoing account are intended to mirror the kinds of circumstances in which questions of justice arise and the constraints which having a morality would impose upon persons so situated. In this way one can see how the acceptance of the principles of justice might come about, for given all these conditions as described, it would be natural if the two principles of justice were to be acknowledged. Since there is no way for anyone to win special advantage for himself, each might consider it reasonable to acknowledge equality as an initial principle. There is, however, no reason why they should regard this position as final; for if there are inequalities which satisfy the second principle, the immediate gain which equality would allow can be considered as intelligently invested in view of its future return. If, as is quite likely, these inequalities work as incentives to draw out better efforts, the members of this society may look upon them as concessions to human nature: they, like us, may think that people ideally should want to serve one another. But as they are mutually self-interested, their acceptance of these inequalities is merely the acceptance of the relations in which they actually stand, and a recognition of the motives which lead them to engage in their common practices. *They* have no title to complain of one another. And so provided that the conditions of the principle are met, there is no reason why they should not allow such inequalities. Indeed, it would be short-sighted of them to do so, and could result, in most cases, only from their being dejected by the bare knowledge, or perception, that others are better situated. Each person will, however, insist on an advantage to himself, and so on a common advantage, for none is willing to sacrifice anything for the others.

* * *

A Theory of Justice

It will be recalled that the general conception of justice as fairness requires that all primary social goods be distributed equally unless an unequal distribution would be to everyone's advantage. No restrictions are placed on exchanges of these goods and therefore a lesser liberty can be compensated for by greater social and economic benefits. Now looking at the situation from the standpoint of one person selected arbitrarily, there is no way for him to win special advantages for himself. Nor, on the other hand, are there grounds for his acquiescing in special disadvantages. Since it is not reasonable for him to expect more than an equal share in the division of social goods, and since it is not rational for him to agree to less, the sensible thing for him to do is to acknowledge as the first principle of justice one requiring an equal distribution. Indeed, this principle is so obvious that we would expect it to occur to anyone immediately.

Thus, the parties start with a principle establishing equal liberty for all, including equality of opportunity, as well as an equal distribution of income and wealth. But there is no reason why this acknowledgment should be final. If there are inequalities in the basic structure that work to make everyone better off in comparison with the benchmark of initial equality, why not permit them? The immediate gain which a greater equality might allow can be regarded as intelligently invested in view of its future return. If, for example, these inequalities set up various incentives which succeed in eliciting more productive efforts, a person in the original position may look upon them as necessary to cover the costs of training and to encourage effective performance. One might think that ideally individuals should want to serve one another. But since the parties are assumed not to take an interest in one another's interests, their acceptance of these inequalities is only the acceptance of the relations in which men stand in the circumstances of justice. They have no grounds for complaining of one another's motives. A person in the original position would, therefore, concede the justice of these inequalities. Indeed, it would be shortsighted of him not to do so. He would hesitate to agree to these regularities only if he would be dejected by the bare knowledge or perception that others were better situated; and I have assumed that the parties decide as if they are not moved by envy. In order to make the principle regulating inequalities determinate, one looks at the system from the standpoint of the least advantaged representative man. Inequalities are permissible when they maximize, or at least all contribute to, the long-term expectations of the least fortunate group in society.

* * *

It seems clear from these remarks that the two principles are at least a plausible conception of justice. The question, though, is how one is to argue for them more systematically. Now there are several things to do. One can work out their consequences for institutions and note their implications for fundamental social policy. In this way they are tested by a comparison with our considered judgments of justice. Part II is devoted to this. But one can also try to find arguments in their favor that are decisive from the standpoint of the original position. In order to see how this might be done, it is useful as a heuristic device to think of the two principles as the maximin solution to the problem of social justice. There is an analogy between the two principles and the maximin rule for choice under uncertainty. This is evident from the fact that the two principles are those a person would choose for the design of a society in which his enemy is to assign him his place. The maximin rule tells us to rank alternatives by their worst possible outcomes: we are to adopt the alternative the worst outcome of which is superior to the worst outcomes of the others. The persons in the original position do not, of course, assume that their initial place in society is decided by a malevolent opponent. As I note below, they should not reason from false premises. The veil of ignorance does not violate this idea, since an absence of information is not misinformation. But that the two principles of justice would be chosen if the parties were forced to protect themselves against such a contingency explains the sense in which this conception is the maximin solution. And this analogy suggests that if the original position has been described so that it is rational for the parties to adopt the conservative attitude expressed by this rule, a conclusive argument can indeed be constructed for these principles. Clearly the maximin rule is not, in general, a suitable guide for choices under uncertainty. But it is attractive in situations marked by certain special features. My aim, then, is to show that a good case can be made for the two principles based on the fact that the original position manifests these features to the fullest possible degree, carrying them to the limit, so to speak.

* * *

Now there appear to be three chief features of situations that give plausibility to this unusual rule. First, since the rule takes no account of the likelihoods of the possible circumstances, there must be some reason for sharply discounting estimates of these probabilities. Offhand, the most natural rule of choice would seem to be to compute the expectation of monetary gain for each decision and then to adopt the course of action with the highest prospect.

* * *

Thus it must be, for example, that the situation is one in which a knowledge of likelihoods is impossible, or at best extremely insecure. In this case it is unreasonable not to be skeptical of probabilistic calculations unless there is no other way out, particularly if the decision is a fundamental one that needs to be justified to others.

The second feature that suggests the maximin rule is the following: the person choosing has a conception of the good such that he cares very little, if anything, for what he might gain above the minimum stipend that he can, in fact, be sure of by following the maximin rule. It is not worthwhile for him to take a chance for the sake of a further advantage, especially when it may turn out that he loses much that is important to him. This last provision brings in the third feature, namely, that the rejected alternatives have outcomes that one can hardly accept. The situation involves grave risks. Of course these features work most effectively in combination. The paradigm situation for following the maximin rule is when all three features are realized to the highest degree. This rule does not, then, generally apply, nor of course is it self-evident. Rather, it is a maxim, a rule of thumb, that comes into its own in special circumstances. Its application depends upon the qualitative structure of the possible gains and losses in relation to one's conception of the good, all this against a background in which it is reasonable to discount conjectural estimates of likelihoods.

* * *

Finally, the third feature holds if we can assume that other conceptions of justice may lead to institutions that the parties would find intolerable. For example, it has sometimes been held that under some conditions the utility principle (in either form) justifies, if not slavery or serfdom, at any rate serious infractions of liberty for the sake of greater social benefits. We need not consider here the truth of this claim, or the likelihood that the requisite conditions obtain. For the moment, this contention is only to illustrate the way in which conceptions of justice may allow for outcomes which the parties may not be able to accept. And having the ready alternative of the two principles of justice which secure a satisfactory minimum, it seems unwise, if not irrational, for them to take a chance that these outcomes are not realized.

Robert Nozick, from *Anarchy, State and Utopia* (1938–)

Robert Nozick defends an "entitlement theory" of justice, in conscientious contrast to Rawls's distributive "end-state" theory of justice.

———

The minimal state is the most extensive state that can be justified. Any state more extensive violates people's rights. Yet many persons have put forth reasons purporting to justify a more extensive state.

* * *

In this chapter we consider the claim that a more extensive state is justified, because necesary (or the best instrument) to achieve distributive justice.

* * *

The term "distributive justice" is not a neutral one. Hearing the term "distribution," most people presume that some thing or mechanism uses some principle or criterion to give out a supply of things. Into this process of distributing shares some error may have crept. So it is an open question, at least, whether *re*distribution should take place; whether we should do again what has already been done once, though poorly. However, we are not in the position of children who have been given portions of pie by someone who now makes last minute adjustments to rectify careless cutting. There is no *central* distribution, no person or group entitled to control all the resources, jointly deciding how they are to be doled out. What each person gets, he gets from others who give to him in exchange for something, or as a gift. In a free society, diverse persons control different resources, and new holdings arise out of the voluntary exchanges and actions of persons. There is no more a distributing or distribution of shares than there is a distributing of mates in a society in which persons choose whom they shall marry. The total result is the product of many individual decisions which the different individuals involved are entitled to make. Some uses of the term "distribution," it is true, do not imply a previous distributing appropriately judged by some criterion (for example, "probability distribution"); nevertheless, despite the title of this chapter, it would be best to use a terminology that clearly is neutral. We shall speak of people's holdings; a principle of justice in holdings describes (part of) what justice tells us (requires) about holdings.

Section I

The Entitlement Theory

The subject of justice in holdings consists of three major topics. The first is the *original acquisition of holdings*, the appropriation of unheld things. This includes the issues of how unheld things may come to be held, the process, or processes, by which unheld things may come to be held, the things that may come to be held by these processes, the extent of what comes to be held by a particular process, and so on. We shall refer to the complicated truth about this topic, which we shall not formulate here, as the principle of justice in acquisition. The second topic concerns the *transfer of holdings* from one person to another. By what processes may a person transfer holdings to another? How may a person acquire a holding from another who holds it? Under this topic come general descriptions of voluntary exchange, and gift and (on the other hand) fraud, as well as reference to particular conventional details fixed upon in a given society. The complicated truth about this subject (with placeholders for conventional details) we shall call the principle of justice in transfer. (And we shall suppose it also includes principles governing how a person may divest himself of a holding, passing it into an unheld state.)

If the world were wholly just, the following inductive definition would exhaustively cover the subject of justice in holdings.

> 1. A person who acquires a holding in accordance with the principle of justice in acquisition is entitled to that holding.
> 2. A person who acquires a holding in accordance with the principle of justice in transfer, from someone else entitled to the holding, is entitled to the holding.
> 3. No one is entitled to a holding except by (repeated) applications of 1 and 2.

The complete principle of distributive justice would say simply that a distribution is just if everyone is entitled to the holdings they possess under the distribution.

A distribution is just if it arises from another just distribution by legitimate means. The legitimate means of moving from one distribution to another are specified by the principle of justice in transfer. The legitimate first "moves" are specified by the principle of justice in acquisition. Whatever arises from a just situation by just steps is itself just. The means of change specified by the principle of justice in transfer preserve justice. As correct rules of inference are truth-preserving, and any conclusion deduced via repeated application of such rules from only true premises is itself true, so the means of transition from one situation to another specified by the principle of justice in transfer are justice-preserving, and any situation actually arising from repeated transitions in accordance with the principle from a just situation is itself just. The parallel between justice-preserving transformations and truth-preserving transformations illuminates where it fails as well as where it holds. That a conclusion could have been deduced by truth-preserving means from premises that are true suffices to show its truth. That from a just situation a situation *could* have arisen via justice-preserving means does *not* suffice to show its justice. The fact that a thief's victims voluntarily *could* have presented him with gifts does not entitle the thief to his ill-gotten gains. Justice in holdings is historical; it depends upon what actually has happened. We shall return to this point later.

Not all actual situations are generated in accordance with the two principles of justice in holdings: the principle of justice in acquisition and the principle of justice in transfer. Some people steal from others, or defraud them, or enslave them, seizing their product and preventing them from living as they choose, or forcibly exclude others from competing in exchanges. None of these are permissible modes of transition from one situation to another. And some persons acquire holdings by means not sanctioned by the principle of justice in aquisition. The existence of past injustice (previous violations of the first two principles of justice in holdings) raises the third major topic under justice in holdings: the rectification of injustice in holdings. If past injustice has shaped present holdings in various ways, some identifiable and some not, what now, if anything, ought to be done to rectify these injustices? What obligations do the performers of injustice have toward those whose position is worse than it would have been had the injustice not been done? Or, than it would have been had compensation been paid promptly? How, if at all, do things change if the beneficiaries and those made worse off are not the direct parties in the act of injustice, but, for example, their descendants? Is an injustice done to someone whose holding was itself based upon an unrectified injustice? How far back must one go in wiping clean the historical slate of injustices? What may victims of injustice permissibly do in order to rectify the injustices being done to them, including the many injustices done by persons acting through their government? I do not know of a thorough or theoretically sophisticated treatment of such issues. Idealizing greatly, let us suppose theoretical investigation will produce a principle of rectification. This principle uses historical information about previous situations and injustices done in them (as defined by the first two principles of justice and rights against interference), and information about the actual course of events that flowed from these injustices, until the present, and it yields a description (or descriptions) of holdings in the society. The principle of rectification presumably will make use of its best estimate of subjunctive information about what would have occurred (or a probability distribution over what might have occurred, using the expected value) if the injustice had not taken place. If the actual description of holdings turns out not to be one of the descriptions yielded by the principle, then one of the descriptions yielded must be realized.

The general outlines of the theory of justice in holdings are that the holdings of a person are just if he is entitled to them by the principles of justice in acquisition and transfer, or by the principle of rectification of injustice (as specified by the first two principles). If each person's holdings are just, then the total set (distribution) of holdings is just.

Historical Principles and End-Result Principles

The general outlines of the entitlement theory illuminate the nature and defects of other conceptions of distributive justice. The entitlement theory of justice in distribution is *historical*; whether a distribution is just depends upon how it came about. In contrast, *current time-slice principles* of justice hold that the justice of a distribution is determined by how things are distributed (who has what) as judged by some *structural* principle(s) of just distribution. A utilitarian who judges between any two distributions by seeing which has the greater sum of utility and, if the sums tie, applies some fixed equality criterion to choose the more equal distribution, would hold a current time-slice principle of justice. As would someone who had a fixed schedule of trade-offs between the sum of happiness and equality. According to a current time-slice principle, all that needs to be looked at, in judging the justice of a distribution, is who ends up with what; in comparing any two distributions one need look only at the matrix presenting the distributions. No further information need be fed into a principle of justice. It is a consequence of such principles of justice that any two structurally identical distributions are equally just. (Two distributions are structurally identical if they present the same profile, but perhaps have different persons occupying the particular slots. My having ten and your having five, and my having five and your having ten are structurally identical distributions.) Welfare economics is the theory of current time-slice principles of justice. The subject is conceived as operating on matrices representing only current information about distribution. This, as well as some of the usual conditions (for example, the choice of distribution is invariant under relabeling of columns), guarantees that welfare economics will be a current time-slice theory, with all of its inadequacies.

Most persons do not accept current time-slice principles as constituting the whole story about distributive shares. They think it relevant in assessing the justice of a situation to consider not only the distribution it embodies, but also how that distribution came about. If some persons are in prison for murder or war crimes, we do not say that to assess the justice of the distribution in the society we must look only at what this person has, and that person has, and that person has, . . . at the current time. We think it relevant to ask whether someone did something so that he *deserved* to be punished, deserved to have a lower share. Most will agree to the relevance of further information with regard to punishments and penalties. Consider also desired things. One traditional socialist view is that workers are entitled to the product and full fruits of their labor; they have earned it; a distribution is unjust if it does not give the workers what they are entitled to. Such entitlements are based upon some past history. No socialist holding this view would find it comforting to be told that because the actual distribution A happens to coincide structurally with the one he desires D, A therefore is no less just than D; it differs only in that the "parasitic" owners of capital receive under A what the workers are entitled to under D, and the workers receive under A what the owners are entitled to under D, namely very little. This socialist rightly, in my view, holds onto the notions of earning, producing, entitlement, desert, and so forth, and he rejects current time-slice principles that look only to the structure of the resulting set of holdings. (The set of holdings resulting from what? Isn't it implausible that how holdings are produced and come to exist has no effect at all on who should hold what?) His mistake lies in his view of what entitlements arise out of what sorts of productive processes.

We construe the position we discuss too narrowly by speaking of *current* time-slice principles. Nothing is changed if structural principles operate upon a time sequence of current time-slice profiles and, for example, give someone more now to counterbalance the less he has had earlier. A utilitarian or an egalitarian or any mixture of the two over time will inherit the difficulties of his more myopic comrades. He is not helped by the fact that *some* of the information others consider relevant in assessing a distribution is reflected, unrecoverably, in past matrices. Henceforth, we shall refer to such unhistorical principles of distributive justice, including the current time-slice principles, as *end-result principles* or *end-state principles.*

In contrast to end-result principles of justice, *historical principles* of justice hold that past circumstances or actions of people can create differential entitlements or differential deserts to things. An injustice can be worked by moving from one distribution to another structurally identical one, for the second, in profile the same, may violate people's entitlements or deserts; it may not fit the actual history.

Patterning

The entitlement principles of justice in holdings that we have sketched are historical principles of justice. To better understand their precise character, we shall distinguish them from another subclass of the historical principles. Consider, as an example, the principle of distribution according to moral merit. This principle requires that total distributive shares vary directly with moral merit; no person should have a greater share than anyone whose moral merit is greater. (If moral merit could be not merely ordered but measured on an interval or ratio scale, stronger principles could be formulated.) Or consider the principle that results by substituting "usefulness to society" for "moral merit" in the previous principle. Or instead of "distribute according to moral merit," or "distribute according to usefulness to society," we might consider "distribute according to the weighted sum of moral merit, usefulness to society, and need," with the weights of the different dimensions equal. Let us call a principle of distribution *patterned* if it specifies that a distribution is to vary along with some natural dimension, weighted sum of natural dimensions, or lexicographic ordering of natural dimensions. And let us say a distribution is patterned if it accords with some patterned principle.

Almost every suggested principle of distributive justice is patterned: to each according to his moral merit, or needs, or marginal product, or how hard he tries, or the weighted sum of the foregoing, and so on. The principle of entitlement we have sketched is *not* patterned. There is no one natural dimension or weighted sum or combination of a small number of natural dimensions that yields the distributions generated in accordance with the principle of entitlement. The set of holdings that results when some persons receive their marginal products, others win at gambling, others receive a share of their mate's income, others receive gifts from admirers, others receive returns on investment, others make for themselves much of what they have, others find things, and so on, will not be patterned. Heavy strands of patterns will run through it; significant portions of the variance in holdings will be accounted for by pattern-variables. If most people most of the time choose to transfer some of their entitlements to others only in exchange for something from them, then a large part of what many people hold will vary with what they held that others wanted. More details are provided by the theory of marginal productivity. But gifts to relatives, charitable donations, bequests to children, and the like, are not best conceived, in the first instance, in this manner. Ignoring the strands of pattern, let us suppose for the moment that a distribution actually arrived at by the operation of the principle of entitlement is random with respect to any pattern. Though the resulting set of holdings will be unpatterned, it will not be incomprehensible, for it can be seen as arising from the operation of a small number of principles. These principles specify how an initial distribution may arise (the principle of acquisition of holdings) and how distributions may be transformed into others (the principle of transfer of holdings). The process whereby the set of holdings is generated will be intelligible, though the set of holdings itself that results from this process will be unpatterned.

The writings of F. A. Hayek focus less than is usually done upon what patterning distributive justice requires. Hayek argues that we cannot know enough about each person's situation to distribute to each according to his moral merit (but would justice demand we do so if we did have this knowledge?); and he goes on to say, "our objection is against all attempts to impress upon society a deliberately chosen pattern of distribution, whether it be an order of equality or of inequality." However, Hayek concludes that in a free society there will be distribution in accordance with value rather than moral merit; that is, in accordance with the perceived value of a person's actions and services to others. Despite his rejection of a patterned conception of distributive justice, Hayek himself suggests a pattern he thinks justifiable: distribution in accordance with the perceived benefits given to others, leaving room for the complaint that a free society does not realize exactly this pattern. Stating this patterned strand of a free capitalist society more precisely, we get "To each according to how much he benefits others who have the resources for benefiting those who benefit them." This will seem arbitrary unless some acceptable initial set of holdings is specified, or unless it is held that the operation of the system over time washes out any significant effects from the initial set of holdings. As an example of the latter, if almost anyone would have bought a car from Henry Ford, the suppositon that it was an arbitrary matter who held the money then (and so bought) would not place Henry Ford's earnings under a cloud. In any event, *his* coming to hold it is not arbitrary. Distribution according to benefits to others *is* a major patterned strand in a free capitalist society, as Hayek correctly points out, but it is only a strand and does not constitute the whole pattern of a system of entitlements (namely, inheritance, gifts for arbitrary reasons, charity, and so on) or a standard that one should insist a society fit. Will people tolerate for long a system yielding distributions that they believe are unpatterned? No doubt people will not long accept a distribution they believe is *unjust*. People want their society to be and to look just. But must the look of justice reside in a resulting pattern rather than in the underlying generating principles? We are in no position to conclude that the inhabitants of a society embodying an entitlement conception of justice in holdings will find it unacceptable. Still, it must be granted that were people's reasons for transferring some of their holdings to others always irrational or arbitrary, we would find this disturbing. (Suppose people always determined what holdings they would transfer, and to whom, by using a random device.) We feel more comfortable upholding the justice of an entitlement system if most of the transfers under it are done for reasons. This does not mean necessarily that all deserve what holdings they receive. It means only that there is a purpose or point to someone's transferring a holding to one person rather than to another; that usually we can see what the transferrer thinks he's gaining, what cause he thinks he's serving, what goals he thinks he's helping to achieve, and so forth. Since in a capitalist society people often transfer holdings to others in accordance with how much they perceive these others benefiting them, the fabric constituted by the individual transactions and transfers is largely reasonable and intelligible. (Gifts to loved ones, bequests to children, charity to the needy also are nonarbitrary components of the fabric.) In stressing the large strand of distribution in accordance with benefits to others, Hayek shows the point of many transfers, and so shows that the system of transfer of entitlements is not just spinning its gears aimlessly. The system of entitlements is defensible when constituted by the individual aims of individual transactions. No overarching aim is needed, no distributional pattern is required.

To think that the task of a theory of distributive justice is to fill in the blank in "to each according to his ———" is to be predisposed to search for a pattern; and the separate treatment of "from each according to his ———" treats production and distribution as two separate and independent issues. On an entitlement view these are *not* two separate questions. Whoever makes something, having bought or contracted for all other held resources used in the process (transferring some of his holdings for these cooperating factors), is entitled to it. The situation is *not* one of something's getting made, and there being an open question of who is to get it. Things come into the world already attached to people having entitlements over them. From the point of view of the historical entitlement conception of justice in holdings, those who start afresh to complete "to each according to his ———" treat objects as if they appeared from nowhere, out of nothing. A complete theory of justice might cover this limit case as well; perhaps here is a use for the usual conceptions of distributive justice.

So entrenched are maxims of the usual form that perhaps we should present the entitlement conception as a competitor. Ignoring acquisition and rectification, we might say:

> From each according to what he chooses to do, to each according to what he makes for himself (perhaps with the contracted aid of others) and what others choose to do for him and choose to give him of what they've been given previously (under this maxim) and haven't yet expended or transferred.

This, the discerning reader will have noticed, has its defects as a slogan. So as a summary and great simplification (and not as a maxim with any independent meaning) we have:

> *From each as they choose, to each as they are chosen.*

Section II

Rawls holds, as we have seen, that

> since everyone's well-being depends upon a scheme of cooperation without which no one could have a satisfactory life, the division of advantages should be such as to draw forth the willing cooperation of everyone taking part in it, including those less well situated. Yet this can be expected only if reasonable terms are proposed. The two principles mentioned seem to be a fair agreement on the basis of which those better endowed or more fortunate in their social position . . . could expect the willing cooperation of others when some workable scheme is a necessary condition of the welfare of all.

No doubt, the difference principle presents terms on the basis of which those less well endowed would be willing to cooperate. (What *better* terms could they propose for themselves?) But is this a fair agreement on the basis of which those *worse* endowed could expect the *willing* cooperation of others?

* * *

Rawls would have us imagine the worse-endowed persons say something like the following: "Look, better endowed: you gain by cooperating with us. If you want our cooperation you'll have to accept reasonable terms. We suggest these terms: We'll cooperate with you only if we get *as much as possible.* That is, the terms of our cooperation should give us that maximal share such that, if it was tried to give us more, we'd end up with less." How generous these proposed terms are might be seen by imagining that the better endowed make the almost symmetrical opposite proposal: "Look, worse endowed: you gain by cooperating with *us.* If you want our cooperation you'll have to accept reasonable terms. We propose these terms: We'll cooperate with you so long as *we* get as much as possible. That is, the terms of our cooperation should give us the maximal share such that, if it was tried to give us more, we'd end up with less." If these terms seem outrageous, as they are, why don't the terms proposed by those worse endowed seem the same? Why shouldn't the better endowed treat this latter proposal as beneath consideration, supposing someone to have the nerve explicitly to state it?

* * *

How can it have been supposed that these terms offered by the less well endowed are fair? Imagine a social pie somehow appearing so that *no one* has any claim at all on any portion of it, no one has any more of a claim than any other person; yet there must be unanimous agreement on how it is to be divided. Undoubtedly, apart from threats or holdouts in bargaining, an equal distribution would be suggested and found plausible as a solution. (It is, in Schelling's sense, a focal point solution.) If *somehow* the size of the pie wasn't fixed, and it was realized that pursuing an equal distribution somehow would lead to a smaller total pie than otherwise might occur, the people might well agree to an unequal distribution which raised the size of the least share. But in any actual situation, wouldn't this realization reveal something about differential claims on parts of the pie? Who is it that could make the pie larger, and would do it if given a larger share, but not if given an equal share under the scheme of equal distribution? To whom is an incentive to be provided to make this larger contribution? (There's no talk here of inextricably entangled joint product; it's known *to whom* incentives are to be offered, or at least to whom a bonus is to be paid after the fact.) Why doesn't this identifiable differential contribution lead to some differential entitlement?

If things fell from heaven like manna, and no one had any special entitlement to any portion of it, and no manna would fall unless all agreed to a particular distribution, and somehow the quantity varied depending on the distribution, then it is plausible to claim that persons placed so they couldn't make threats, or hold out for specially large shares, would agree to the difference principle rule of distribution. But is *this* the appropriate model for thinking about how the things people produce are to be distributed? Why think the same results should obtain for situations where there *are* differential entitlements as for situations where there are not?

A procedure that founds principles of distributive justice on what rational persons who know nothing about themselves or their histories would agree to *guarantees that end-state principles of justice will be taken as fundamental.* Perhaps some historical principles of justice are derivable from end-state principles, as the utilitarian tries to derive individual rights, prohibitions on punishing the innocent, and so forth, from *his* end-state principle; perhaps such arguments can be constructed even for the entitlement principle. But no historical principle, it seems, could be agreed to in the first instance by the participants in Rawls' original position. For people meeting together behind a veil of ignorance to decide who gets what, knowing nothing about any special entitlements people may have, will treat anything to be distributed as manna from heaven.

Alasdair MacIntyre,
from *After Virtue* (1929–)

Alasdair MacIntyre rejects the classical "liberal" position in which both Rawls and Nozick make their stand and develop their theories. In the selection that follows, he famously juxtaposes Rawls and Nozick and emphasizes their similarities as well as their differences and explains why the dispute between them is, in their own terms, irresolvable.

When Aristotle praised justice as the first virtue of political life, he did so in such a way as to suggest that a community which lacks practical agreement on a conception of justice must also lack the necessary basis for political community. But the lack of such a basis must therefore threaten our own society. For the outcome of that history, some aspects of which I sketched in the preceding chapter, has not only been an inability to agree upon a catalogue of the virtues and an even more fundamental inability to agree upon the relative importance of the virtue concepts within a moral scheme in which notions of rights and of utility also have a key place. It has also been an inability to agree upon the content and character of particular virtues. For since a virtue is now generally understood as a disposition or sentiment which will produce in us obedience to certain rules, agreement on what the relevant rules are to be is always a prerequisite for agreement upon the nature and content of a particular virtue. But this prior agreement in rules is, as I have emphasized in the earlier part of this book, something which our individualist culture is unable to secure. Nowhere is this more marked and nowhere are the consequences more threatening than in the case of justice. Everyday life is pervaded by them and basic controversies cannot therefore be rationally resolved. Consider one such controversy, endemic in the politics of the United States today—I present it in the form of a debate between two ideal-typical characters unimaginatively named 'A' and 'B'.

A, who may own a store or be a police officer or a construction worker, has struggled to save enough from his earnings to buy a small house, to send his children to the local college, to pay for some special type of medical care for his parents. He now finds all of his projects threatened by rising taxes. He regards this threat to his projects as *unjust*; he claims to have a right to what he has earned and that nobody else has a right to take away what he acquired legitimately and to which he has a just title. He intends to vote for candidates for political office who will defend his property, his projects *and* his conception of justice.

B, who may be a member of one of the liberal professions, or a social worker, or someone with inherited wealth, is impressed with the arbitrariness of the inequalities in the distribution of wealth, income and opportunity. He is, if anything, even more impressed with the inability of the poor and the deprived to do very much about their own condition as a result of inequalities in the distribution of power. He regards both these types of inequality as *unjust* and as constantly engendering further injustice. He believes more generally that all inequality stands in need of justification and that the only possible justification for inequality is to improve the condition of the poor and the deprived—by, for example, fostering economic growth. He draws the conclusion that in present circumstances redistributive taxation which will finance welfare and the social services is what justice demands. He intends to vote for candidates for political office who will defend redistributive taxation *and* his conception of justice.

It is clear that in the actual circumstances of our social and political order A and B are going to disagree about politics and politicians. But *must* they so disagree? The answer seems to be that under certain types of economic condition their disagreement need not manifest itself at the level of political conflict. If A and B belong to a society where economic resources are such, or are at least believed to be such, that B's public redistributive projects can be carried through at least to a certain point without threatening A's private life-plan projects, A and B might for some time vote for the same politicians and policies. Indeed they might on occasion be one and the same person. But if it is, or comes to be, the case that economic circumstances are such that either A's projects must be sacrificed to B's or *vice versa*, it at once becomes clear that A and B have views of justice which are not only logically incompatible with each other but which invoke considerations which are incommensurable with those advanced by the adversary party.

The logical incompatibility is not difficult to identify. A holds that principles of just acquisition and entitlement set limits to redistributive possibilities. If the outcome of the application of the principles of just acquisition and entitlement is gross inequality, the toleration of such inequality is a price that has to be paid for justice. B holds that principles of just distribution set limits to legitimate acquisition and entitlement. If the outcome of the application of the principles of just distribution is interference—by means of taxation or such devices as eminent domain—with what has up till now been regarded in this social order as legitimate acquisition and entitlement, the toleration of such interference is a price that has to be paid for justice. We may note in passing—it will not be unimportant later—that in the case of both A's principle and B's principle the price for one person or group of persons receiving justice is always paid by someone else. Thus different identifiable social groups have an interest in the acceptance of one of the principles and the rejection of the other. Neither principle is socially or politically neutral.

Moreover it is not simply that A and B advance principles which produce incompatible practical conclusions. The type of concept in terms of which each frames his claim is so different from that of the other that the question of how and whether the dispute between them may be rationally settled begins to pose difficulties. For A aspires to ground the notion of justice in some account of what and how a given person is entitled to in virtue of what he has acquired and earned; B aspires to ground the notion of justice in some account of the equality of the claims of each person in respect of basic needs and of the means to meet such needs. Confronted by a given piece of property or resource, A will be apt to claim that it is justly his because he owns it—he acquired it legitimately, he earned it; B will be apt to claim that it justly ought to be someone else's, because they need it much more, and if they do not have it, their basic needs will not be met. But our pluralist culture possesses no method of weighing, no rational criterion for deciding between claims based on legitimate entitlement against claims based on need. Thus these two types of claim are indeed, as I suggested, incommensurable, and the metaphor of 'weighing' moral claims is not just inappropriate but misleading.

It is at this point that recent analytical moral philosophy makes important claims. For it aspires to provide rational principles to which appeal may be made by contending parties with conflicting interests. And the two most distinguished recent attempts to carry through this project have a special relevance for the argument between A and B. For Robert Nozick's account of justice (1974) is at least to some large degree a rational articulation of key elements in A's position, while John Rawls's account (1971) is in the same way a rational articulation of key elements in B's position. Thus if the philosophical considerations which either Rawls or Nozick urge upon us turn out to be rationally compelling, the argument between A and B will have been rationally settled one way or another and my own characterization of the dispute will in consequence turn out to be quite false.

* * *

Many critics of Rawls have focused their attention on the ways in which Rawls derives his principles of justice from his statement of the initial position of the rational agent 'situated behind the veil of ignorance'. Such critics have made a number of telling points, but I do not intend to dwell on them, if only because I take it not only that a rational agent in *some such* situation as that of the veil of ignorance would indeed choose *some such* principles of justice as Rawls claims, but also that it is *only* a rational agent in such a situation who would choose such principles. Later in my argument this point will become important.

* * *

What I want to argue is threefold: first, that the incompatibility of Rawls's and Nozick's accounts does up to a point genuinely mirror the incompatibility of A's position with B's, and that to this extent at least Rawls and Nozick successfully articulate at the level of moral philosophy the disagreement between such ordinary non-philosophical citizens as A and B; but that Rawls and Nozick also reproduce the very same type of incompatibility and incommensurability at the level of philosophical argument that made A's and B's debate unsettlable at the level of social conflict; and secondly, that there is nonetheless an element in the position of both A and B which neither Rawls's account nor Nozick's captures, an element which survives from that older classical tradition in which the virtues were central. When we reflect on both these points, a third emerges: namely, that in their conjunction we have an important clue to the social presuppositions which Rawls and Nozick to some degree share.

Rawls makes primary what is in effect a principle of equality with respect to needs. His conception of 'the worst off' sector of the community is a conception of those whose needs are gravest in respect of income, wealth and other goods. Nozick makes primary what is a principle of equality with respect to entitlement. For Rawls how those who are now in grave need come to be in grave need is irrelevant; justice is made into a matter of present patterns of distribution to which the past is irrelevant. For Nozick only evidence about what has been legitimately acquired in the past is relevant; present patterns of distribution in themselves must be irrelevant to *justice* (although not perhaps to kindness or generosity). To say even this much makes it clear how close Rawls is to B and how close Nozick is to A. For A appealed against distributive canons to a justice of entitlement, and B appealed against canons of entitlement to a justice which regards needs. Yet it is also at once clear not only that Rawls's priorities are incompatible with Nozick's in a way parallel to that in which B's position is incompatible with A's, but also that Rawls's position is incommensurable with Nozick's in a way similarly parallel to that in which B's is incommensurable with A's. For how can a claim that gives priority to equality of needs be rationally weighed against one which gives priority to entitlements? If Rawls were to argue that anyone *behind the veil of ignorance*, who knew neither whether and how his needs would be met nor what his entitlements would be, ought rationally to prefer a principle which respects needs to one which respects entitlements, invoking perhaps principles of rational decision theory to do so, the immediate answer must be not only that *we* are *never* behind such a veil of ignorance, but also that this leaves unimpugned Nozick's premise about inalienable rights. And if Nozick were to argue that any distributive principle, if enforced, could violate a freedom to which everyone of us is entitled—as he does indeed argue—the immediate answer must be that in so interpreting the inviolability of basic rights he begs the question in favor of his own argument and leaves unimpugned Rawls's premises.

Nonetheless there is something important, if negative, which Rawls's account shares with Nozick's. Neither of them make any reference to *desert* in their account of justice, nor could they consistently do so. And yet both A and B did make such a reference—and it is imperative here to notice that 'A' and 'B' are not the names of mere arbitrary constructions of my own; their arguments faithfully reproduce, for example, a good deal of what was actually said in recent fiscal debates in California, New Jersey and elsewhere. What A complains of on his own behalf is not merely that he is entitled to what he has earned, but that he *deserves* it in virtue of his life of hard work; what B complains of on behalf of the poor and deprived is that their poverty and deprivation is *undeserved* and therefore unwarranted. And it seems clear that in the case of the real-life counterparts of A and B it is the reference to desert which makes them feel strongly that what they are complaining about is injustice, rather than some other kind of wrong or harm.

Neither Rawls's account nor Nozick's allows this central place, or indeed any kind of place, for desert in claims about justice and injustice. Rawls (p. 310) allows that common sense views of justice connect it with desert, but argues first that we do not know what anyone deserves until we have already formulated the rules of justice (and hence we cannot base our understanding of justice upon desert), and secondly that when we have formulated the rules of justice it turns out that it is not desert that is in question anyway, but only legitimate expectations. He also argues that to attempt to apply notions of desert would be impracticable—the ghost of Hume walks in his pages at this point.

Nozick is less explicit, but his scheme of justice being based exclusively on entitlements can allow no place for desert. He goes at one point discuss the possibility of a principle for the rectification of injustice, but what he writes on that point is so tentative and cryptic that it affords no guidance for amending his general viewpoint. It is in any case clear that for both Nozick and Rawls a society is composed of individuals, each with his or her own interest, who then have to come together and formulate common rules of life. In Nozick's case there is the additional negative constraint of a set of basic rights. In Rawls's case the only constraints are those that a prudent rationality would impose. Individuals are thus in both accounts primary and society secondary, and the identification of individual interests is prior to, and independent of, the construction of any moral or social bonds between them. But we have already seen that the notion of desert is at home only in the context of a community whose primary bond is a shared understanding both of the good for man and of the good of that community and where individuals identify their primary interests with reference to those goods. Rawls explicitly makes it a presupposition of his view that we must expect to disagree with others about what the good life for man is and must therefore exclude any understanding of it that we may have from our formulation of the principles of justice. Only those goods in which everyone, whatever their view of the good life, takes an interest are to be admitted to consideration. In Nozick's argument too, the concept of community required for the notion of desert to have application is simply absent. To understand this is to clarify two further points.

The first concerns the shared social presuppositions of Rawls and Nozick. It is, from both standpoints, as though we had been shipwrecked on an uninhabited island with a group of other individuals, each of whom is a stranger to me and to all the others. What have to be worked out are rules which will safeguard each one of us maximally in such a situation. Nozick's premise concerning rights introduces a strong set of constraints; we do know that certain types of interference with each other are absolutely prohibited. But there is a limit to the bonds between us, a limit set by our private and competing interests. This individualistic view has of course, as I noticed earlier, a distinguished ancestry: Hobbes, Locke (whose views Nozick treats with great respect), Machiavelli and others. And it contains within itself a certain note of realism about modern society; modern society is indeed often, at least in surface appearance, nothing but a collection of strangers, each pursuing his or her own interests under minimal constraints. We still of course, even in modern society, find it difficult to think of families, colleges and other genuine communities in this way; but even our thinking about those is now invaded to an increasing degree by individualist conceptions, especially in the law courts. Thus Rawls and Nozick articulate with great power a shared view which envisages entry into social life as—at least ideally—the voluntary act of at least potentially rational individuals with prior interests who have to ask the question 'What kind of social contract with others is it reasonable for me to enter into?' Not surprisingly it is a consequence of this that their views exclude any account of human community in which the notion of desert in relation to contributions to the common tasks of that community in pursuing shared goods could provide the basis for judgments about virtue and injustice.

Desert is ruled out too in another way. I have remarked upon how Rawls's distributive principles exclude reference to the past and so to claims to desert based on past actions and sufferings. Nozick too excludes that of the past on which such claims might be based, by making a concern for the legitimacy of entitlements the sole ground for taking an interest in the past in connection with justice. What makes this important is that Nozick's account serves the interest of a particular mythology about the past precisely by what it excludes from view. For central to Nozick's account is the thesis that all legitimate entitlements can be traced to legitimate acts of original acquisition. But, if that is so, there are in fact very few, and in some large areas of the world *no*, legitimate entitlements. The property-owners of the modern world are not the legitimate heirs of the Lockean individuals who performed quasi-Lockean ('quasi' to allow for Nozick's emendations of Locke) acts of original acquisition; they are the inheritors of those who, for example, stole, and used violence to steal the common lands of England from the common people, vast tracts of North America from the American Indian, much of Ireland from the Irish, and Prussia from the original non-German Prussians. This is the historical reality ideologically concealed behind any Lockean thesis. The lack of any principle of rectification is thus not a small side issue for a thesis such as Nozick's; it tends to vitiate the theory as a whole—even if we were to suppress the overwhelming objections to any belief in inalienable human rights.

A and B differ from Rawls and Nozick at the price of inconsistency. Each of them in conjoining either Rawls's principles or Nozick's with an appeal to desert exhibits an adherence to an older, more traditional, more Aristotelian and Christian view of justice. This inconsistency is thus a tribute to the residual power and influence of the tradition, a power and influence with two distinct sources. In the conceptual *mélange* of moral thought and practice today fragments from the tradition—virtue concepts for the most part—are still found alongside characteristically modern and individualist concepts such as those of rights or utility. But the tradition also survives in a much less fragmented, much less distorted form in the lives of certain communities whose historical ties with their past remain strong. So the older moral tradition is discernible in the United States and elsewhere among, for example, some Catholic Irish, some Orthodox Greeks and some Jews of an Orthodox persuasion, all of them communities that inherit their moral tradition not only through their religion, but also from the structure of the peasant villages and households which their immediate ancestors inhabited on the margins of modern Europe. Moreover it would be wrong to conclude from the stress that I have laid on the medieval background that Protestantism did not in some areas become the bearer of this very same moral tradition; in Scotland, for example, Aristotle's *Nicomachean Ethics* and *Politics* were the secular moral texts in the universities, coexisting happily with a Calvinist theology which was often elsewhere hostile to them, until 1690 and after. And there are today both black and white Protestant communities in the United States, especially perhaps those in or from the South, who will recognize in the tradition of the virtues a key part of their own cultural inheritance.

Even however in such communities the need to enter into public debate enforces participation in the cultural *mélange* in the search for a common stock of concepts and norms which all may employ and to which all may appeal. Consequently the allegiance of such marginal communities to the tradition is constantly in danger of being eroded, and this in search of what, if my argument is correct, is a chimaera. For what analysis of A's and B's position reveals once again is that we have all too many disparate and rival moral concepts, in this case rival and disparate concepts of justice, and that the moral resources of the culture allow us no way of settling the issue between them rationally. Moral philosophy, as it is dominantly understood, reflects the debates and disagreements of the culture so faithfully that its controversies turn out to be unsettlable in just the way that the political and moral debates themselves are.

It follows that our society cannot hope to achieve moral consensus.

* * *

This does not mean that there are not many tasks only to be performed in and through government which still require performing: the rule of law, so far as it is possible in a modern state, has to be vindicated, injustice and unwarranted suffering have to be dealt with, generosity has to be exercised, and liberty has to be defended, in ways that are sometimes only possible through the use of governmental institutions. But each particular task, each particular responsibility has to be evaluated on its own merits. Modern systematic politics, whether liberal, conservative, radical or socialist, simply has to be rejected from a standpoint that owes genuine allegiance to the tradition of the virtues; for modern politics itself expresses in its institutional forms a systematic rejection of that tradition.

Thomas Nagel, "Rawls on Justice" (1937-)

Thomas Nagel was one of the first and one of the most thoughtful enthusiastic reviewers of Rawls's *Theory of Justice*. He had serious reservations about the book, nevertheless, anticipating much subsequent controversy. In particular, he found Rawls's seemingly innocuous presuppositions not so innocuous at all, and the two principles not nearly so obvious, even from a sympathetic point of view.

A Theory of Justice is a rich, complicated, and fundamental work. It offers an elaborate set of arguments and provides many issues for discussion. This review will focus on its contribution to the more abstract portions of ethical theory.

The book contains three elements. One is a vision of men and society as they should be. Another is a conception of moral theory. The third is a construction that attempts to derive principles expressive of the vision, in accordance with methods that reflect the conception of moral theory. In that construction Rawls has pursued the contractarian tradition in moral and political philosophy. His version of the social contract, a hypothetical choice situation called the original position, was first presented in 1958 and is here developed in great and explicit detail. The aim is to provide a way of treating the basic problems of social choice, for which no generally recognized methods of precise solution exist, through the proxy of a specially constructed parallel problem of individual choice, which can be solved by the more reliable intuitions and decision procedures of rational prudence.

* * *

Rawls's substantive doctrine is a rather pure form of egalitarian liberalism, whose controversial elements are its egalitarianism, its anti-perfectionism and anti-meritocracy, the primacy it gives to liberty, and the fact that it is more egalitarian about liberty than about other goods. The justice of social institutions is measured not by their tendency to maximize the sum or average of certain advantages, but by their tendency to counteract the natural inequalities deriving from birth, talent, and circumstance, pooling those resources in the service of the common good. The common good is measured in terms of a very restricted, basic set of benefits to individuals: personal and political liberty, economic and social advantages, and self-respect.

The justice of institutions depends on their conformity to two principles. The first requires the greatest equal liberty compatible with a like liberty for all. The second (the difference principle) permits only those inequalities in the distribution of

Reprinted by permission of the author and *The Philosophical Review*.

primary economic and social advantages that benefit everyone, in particular the worst off. Liberty is prior in the sense that it cannot be sacrificed for economic and social advantages, unless they are so scarce or unequal as to prevent the meaningful exercise of equal liberty until material conditions have improved.

The view is firmly opposed to mere equality of opportunity, which allows too much influence to the morally irrelevant contingencies of birth and talent; it is also opposed to counting a society's advanced cultural or intellectual achievements among the gains which can make sacrifice of the more primary goods just. What matters is that everyone be provided with the basic conditions for the realization of his own aims, regardless of the absolute level of achievement that may represent.

When the social and political implications of this view are worked out in detail, as is done in Part Two of the book, it is extremely appealing, but far from self-evident. In considering its theoretical basis, one should therefore ask whether the contractarian approach, realized in terms of the original position, depends on assumptions any less controversial than the substantive conclusions it is adduced to support.

The notion that a contract is the appropriate model for a theory of social justice depends on the view that it is fair to require people to submit to procedures and institutions only if, given the opportunity, they could in some sense have agreed in advance on the principles to which they must submit. That is why Rawls calls the theory "justice as fairness." (Indeed, he believes that a similar contractual basis can be found for the principles of individual morality, yielding a theory of rightness as fairness.) The fundamental attitude toward persons on which justice as fairness depends is a respect for their autonomy or freedom. Since social institutions are simply there and people are born into them, submission cannot be literally voluntary, but (p. 13) "A society satisfying the principles of justice as fairness comes as close as a society can to being a voluntary scheme, for it meets the principles which free and equal persons would assent to under circumstances that are fair."

Before considering whether the original position embodies these conditions, we must ask why respect for the freedom of others, and the desire to make society as near to voluntary as possible, should be taken as the mainspring of the sense of justice. That gives liberty a position of great importance from the very beginning, an importance that it retains in the resulting substantive theory. But we must ask how the respect for autonomy by itself can be expected to yield further results as well.

When one justifies a policy on the ground that the affected parties would have (or even have) agreed to it, much depends on the reasons for their agreement. If it is motivated by ignorance or fear or helplessness or a defective sense of what is reasonable, then actual or possible prior agreement does not sanction anything. In other cases, prior agreement for the right reasons can be obtained or presumed, but it is not the agreement that justifies what has been agreed to, but rather whatever justifies the agreement itself. If, for example, certain principles would be agreed to because they are just, that cannot be what makes them just. In many cases the appeal to hypothetical prior agreement is actually of this character. It is not a final justification, not a mark of respect for autonomy, but merely a way of recalling someone to the kind of *moral* judgment he would make in the absence of distorting influences derived from his special situation.

Actual or presumable consent can be the *source* of a justification only if it is already accepted that the affected parties are to be treated as certain reasons would incline each of them to want to be treated. The circumstances of consent are designed to bring those reasons into operation, suppressing irrelevant considerations, and the fact that the choice would have been made becomes a further reason for adhering to the result.

When the interests of the parties do not naturally coincide, a version of consent may still be preserved if they are able to agree in advance on a procedure for settling conflicts. They may agree unanimously that the procedure treats them equally in relevant respects, though they would not be able to agree in advance to any of the particular distributions of advantages that it might yield. (An example would be a lottery to determine the recipient of some indivisible benefit.)

For the result of such a choice to be morally acceptable, two things must be true: (*a*) the choice must be unanimous; (*b*) the circumstances that make unanimity possible must not undermine the equality of the parties in other respects. Presumably they must be deprived of some knowledge (for example, of who will win the lottery) in order to reach agreement, but it is essential that they not be unequally deprived (as would be the case, for example, if they agreed to submit a dispute to an arbitrator who, unknown to any of them, was extremely biased).

The more disparate the conflicting interests to be balanced, however, the more information the parties must be deprived of to insure unanimity, and doubts begin to arise whether any procedure can be relied on to treat everyone equally in respect of the relevant interests. There is then a real question whether hypothetical choice under conditions of ignorance, as a representation of consent, can by itself provide a moral justification for outcomes that could not be unanimously agreed to if they were known in advance.

* * *

I do not believe that the assumptions of the original position are either weak or innocuous or uncontroversial. In fact, the situation thus constructed may not be fair. Rawls says that the aim of the veil of ignorance is "to rule out those principles that it would be rational to propose for acceptance, however little the chance of success, only if one knew certain things that are irrelevant from the standpoint of justice" (p. 18). Let us grant that the parties should be equal and should not be in possession of information which would lead them to seek advantages on morally irrelevant grounds like race, sex, parentage, or natural endowments. But they are deprived also of knowledge of their particular conception of the good. It seems odd to regard that as morally irrelevant from the standpoint of justice. If someone favors certain principles because of his conception of the good, he will not be seeking special advantages for himself so long as he does not know who in the society he is. Rather he will be opting for principles that advance the good for everyone, as defined by that conception. (I assume a conception of the good is just that, and not simply a system of tastes or preferences.) Yet Rawls appears to believe that it would be as unfair to permit people to press for the realization of their conception of the good as to permit them to press for the advantage of their social class.

It is true that men's different conceptions of the good divide them and produce conflict, so allowing this knowledge to the parties in the original position would prevent unanimity. Rawls concludes that the information must be suppressed and a common idea substituted which will permit agreement without selecting any particular conception of the good. This is achieved by means of the class of primary goods that it is supposedly rational to want whatever else one wants. Another possible conclusion, however, is that the model of the original position will not work because in order to secure spontaneous unanimity and avoid the necessity of bargaining one must suppress information that is morally relevant, and moreover suppress it in a way that does not treat the parties equally.

What Rawls wishes to do, by using the notion of primary goods, is to provide an Archimedean point, as he calls it, from which choice is possible without unfairness to any of the fuller conceptions of the good that lead people to differ. A *theory* of the good is presupposed, but it is ostensibly neutral between divergent particular conceptions, and supplies a least common denominator on which a choice in the original position can be based without unfairness to any of the parties. Only later, when the principles of justice have been reached on this basis, will it be possible to rule out certain particular interests or aims as illegitimate because they are unjust. It is a fundamental feature of Rawls's conception of the fairness of the original position that it should not permit the choice of principles of justice to depend on a particular conception of the good over which the parties may differ.

The construction does not, I think, accomplish this, and there are reasons to believe that it cannot be successfully carried out. Any hypothetical choice situation which requires agreement among the parties will have to impose strong restrictions on the grounds of choice, and these restrictions can be justified only in terms of a conception of the good. It is one of those cases in which there is no neutrality to be had, because neutrality needs as much justification as any other position.

Rawls's minimal conception of the good does not amount to a weak assumption: it depends on a strong assumption of the sufficiency of that reduced conception for the purposes of justice. The refusal to rank particular conceptions of the good implies a very marked tolerance for individual inclinations. Rawls is opposed not only to teleological conceptions according to which justice requires adherence to the principles that will maximize the good. He is also opposed to the natural position that even in a nonteleological theory what is just must depend on what is good, at least to the extent that a correct conception of the good must be used in determining what counts as an advantage and what as a disadvantage, and how much, for purposes of distribution and compensation. I interpret him as saying that the principles of justice are objective and interpersonally recognizable in a way that conceptions of the good are not. The refusal to rank individual conceptions and the reliance on primary goods are intended to insure this objectivity.

Objectivity may not be so easily achieved. The suppression of knowledge required to achieve unanimity is not equally fair to all the parties, because the primary goods are not equally valuable in pursuit of all conceptions of the good. They will serve to advance many different individual life plans (some more efficiently than others), but they are less useful in implementing views that hold a good life to be readily achievable only in certain well-defined types of social structure, or only in a society that works concertedly for the realization of certain higher human capacities and the suppression of baser ones, or only given certain types of economic relations among men. The model contains a strong individualistic bias, which is further strengthened by the motivational assumptions of mutual disinterest and absence of envy. These assumptions have the effect of discounting the claims of conceptions of the good that depend heavily on the relation between one's own position and that of others (though Rawls is prepared to allow such considerations to enter in so far as they affect self-esteem). The original position seems to presuppose not just a neutral theory of the good, but a liberal, individualistic conception according to which the best that can be wished for someone is the unimpeded pursuit of his own path, provided it does not interfere with the rights of others. The view is persuasively developed in the later portions of the book, but without a sense of its controversial character.

Among different life plans of this general type the construction is neutral. But given that many conceptions of the good do not fit into the individualistic pattern, how can this be described as a fair choice situation for principles of justice? Why should parties in the original position be prepared to commit themselves to principles that may frustrate or contravene their deepest convictions, just because they are deprived of the knowledge of those convictions?

There does not seem to be any way of redesigning the original position to do away with a restrictive assumption of this kind. One might think it would be an improvement to allow the parties full information about everyone's preferences and conception of the good, merely depriving them of the knowledge of who they were. But this, as Rawls points out (pp. 173–174), would yield no result at all. For either the parties would retain their conceptions of the good and, choosing from different points of view, would not reach unanimity, or else they would possess no aims of their own and would be asked to choose in terms of the aims of all the people they might be—an unintelligible request which provides no basis for a unified choice, in the absence of a dominant conception. The reduction to a common ground of choice is therefore essential for the model to operate at all, and the selection of that ground inevitably represents a strong assumption.

Let us now turn to the argument leading to the choice of the two principles in the original position as constructed. The core of this argument appears in Sections 26–29, intertwined with an argument against the choice of the principle of average utility. Rawls has gone to some lengths to defend his controversial claim that in the original position it is rational to adopt the maximin rule which leads one to choose principles that favor the bottom of the social hierarchy, instead of accepting a greater risk at the bottom in return for the possibility of greater benefits at the top (as might be prudentially rational if one had an equal chance of being anyone in the society).

Rawls states (p. 154) that three conditions which make maximin plausible hold in the original position to a high degree. (1) "There must be some reason for sharply discounting estimates of . . . probabilities." (2) "The person choosing has a conception of the good such that he cares very little, if anything, for what he might gain above the minimum stipend that he can, in fact, be sure of by following the maximin rule." (3) "The rejected alternatives have outcomes that one can hardly accept." Let us consider these in turn.

The first condition is very important, and the claim that it holds in the original position is not based simply on a general rejection of the principle of insufficient reason (that is, the principle that where probabilities are unknown they should be regarded as equal). For one could characterize the original position in such a way that the parties would be prudentially rational to choose as if they had an equal chance of being anyone in the society, and the problem is to see why this would be an inappropriate representation of the grounds for a choice of principles.

One factor mentioned by Rawls is that the subject matter of the choice is extremely serious, since it involves institutions that will determine the total life prospects for the parties and those close to them. It is not just a choice of alternatives for a single occasion. Now this would be a reason for a conservative choice even if one knew the relative probabilities of different outcomes. It would be irresponsible to accept even a small risk of dreadful life prospects for oneself and one's descendants in exchange for a good chance of wealth or power. But what is needed is an account of why probabilities should be totally discounted, and not just with regard to the most unacceptable outcomes. The difference principle, for example, is supposed to apply at all levels of social development, so it is not justified merely by the desire to avoid grave risks. The fact that total life prospects are involved does not seem an adequate explanation. There must be some reason against allowing probabilities (proportional, for instance, to the number of persons in each social position) to enter into the choice of distributions above an acceptable minimum. Let me stress that I am posing a question not about decision theory but about the design of the original position and the comprehensiveness of the veil of ignorance. Why should it be thought that a just solution will be reached only if these considerations are suppressed?

Their suppression is justified, I think, only on the assumption that the proportions of people in various social positions are regarded as morally irrelevant, and this must be because it is not thought acceptable to sum advantages and disadvantages over persons, so that a loss for some is compensated by a gain for others. This aspect of the design of the original position appears, therefore, to be motivated by the wish to avoid extending to society as a whole the principle of rational choice for one man. Now this is supposed to be one of the *conclusions* of the contract approach, not one of its presuppositions. Yet the constraints on choice in Rawls's version of the original position are designed to rule out the possibility of such an extension, by requiring that probabilities be discounted. I can see no way to avoid presupposing some definite view on this matter in the design of a contract situation. If that is true, then a contract approach cannot give any particular view very much support.

Consider next the second condition. Keeping in mind that the parties in the original position do not know the stage of development of their society, and therefore do not know what minimum will be guaranteed by a maximin strategy, it is difficult to understand how an individual can know that he "cares very little, if anything, for what he might gain above the minimum." The explanation Rawls offers (p. 156) seems weak. Even if parties in the original position accept the priority of liberty, and even if the veil of ignorance leaves them with skeletal conception of the good, it seems impossible that they should care very little for increases in primary economic and social goods above what the difference principle guarantees at any given stage of social development.

Finally, the third condition, that one should rule out certain possibilities as unacceptable, is certainly a ground for requiring a social minimum and the priority of basic personal liberties, but it is not a ground for adopting the maximin rule in that general form needed to justify the choice of the difference principle. That must rely on stronger egalitarian premises.

Charles Taylor,
from "Distributive Justice" (1931–)

Charles Taylor shares MacIntyre's tradition-oriented conception of justice, and he too complains that liberalism gives too little credit to the all-important notion of "the good."

———————

A vigorous debate is raging today about the nature of distributive justice. The controversy doesn't concern only the criteria or standards of justice, what we would have to do or be to be just; it also touches the issue of what kind of good distributive justice is. Indeed, I would argue that as the debate has progressed, it has become clearer that the solution to the first kind of question presupposes some clarification on the second. In any case, recent extremely interesting works by Michael Walzer and Michael Sandel raise fundamental questions in the second range.

I want to take up both issues in this paper. In the first part, I raise questions about the nature of distributive justice. In the second, I want to look at the actual debates about criteria which now divide our societies.

First, what kind of good, or mode of right, is distributive justice? John Rawls helps us by giving us a formulation of the circumstances of justice. We have separate human beings who are nevertheless collaborating in conditions of moderate scarcity. This distinguishes it from other kinds and contexts of good. For instance, there is a mode of justice which holds between quite independent human beings, not bound together by any society or collaborative arrangement. If two nomadic tribes meet in the desert, very old and long-standing intuitions about justice tell us that it is wrong (unjust) for one to steal the flocks of the other. The principle here is very simple: we have a right to what we have. But this is not a principle of *distributive* justice, which presupposes that people are in a society together or some kind of collaborative arrangement.

Similarly we have to distinguish distributive justice from other kinds of good or right action. If in the above case one of the tribes were starving, the other would, according to a widespread moral tradition, have a natural duty to succor it; and by extension, it is usually held that the starving tribe could legitimately steal from the other if it refused to help. But acting according to this natural duty is not the same thing as acting according to justice—although the demands of natural duty can have moral repercussions on justice, as we see in the second case: the necessity in which the starving tribe finds itself and the refusal of the better off cancel what would otherwise be the injustice of the act of stealing.

From *Philosophical Papers*, Vol. 2: *Philosophy and the Human Sciences* by Charles Taylor. Copyright © 1985 by Charles Taylor. Reprinted with the permission of Cambridge University Press.

Nomadic clans of herders are rather far removed from our predicament. They enter here only as exemplars of man in what is called the state of nature. The basic point is that there is no such thing as distributive justice in the state of nature. Everyone agrees with this truism; but beyond it agreement stops. The really important questions are: In what way do the principles of distributive justice differ from those of justice among independent agents (agents in the state of nature)? And what is it about human society that makes the difference?

This second question is not even recognized as a question by many thinkers. But I want to claim that it is the fundamental one. To argue or reason about distributive justice requires us to give clear formulations to strong and originally inchoate intuitions and to attempt to establish some coherent order among these formulations. In the process, as Rawls points out in his excellent discussion of this question, both formulations and intuitions can undergo alteration, until that limiting stage where they are in "reflective equilibrium."

Now our intuitions about distributive justice are continuous with our basic moral intuitions about human beings as beings who demand a certain respect (to use one moral language among many possible ones, but it is not possible to talk about this subject at all without using *someone's* formulations). It is because people ought to be treated in a certain way, and thus enjoy a status not shared by stones and (some think also) animals, that they ought to be treated *equally* in collaborative situations (I use the term "equally" in the wide sense of Aristotle's *Ethics* V, where it includes also "proportionate equality").

If we introduce the Kantian term "dignity" as a term of art to describe this status that human beings enjoy, then it is plain that there has been widespread disagreement on what human dignity consists in. But I would like to add that this disagreement lies behind the disputes about the nature of distributive justice. We can't really get clear about these disputes without exploring the different notions of human dignity.

Now our notion of human dignity is in turn bound up with a conception of the human good, that is, our answer to the question: What is the good for human beings? What is the good human life? This question, too, is part of the background of a conception of distributive justice. Differences about justice are related to differences about the nature of the good (if I may be permitted this Aristotelian expression). And they are related in particular to a key issue, which is whether and in what way human beings can realize the good alone, or to turn it around the other way, in what way they must be part of society to be human in the full sense or to realize the human good.

The claim I am making could be put in this way: that different principles of distributive justice are related to conceptions of the human good, and in particular to different notions of individuals' dependence on society to realize the good. Thus deep disagreements about justice can be clarified only if we formulate and confront the underlying notions of the individual and society. This is the nature of the argument, and it also underlies the actual disputes we witness in our society.

The above paragraph would be a crashing truism but for two related factors. The first is the tendency of much Anglo-Saxon philosophy to shy away from any exploration of the human subject. Seventeenth-century epistemology started with an unexamined and unexaminable subject—unexaminable because any examination deals with data, and data are on the side of the known, not the knower. In a parallel way, seventeenth-century theory of natural right started with the unexamined subject. Much Anglo-Saxon philosophy seems to want to continue in this direction. Rawls is a partial exception, since he speaks of a Kantian basis for his principles of justice, expressing as they do our nature as free and equal rational beings; but even he doesn't bring this foundation out as an explicit theme. Robert Nozick is an extreme example. He argues from our current conception of individual rights, and reasons as though this conception were sufficient to build our notion of the entire social context. The question is never raised whether the affirmation of these rights is bound up with a notion of human dignity and the human good which may require a quite other context. In short, the question is never raised whether the human being is morally self-sufficient, as Locke thought, or whether perhaps Aristotle is not right about this matter. An argument that abstracts from this conception, and that is insensitive to the social nature of human beings, naturally produces the most bizarre consequences.

I really want to argue here for an Aristotelian way of putting the issue about distributive justice. But Aristotle not only has a substantive view about man as *zoon politikon* which conflicts with Locke. He also has an implicit metaview (a view about what is at stake in the argument). And this view conflicts with that of a philosophical tradition of which Locke is one of the patrons, which wants to make questions of human nature irrelevant to morals and political philosophy, and to start instead with rights. This means that it is, alas, not just platitudinous to restate Aristotle's way of putting the issue. A second reason for reediting Aristotle naturally follows: the precise way in which different notions of distributive justice are related to their foundation in a view of human beings needs to be restated.

The Aristotelian metaview I want to put forward here is that principles of distributive justice are related to some notion of the good which is sustained or realized or sought in the association concerned. We can illustrate this view first with a very un-Aristotelian theory, the atomist view of Locke. For the purposes of this discussion we can describe as "atomist" views of the human good in which it is conceivable for an individual to attain it alone. On these views, in other words, what people derive from association in realizing the good are a set of aids only contingently, even if almost unfailingly, linked to this association; protection against attack from others, for example, or the benefits of higher production. Such benefits always require association, or almost always. But there are imaginable circumstances in which we could enjoy security or a high living standard alone; on a continent in which there were no others, for example, or in a land of paradisiac natural abundance.

A social view of the human good, by contrast, holds that an essential constitutive condition of the search for it is bound up with being in society. Thus if I argue that an individual cannot even be a moral subject, and thus a candidate for the realization of the human good, outside of a community of language and mutual discourse about the good and bad, just and unjust, I am rejecting all atomist views, since what a person derives from society is not some aid in realizing his or her good but the very possibility of being an agent seeking that good. (The Aristotelian resonances in the above sentence are, of course, not coincidental.)

To put the issue in other terms, social views see some form of society as essentially bound up with human dignity, since outside of society the very potentiality to realize that wherein this dignity consists is undermined; whereas atomist views see human dignity as quite independent of society—which is why they have no difficulty ascribing *rights* (as against just the status of being an object of respect) to a person alone, in the state of nature.

* * *

First, any social view sees a certain kind or structure of society as an essential condition of human potentiality, be it the polis or the classless society or the hierarchical society under God and king or any other of the host of social structures that we have seen in history. This structure itself, or order, or type of relation, thus provides the essential background for any principles of distributive justice.

This means, of course, that the structure itself cannot be called into question in the name of distributive justice. Within the bounds of a hierarchical conception of society in which the political order is thought to reflect the order in the universe, it makes no sense to object to the special "status" or "privileges" of king or priest as violations of equality. When this objection is made, it involves a challenge to the entire hierarchical conception.

* * *

The second form of argument that arises in a social perspective concerns the principles of distributive justice themselves, and not just the framework. Granted a certain view of a common good, in the sense of an indivisible good, which a social perspective necessarily offers us, since it sees human beings as realizing their potential only in a certain common structure, it may appear evident that certain people deserve more than others, in the sense that their contribution to this common good is more marked, or more important.

We could put it this way: For this common good of living in a family or a community or whatever, we are all in each other's debt. But we might see the balance of mutual indebtedness as not entirely reciprocal. Some people, those who contribute in some special way to the animation of the community or to the common deliberative life of the society or to the defense of its integrity or whatever, deserve more, because we are more in their debt than vice versa.

This is the perspective of Aristotle's discussion of distributive justice in Books III and IV of the *Politics*. He closely links together what the principle of distribution of offices and honors ought to be and what the common good is which the polis is for.

* * *

What emerges out of the above discussion is above all two major points about the nature and scope of distributive justice, confusion about which bedevils discussions today. The confusions arise from conflating distributive justice with other virtues, so that it is no longer clear just what is being demanded or advocated. This is not a merely "academic" question, as I hope to show; rather the confusions may hide from us important political choices.

1. Aristotle, in distinguishing particular from general justice, points out that the former is a virtue whose opposite is "pleonexia," grasping more than one's share. Criteria of distributive justice are meant to give us the basis for knowing what our share is, and therefore when we are being grasping. But what falls out of the above discussion is that there are two rather different kinds of argument that do this. There

are arguments about the nature of the framework, from considerations of the goods sought and the nature of the agents associated; and on these grounds we sometimes judge that certain distributions are wrong and that those who push for them are "grasping." And then there are arguments about the balance of mutual indebtedness which justify some distributions within the framework and rule out others.

Both of these arguments concern distributive justice in a sense; but in another sense, only the latter do so, because only they tell us how to resolve questions of distribution that can be considered as open and allowable questions in the light of the framework. They help us decide the distributive questions that we can legitimately allow as questions without subverting the good or the very basis of our association. However, the issue of how to use the term is not vital, so long as these two levels are not confused.

2. Both framework questions and the criteria of distribution are derived at least in part from the nature of the association and that of the goods sought in common. But that means that the demands of distributive justice can and will differ across different societies and at different moments in history.

* * *

What all this means is that we have to abandon the search for a single set of principles of distributive justice. A modern society can be seen under different, mutually irreducible perspectives, and consequently can be judged by independent, mutually irreducible principles of distributive justice. Complexity is further compounded when we reflect that there is no single answer to the question of the unit within which people owe each other distributive justice; that even within one model of society, there are different degrees of mutual involvement which create different degrees of mutual obligation. So that we may have to think of justice both between individuals and between communities, and perhaps within communities as well.

If this all means that there may be no such thing as *the* coherent set of principles of distributive justice for a modern society, we should not be distressed. The same plurality emerges in Aristotle's discussion of justice in *Politics* III and IV. Those who adopt a single exclusive principle, Aristotle says, "speak of a part of justice only" (*meros ti tou dikaiou legousi*, 1281a10).

Michael Walzer,
from *Spheres of Justice* (1935–)

Michael Walzer also rejects Rawls's idea of an all-embracing theory of justice, but he does not thereby reject liberalism and Rawls's central concern for equality as such. Rather, he defends a much more "complex" and pluralistic notion of equality, and a conception of justice that is variously defined within different contexts or "spheres" of justice.

Distribution is what social conflict is all about. Marx's heavy emphasis on productive processes should not conceal from us the simple truth that the struggle for control of the means of production is a distributive struggle. Land and capital are at stake, and these are goods that can be shared, divided, exchanged, and endlessly converted. But land and capital are not the only dominant goods; it is possible (it has historically been possible) to come to them by way of other goods—military or political power, religious office and charisma, and so on. History reveals no single dominant good and no naturally dominant good, but only different kinds of magic and competing bands of magicians.

The claim to monopolize a dominant good—when worked up for public purposes—constitutes an ideology. Its standard form is to connect legitimate possession with some set of personal qualities through the medium of a philosophical principle. So aristocracy, or the rule of the best, is the principle of those who lay claim to breeding and intelligence: they are commonly the monopolists of landed wealth and familial reputation. Divine supremacy is the principle of those who claim to know the word of God: they are the monopolists of grace and office. Meritocracy, or the career open to talents, is the principle of those who claim to be talented: they are most often the monopolists of education. Free exchange is the principle of those who are ready, or who tell us they are ready, to put their money at risk: they are the monopolists of movable wealth. These groups—and others, too, similarly marked off by their principles and possessions—compete with one another, struggling for supremacy. One group wins, and then a different one; or coalitions are worked out, and supremacy is uneasily shared. There is no final victory, nor should there be. But that is not to say tht the claims of the different groups are necessarily wrong, or that the principles they invoke are of no value as distributive criteria; the principles are often exactly right within the limits of a particular sphere. Ideologies are readily corrupted, but their corruption is not the most interesting thing about them.

It is in the study of these struggles that I have sought the guiding thread of my own argument. The struggles have, I think, a paradigmatic form. Some group of men and women—class, caste, strata, estate, alliance, or social formation—comes to enjoy a monopoly or a near monopoly of some dominant good; or, a coalition of groups comes to enjoy, and so on. This dominant good is more or less systematically converted into all sorts of other things—opportunities, powers, and reputations. So wealth is seized by the strong, honor by the wellborn, office by the well educated. Perhaps the ideology that justifies the seizure is widely believed to be true. But resentment and resistance are (almost) as pervasive as belief. There are always some people, and after a time there are a great many, who think the seizure is not justice but usurpation. The ruling group does not possess, or does not uniquely possess, the qualities it claims; the conversion process violates the common understanding of the goods at stake. Social conflict is intermittent, or it is endemic; at some point, counterclaims are put forward. Though these are of many different sorts, three general sorts are especially important:

> 1. The claim that the dominant good, whatever it is, should be redistributed so that it can be equally or at least more widely shared: this amounts to saying that monopoly is unjust.
> 2. The claim that the way should be opened for the autonomous distribution of all social goods: this amounts to saying that dominance is unjust.
> 3. The claim that some new good, monopolized by some new group, should replace the currently dominant good: this amounts to saying that the existing pattern of dominance and monopoly is unjust.

The third claim is, in Marx's view, the model of every revolutionary ideology—except, perhaps, the proletarian or last ideology. Thus, the French Revolution in Marxist theory: the dominance of noble birth and blood and of feudal landholding is ended, and bourgeois wealth is established in its stead. The original situation is reproduced with different subjects and objects (this is never unimportant), and then the class war is immediately renewed. It is not my purpose here to endorse or to criticize Marx's view. I suspect, in fact, that there is something of all three claims in every revolutionary ideology, but that, too, is not a position that I shall try to defend here. Whatever its sociological significance, the third claim is not philosophically interesting—unless one believes that there is a naturally dominant good, such that its possessors could legitimately claim to rule the rest of us. In a sense, Marx believed exactly that. The means of production is the dominant good throughout history, and Marxism is a historicist doctrine insofar as it suggests that whoever controls the prevailing means legitimately rules. After the communist revolution, we shall all control the means of production: at that point, the third claim collapses into the first. Meanwhile, Marx's model is a program for ongoing distributive struggle. It will matter, of course, who wins at this or that moment, but we won't know why or how it matters if we attend only to the successive assertions of dominance and monopoly.

Simple Equality

It is with the first two claims that I shall be concerned, and ultimately with the second alone, for that one seems to me to capture best the plurality of social meanings and the real complexity of distributive systems. But the first is the more common among philosophers; it matches their own search for unity and singularity; and I shall need to explain its difficulties at some length.

Men and women who make the first claim challenge the monopoly but not the dominance of a particular social good. This is also a challenge to monopoly in general; for if wealth, for example, is dominant and widely shared, no other good can possibly be monopolized. Imagine a society in which everything is up for sale and every citizen has as much money as every other. I shall call this the "regime of simple equality." Equality is multiplied through the conversion process, until it extends across the full range of social goods. The regime of simple equality won't last for long, because the further progress of conversion, free exchange in the market, is certain to bring inequalities in its train. If one wanted to sustain simple equality over time, one would require a "monetary law" like the agrarian laws of ancient times or the Hebrew sabbatical, providing for a periodic return to the original condition. Only a centralized and activist state would be strong enough to force such a return; and it isn't clear that state officials would actually be able or willing to do that, if money were the dominant good. In any case, the original condition is unstable in another way. It's not only that monopoly will reappear, but also that dominance will disappear.

In practice, breaking the monopoly of money neutralizes its dominance. Other goods come into play, and inequality takes on new forms. Consider again the regime of simple equality. Everything is up for sale, and everyone has the same amount of money. So everyone has, say, an equal ability to buy an education for his children. Some do that, and others don't. It turns out to be a good investment: other social goods are, increasingly, offered for sale only to people with educational certificates. Soon everyone invests in education; or, more likely, the purchase is universalized through the tax system. But then the school is turned into a competitive world within which money is no longer dominant. Natural talent or family upbringing or skill in writing examinations is dominant instead, and educational success and certification are monopolized by some new group. Let's call them (what they call themselves) the "group of the talented." Eventually the members of this group claim that the good they control should be dominant outside the school: offices, titles, prerogatives, wealth too, should all be possessed by themselves. This is the career open to talents, equal opportunity, and so on. This is what fairness requires; talent will out; and in any case, talented men and women will enlarge the resources available to everyone else. So Michael Young's meritocracy is born, with all its attendant inequalities.

What should we do now? It is possible to set limits to the new conversion patterns, to recognize but constrain the monopoly power of the talented. I take this to be the purpose of John Rawls's difference principle, according to which inequalities are justified only if they are designed to bring, and actually do bring, the greatest possible benefit to the least advantaged social class. More specifically, the difference principle is a constraint imposed on talented men and women, once the monopoly of wealth has been broken. It works in this way: Imagine a surgeon who claims more than his equal share of wealth on the basis of the skills he has learned and the certificates he has won in the harsh competitive struggles of college and medical school. We will grant the claim if, and only if, granting it is beneficial in the stipulated ways. At the same time, we will act to limit and regulate the sale of surgery—that is, the direct conversion of surgical skill into wealth.

This regulation will necessarily be the work of the state, just as monetary laws and agrarian laws are the work of the state. Simple equality would require continual state intervention to break up or constrain incipient monopolies and to repress new forms of dominance. But then state power itself will become the central object of competitive struggles. Groups of men and women will seek to monopolize and then to use the state in order to consolidate their control of other social goods. Or, the state will be monopolized by its own agents in accordance with the iron law of oligarchy. Politics is always the most direct path to dominance, and political power (rather than the means of production) is probably the most important, and certainly the most dangerous, good in human history. Hence the need to constrain the agents of constraint, to establish constitutional checks and balances. These are limits imposed on political monopoly, and they are all the more important once the various social and economic monopolies have been broken.

One way of limiting political power is to distribute it widely. This may not work, given the well-canvassed dangers of majority tyranny; but these dangers are probably less acute than they are often made out to be. The greater danger of democratic government is that it will be weak to cope with re-emerging monopolies in society at large, with the social strength of plutocrats, bureaucrats, technocrats, meritocrats, and so on. In theory, political power is the dominant good in a democracy, and it is convertible in any way the citizens choose. But in practice, again, breaking the monopoly of power neutralizes its dominance. Political power cannot be widely shared without being subjected to the pull of all the other goods that the citizens already have or hope to have. Hence democracy is, as Marx recognized, essentially a reflective system, mirroring the prevailing and emerging distribution of social goods. Democratic decision making will be shaped by the cultural conceptions that determine or underwrite the new monopolies. To prevail against these monopolies, power will have to be centralized, perhaps itself monopolized. Once again, the state must be very powerful if it is to fulfill the purposes assigned to it by the difference principle or by any similarly interventionist rule.

Still, the regime of simple equality might work. One can imagine a more or less stable tension between emerging monopolies and political constraints, between the claim to privilege put forward by the talented, say, and the enforcement of the difference principle, and then between the agents of enforcement and the democratic constitution. But I suspect that difficulties will recur, and that at many points in time the only remedy for private privilege will be statism, and the only escape from statism will be private privilege. We will mobilize power to check monopoly, then look for some way of checking the power we have mobilized. But there is no way that doesn't open opportunities for strategically placed men and women to seize and exploit important social goods.

These problems derive from treating monopoly, and not dominance, as the central issue in distributive justice. It is not difficult, of course, to understand why philosophers (and political activists, too) have focused on monopoly. The distributive struggles of the modern age begin with a war against the aristocracy's singular hold on land, office, and honor. This seems an especially pernicious monopoly because it rests upon birth and blood, with which the individual has nothing to do, rather than upon wealth, or power, or education, all of which—at least in principle—can be earned. And when every man and woman becomes, as it were, a smallholder in the sphere of birth and blood, an important victory is indeed won. Birthright ceases to be a dominant good; henceforth, it purchases very little; wealth, power, and education come to the fore. With regard to these latter goods, however, simple equality cannot be sustained at all, or it can only be sustained subject to the vicissitudes I have just described. Within their own spheres, as they are currently understood, these three tend to generate natural monopolies that can be repressed only if state power is itself dominant and if it is monopolized by officials committed to the repression. But there is, I think, another path to another kind of equality.

Tyranny and Complex Equality

I want to argue that we should focus on the reduction of dominance—not, or not primarily, on the break-up or the constraint of monopoly. We should consider what it might mean to narrow the range within which particular goods are convertible and to vindicate the autonomy of distributive spheres. But this line of argument, though it is not uncommon historically, has never fully emerged in philosophical writing. Philosophers have tended to criticize (or to justify) existing or emerging monopolies of wealth, power, and education. Or, they have criticized (or justified) particular conversions—of wealth into education or of office into wealth. And all this, most often, in the name of some radically simplified distributive system. The critique of dominance will suggest instead a way of reshaping and then living with the actual complexity of distributions.

Imagine now a society in which different social goods are monopolistically held—as they are in fact and always will be, barring continual state intervention— but in which no particular good is generally convertible. As I go along, I shall try to define the precise limits on convertibility, but for now the general description will suffice. This is a complex egalitarian society. Though there will be many small inequalities, inequality will not be multiplied through the conversion process. Nor will it be summed across different goods, because the autonomy of distributions will tend to produce a variety of local monopolies, held by different groups of men and women. I don't want to claim that complex equality would necessarily be more stable than simple equality, but I am inclined to think that it would open the way for more diffused and particularized forms of social conflict. And the resistance to convertibility would be maintained, in large degree, by ordinary men and women within their own spheres of competence and control, without large-scale state action.

This is, I think, an attractive picture, but I have not yet explained just why it is attractive. The argument for complex equality begins from our understanding—I mean, our actual, concrete, positive, and particular understanding—of the various social goods. And then it moves on to an account of the way we relate to one another through those goods. Simple equality is a simple distributive condition, so that if I have fourteen hats and you have fourteen hats, we are equal. And it is all to the good if hats are dominant, for then our equality is extended through all the spheres of social life. On the view that I shall take here, however, we simply have the same number of hats, and it is unlikely that hats will be dominant for long. Equality is a complex relation of persons, mediated by the goods we make, share, and divide among ourselves; it is not an identity of possessions. It requires then, a diversity of distributive criteria that mirrors the diversity of social goods.

The argument of complex equality has been beautifully put by Pascal in one of his *Pensées*.

> The nature of tyranny is to desire power over the whole world and outside its own sphere.
>
> There are different companies—the strong, the handsome, the intelligent, the devout—and each man reigns in his own, not elsewhere. But sometimes they meet, and the strong and the handsome fight for mastery—foolishly, for their mastery is of different kinds. They misunderstand one another, and make the mistake of each aiming at universal dominion. Nothing can win this, not even strength, for it is powerless in the kingdom of the wise. . . .
>
> *Tyranny.* The following statements, therefore, are false and tyrannical: "Because I am handsome, so I should command respect." "I am strong, therefore men should love me. . . ." "I am . . . et cetera."
>
> Tyranny is the wish to obtain by one means what can only be had by another. We owe different duties to different qualities: love is the proper response to charm, fear to strength, and belief to learning.

Marx made a similar argument in his early manuscripts; perhaps he had this *pensée* in mind:

> Let us assume man to be man, and his relation to the world to be a human one. Then love can only be exchanged for love, trust for trust, etc. If you wish to enjoy art you must be an artistically cultivated person; if you wish to influence other people, you must be a person who really has a stimulating and encouraging effect upon others. . . . If you love without evoking love in return, i.e., if you are not able, by the manifestation of yourself as a loving person, to make yourself a beloved person—then your love is impotent and a misfortune.

These are not easy arguments, and most of my book is simply an exposition of their meaning. But here I shall attempt something more simple and schematic: a translation of the arguments into the terms I have already been using.

The first claim of Pascal and Marx is that personal qualities and social goods have their own spheres of operation, where they work their effects freely, spontaneously, and legitimately. There are ready or natural conversions that follow from, and are intuitively plausible because of, the social meaning of particular goods. The appeal is to our ordinary understanding and, at the same time, against our common acquiescence in illegitimate conversion patterns. Or, it is an appeal from our acquiescence to our resentment. There is something wrong, Pascal suggests, with the conversion of strength into belief. In political terms, Pascal means that no ruler can rightly command my opinions merely because of the power he wields. Nor can he, Marx adds, rightly claim to influence my actions: if a ruler wants to do that, he must be persuasive, helpful, encouraging, and so on. These arguments depend for their force on some shared understanding of knowledge, influence, and power. Social goods have social meanings, and we find our way to distributive justice through an interpretation of those meanings. We search for principles internal to each distributive sphere.

The second claim is that the disregard of these principles is tyranny. To convert one good into another, when there is no intrinsic connection between the two, is to invade the sphere where another company of men and women properly rules. Monopoly is not inappropriate within the spheres. There is nothing wrong, for example, with the grip that persuasive and helpful men and women (politicians) establish on political power. But the use of political power to gain access to other goods is a tyrannical use. Thus, an old description of tyranny is generalized: princes become tyrants, according to medieval writers, when they seize the property or invade the family of their subjects. In political life—but more widely, too—the dominance of goods makes for the domination of people.

The regime of complex equality is the opposite of tyranny. It establishes a set of relationships such that domination is impossible. In formal terms, complex equality means that no citizen's standing in one sphere or with regard to one social good can be undercut by his standing in some other sphere, with regard to some other good. Thus, citizen X may be chosen over citizen Y for political office, and then the two of them will be unequal in the sphere of politics. But they will not be unequal generally so long as X's office gives him no advantages over Y in any other sphere—superior medical care, access to better schools for his children, entrepreneurial opportunities, and so on. So long as office is not a dominant good, is not generally convertible, office holders will stand, or at least can stand, in a relation of equality to the men and women they govern.

But what if dominance were eliminated, the autonomy of the spheres established—and the same people were successful in one sphere after another, triumphant in every company, piling up goods without the need for illegitimate conversions? This would certainly make for an inegalitarian society, but it would also suggest in the strongest way that a society of equals was not a lively possibility. I doubt that any egalitarian argument could survive in the face of such evidence. Here is a person whom we have freely chosen (without reference to his family ties or personal wealth) as our political representative. He is also a bold and inventive entrepreneur. When he was younger, he studied science, scored amazingly high grades in every exam, and made important discoveries. In war, he is surpassingly brave and wins the highest honors. Himself compassionate and compelling, he is loved by all who know him. Are there such people? Maybe so, but I have my doubts. We tell stories like the one I have just told, but the stories are fictions, the conversion of power or money or academic talent into legendary fame. In any case, there aren't enough such people to constitute a ruling class and dominate the rest of us. Nor can they be successful in every distributive sphere, for there are some spheres to which the idea of success doesn't pertain. Nor are their children likely, under conditions of complex equality, to inherit their success. By and large, the most accomplished politicians, entrepreneurs, scientists, soldiers, and lovers will be different people; and so long as the goods they possess don't bring other goods in train, we have no reason to fear their accomplishments.

The critique of dominance and domination points toward an open-ended distributive principle. *No social good* x *should be distributed to men and women who possess some other good* y *merely because they possess* y *and without regard to the meaning of* x. This is a principle that has probably been reiterated, at one time of another, for every y that has ever been dominant. But it has not often been stated in general terms. Pascal and Marx have suggested the application of the principle against all possible y's.

* * *

The purpose of the principle is to focus our attention; it doesn't determine the shares or the division. The principle directs us to study the meaning of social goods, to examine the different distributive spheres from the inside.

Elizabeth Wolgast, from *The Grammar of Justice* (1929–)

Elizabeth Wolgast rejects not only the "social atomism" that underlies all of the classical liberal theories, but the very basic idea that justice is as such amenable and accessible to an analysis or "theory" of the sort attempted by Rawls, Nozick, and so many others ever since Plato. She suggests that the "grammar" of justice does not allow for such an analysis, but rather shows the primacy of the concept of *in*justice.

Why Justice Isn't an Ideal

We imagine that justice is an ideal or standard from which injustice departs. That we have such a standard seems necessary, for how else will we recognize a case of injustice when we come upon it? We need justice the way we need a pattern or standard that something can fail to fit. Though the idea is natural, I argue that it's mistaken. And from this mistake others flow, such as antinomies that have no rational solution in the face of morality's demand for one. My approach here involves looking not at the concept by itself but more broadly at its grammar, at the ways and the contexts in which justice is invoked.

I

It seems at first glance reasonable that injustice should be a violation of just rules or a just system, that injustice is a departure and goes contrary to justice, whatever it is. Justice is the prior notion, and it connects with some positive vision. Thus we have the image of the scales that are originally in balance, and imbalance comes as a departure from the original equilibrium. Therefore to redress injustice we must tip the scales back into symmetry.

How do we get the tipping done? By punishing criminals, fining civil offenders, granting claims to damages, and exiling traitors. Public justice, Kant writes, "is just the principle of equality, by which the pointer of the scale of justice is made to incline no more to the one side than the other." The language of debt and repayment goes along naturally with the image of the scales, for both images suggest returning things to some prior condition of rest, or, as Plato would say, of harmony. When that point is reached, nothing further needs to be done. The condition of justice is one of stasis: the account books are in balance and can be closed. Injustice is an upsetting of the balance, which restlessly demands adjustment, while payment restores things to the state of satisfaction and rest.

Yet notoriously the metaphor fails at just this point, that is, the return of things to their original equitable state. For when a criminal is punished—having committed rape or burglary or murder, say—things are *not* returned to their original state. Even though a theft is recompensed, the injury and wrong done cannot be undone; they become a permanent part of the universe, never to be erased. Thus the condition of stasis, defined as the condition in which the wrong is essentially corrected, is unattainable. The debt cannot be repaid.

If there were a debt it could be repaid; and scales out of balance can be brought into balance again. The problem with our reasoning here is a problem with the images we use. The images of debt and scales are unreliable guides to show us the grammar of justice, to show how that concept works. The point at which both fail is the identification of justice with a logically prior, harmonious state of affairs, a state to which theoretically we could return by redressing injustices in an appropriate way.

The view that wrong is unexpungible is defended eloquently by Fyodor Dostoyevsky in *The Brothers Karamazov*. Ivan asks young Alyosha, who plans to enter the priesthood, how there can be so much cruelty in the universe if a good God created it. Take the mistreatment of very young children; you cannot argue that they deserve what they are made to suffer, since they have had no opportunity to sin: they are innocent. Yet children are mistreated in terrible ways in every culture. How do you atone for such acts? If you suppose justice to be a state of harmony, the sufferings of these children "must be atoned for, or there can be no harmony. But how? How are you going to atone for them? Is it possible? By their being avenged? But what do I care for avenging them? What do I care for a hell for oppressors? What good can hell do, since those children have already been tortured? . . . I don't want more suffering" (Fyodor Dostoyevsky, *The Brothers Karamazov*). A state of harmony looks like a state unspoiled by wrong or perhaps one where wrong has been atoned for. But atonement may be impossible; wrong cannot be undone. Therefore, once a wrong is committed, harmony or stasis must be forever unattainable. But in that case, what can we say of achieving justice? If justice *were* the state of wrong expunged, its attainment appears to be logically impossible.

This is the puzzle we are driven to by supposing that justice is a state of harmony, an "original state," morally speaking, from which injustice digresses. That idea is what leads us to think in terms of scales and debts.

From a psychological and social point of view, one can find support for the argument that justice is an original state in the fact that wrongdoing takes place against a background of decency. Only in a setting where people normally treat one another in respectful ways do hurtful and unjust actions stand out and shock us. Injustice is then seen as diverging from these norms. I don't want to quarrel with this claim. What I contest is the idea that injustice is defined and made recognizable by some positive vision of justice, that the conception of justice is primary.

If wrong is ineradicable, then justice is a poor virtue. What might be preferable would be a state where the need for justice was absent; then the scales would be in balance, for they would be undisturbed. However, we need to look more carefully at the assumptions behind this picture.

II

Instead of fastening our attention on justice the substantive, let us examine some of the contexts where justice is invoked, that is, complaints against injustice. In the face of wrong, justice is demanded and cried out for, and with passion and intensity. "We must have justice!" and "Justice must be done!" are its expressions, and they characteristically have imperative force as well as urgency.

A demand for justice is generally a demand for action in the face of wrongdoing. It is a demand for counteraction, some kind of "corrective." The implication is plain: something must be done because to do nothing would be to accept the wrong and thus to sanction it. To tolerate the wrong is to associate oneself with it. This much is part of our everyday moral understanding.

The need to dissociate oneself from wrong pervades retributive theories of punishment, and is reflected in Kant's remark that a community that doesn't punish its offenders bears some of the guilt on its own head. But what does the demand for justice tell one to *do*? Kant and other retributivists held that what is demanded is punishment of offenders, for only in this way can the innocent dissociate themselves from the wrong and bring about moral restitution. But how can this account be squared with the consequence that nothing can restore the harmony? And what kind of punitive action would restore it in their eyes? We are left in the dark.

If there *were* an ideal of justice which stood as the original standard for denouncing injustice, then the demand might be understood as a demand to move as close as possible back toward the ideal. Even though only small and partial steps could be taken, the route itself would lie clear, for the ideal of justice would define both the wrong and the degree of wrong, would show whether the departure from it was greater or lesser. But, as I argue, we have reason to doubt that there is such a state or ideal. And then how are we to understand the concept of a just response to wrong?

Let us go back one step. The demand for justice appears to be a demand for action, and the necessity for acting lies somehow in the fact that inaction signifies an acceptance of the wrong, which is to say toleration of it. So part of what is demanded is that one go on record as opposing a wrong and expressing abhorrence of it.

Such expression may take a variety of forms, none of them dictated by the wrong itself. In some cultures it may involve ritual cleansing and repentance, with a public ceremony; or the wrongdoer may be publicly condemned; in others the rule will be physical incarceration and punishment. So in its form the requirement seems exceedingly loose. The commission of a wrong, by its description, does not imply that any action of a particular description will be *the* appropriate response to it.

I think that this fact has sometimes been taken to mean that justice and injustice are entirely relative, even conventional, changing from culture to culture and impossible to pin down. This inference stems, however, from an attempt to see justice either as an ideal or as mere convention, while I suggest that neither of these options is right, that in setting these alternatives we misread the way the concept works. Justice is essentially and grammatically "unwilling," as Edmond Cahn put it, "to be captured in a formula" while it remains "a word of magic evocations."

Michael Sandel, from *Liberalism and the Limits of Justice* (1953–)

In what follows, Michael Sandel wraps up his argument against Rawls's theory. Having rejected the stark individualism of the "original position" (see selection in Part II), he now concludes that it is the concept of "community" that both explains and limits justice.

Liberalism and the Limits of Justice

For justice to be the first virtue, certain things must be true of us. We must be creatures of a certain kind, related to human circumstance in a certain way. We must stand at a certain distance from our circumstance, whether as transcendental subject in the case of Kant, or as essentially unencumbered subject of possession in the case of Rawls. Either way, we must regard ourselves as independent: independent from the interests and attachments we may have at any moment, never identified by our aims but always capable of standing back to survey and assess and possibly to revise them (Rawls 1979: 7; 1980: 544–5).

Deontology's Liberating Project

Bound up with the notion of an independent self is a vision of the moral universe this self must inhabit. Unlike classical Greek and medieval Christian conceptions, the universe of the deontological ethic is a place devoid of inherent meaning, a world 'disenchanted' in Max Weber's phrase, a world without an objective moral order. Only in a universe empty of *telos*, such as seventeenth-century science and philosophy affirmed, is it possible to conceive a subject apart from and prior to its purposes and ends. Only a world ungoverned by a purposive order leaves principles of justice open to human construction and conceptions of the good to individual choice. In this the depth of opposition between deontological liberalism and teleological world views most fully appears.

Where neither nature nor cosmos supplies a meaningful order to be grasped or apprehended, it falls to human subjects to constitute meaning on their own. This would explain the prominence of contract theory from Hobbes onward, and the corresponding emphasis on voluntarist as against cognitive ethics culminating in Kant. What can no longer be found remains somehow to be created. Rawls describes his own view in this connection as a version of Kantian 'constructivism'.

The parties to the original position do not agree on what the moral facts are, as if there were already such facts. It is not that, being situated impartially, they have a clear and undistorted view of a prior and independent moral order. Rather (for constructivism), *there is no such order*, and therefore no such facts apart from the procedure as a whole [emphasis added] (1980: 568).

Similarly for Kant, the moral law is not a discovery of theoretical reason but a deliverance of practical reason, the product of pure will. 'The elementary practical concepts have as their foundation the form of a pure will given in reason,' and what makes this will authoritative is that it legislates in a world where meaning has yet to arrive. Practical reason finds its advantage over theoretical reason precisely in this voluntarist faculty, in its capacity to generate practical precepts directly, without recourse to cognition. 'Since in all precepts of the pure will it is only a question of the determination of will,' there is no need for these precepts 'to wait upon intuitions in order to acquire a meaning. This occurs for the noteworthy reason that *they themselves produce the reality of that to which they refer*' [emphasis added] (1788: 67–8).

It is important to recall that, on the deontological view, the notion of a self barren of essential aims and attachments does not imply that we are beings wholly without purpose or incapable of moral ties, but rather that the values and relations we have are the products of choice, the possessions of a self given prior to its ends. It is similar with deontology's universe. Though it rejects the possibility of an objective moral order, this liberalism does not hold that just anything goes. It affirms justice, not nihilism. The notion of a universe empty of intrinsic meaning does not, on the deontological view, imply a world wholly ungoverned by regulative principles, but rather a moral universe inhabited by subjects capable of constituting meaning on their own—as agents of *construction* in case of the right, as agents of *choice* in the case of the good. *Qua* noumenal selves, or parties to the original position, we arrive at principles of justice; *qua* actual, individual selves, we arrive at conceptions of the good. And the principles we construct as noumenal selves constrain (but do not determine) the purposes we choose as individual selves. This reflects the priority of the right over the good.

The deontological universe and the independent self that moves within it, taken together, hold out a liberating vision. Freed from the dictates of nature and the sanction of social roles, the deontological subject is installed as sovereign, cast as the author of the only moral meanings there are. As inhabitants of a world without *telos*, we are free to construct principles of justice unconstrained by an order of value antecedently given. Although the principles of justice are not strictly speaking a matter of choice, the society they define 'comes as close as a society can to being a voluntary scheme' (13), for they arise from a pure will or act of construction not answerable to a prior moral order. And as independent selves, we are free to choose our purposes and ends unconstrained by such an order, or by custom or traditional or inherited status. So long as they are not unjust, our conceptions of the good carry weight, whatever they are, simply in virtue of our having chosen them. We are 'self-originating sources of valid claims' (Rawls 1980: 543).

Now justice is the virtue that embodies deontology's liberating vision and allows it to unfold. It embodies this vision by describing those principles the sovereign subject is said to construct while situated prior to the constitution of all value. It allows the vision to unfold in that, equipped with these principles, the just society regulates each person's choice of ends in a way compatible with a similar liberty for all. Citizens governed by justice are thus enabled to realize deontology's liberating project—to exercise their capacity as 'self-originating sources of valid claims'—as fully as circumstances permit. So the primacy of justice at once expresses and advances the liberating aspirations of the deontological world view and conception of the self.

But the deontological vision is flawed, both within its own terms and more generally as an account of our moral experience. Within its own terms, the deontological self, stripped of all possible constitutive attachments, is less liberated than disempowered. As we have seen, neither the right nor the good admits of the voluntarist derivation deontology requires. As agents of construction we do not really construct (chapter 3), and as agents of choice we do not really choose (chapter 4). What goes on behind the veil of ignorance is not a contract or an agreement but if anything a kind of discovery; and what goes on in 'purely preferential choice' is less a choosing of ends than a matching of pre-existing desires, undifferentiated as to worth, with the best available means of satisfying them. For the parties to the original position, as for the parties to ordinary deliberative rationality, the liberating moment fades before it arrives; the sovereign subject is left at sea in the circumstances it was thought to command.

The moral frailty of the deontological self also appears at the level of first-order principles. Here we found that the independent self, being essentially dispossessed, was too thin to be capable of desert in the ordinary sense (chapter 2). For claims of desert presuppose thickly-constituted selves, beings capable of possession in the constitutive sense, but the deontological self is wholly without possessions of this kind. Acknowledging this lack, Rawls would found entitlements on legitimate expectations instead. If we are incapable of desert, at least we are entitled that institutions honor the expectations to which they give rise.

But the difference principle requires more. It begins with the thought, congenial to the deontological view, that the assets I have are only accidentally mine. But it ends by assuming that these assets are therefore common assets and that society has a prior claim on the fruits of their exercise. This either disempowers the deontological self or denies its independence. Either my prospects are left at the mercy of institutions established for 'prior and independent social ends' (313), ends which may or may not coincide with my own, or I must count myself a member of a community defined in part by those ends, in which case I cease to be unencumbered by constitutive attachments. Either way, the difference principle contradicts the liberating aspiration of the deontological project. We cannot be persons for whom justice is primary and also be persons for whom the difference principle is a principle of justice.

Character, Self-Knowledge, and Friendship

If the deontological ethic fails to redeem its own liberating promise, it also fails plausibly to account for certain indispensable aspects of our moral experience. For deontology insists that we view ourselves as independent selves, independent in the sense that our identity is never tied to our aims and attachments. Given our 'moral power to form, to revise, and rationally to pursue a conception of the good' (Rawls 1980: 544), the continuity of our identity is unproblematically assured. No transformation of my aims and attachments could call into question the person I am, for no such allegiances, however deeply held, could possibly engage my identity to begin with.

But we cannot regard ourselves as independent in this way without great cost to those loyalties and convictions whose moral force consists partly in the fact that living by them is inseparable from understanding ourselves as the particular persons we are—as members of this family or community or nation or people, as bearers of this history, as sons and daughters of that revolution, as citizens of this republic. Allegiances such as these are more than values I happen to have or aims I 'espouse at any given time.' They go beyond the obligations I voluntarily incur and the 'natural duties' I owe to human beings as such. They allow that to some I owe more than justice requires or even permits, not by reason of agreements I have made but instead in virtue of those more or less enduring attachments and commitments which taken together partly define the person I am.

To imagine a person incapable of constitutive attachments such as these is not to conceive an ideally free and rational agent, but to imagine a person wholly without character, without moral depth. For to have character is to know that I move in a history I neither summon nor command, which carries consequences none the less for my choices and conduct. It draws me closer to some and more distant from others; it makes some aims more appropriate, others less so. As a self-interpreting being, I am able to reflect on my history and in this sense to distance myself from it, but the distance is always precarious and provisional, the point of reflection never finally secured outside the history itself. A person with character thus knows that he is complicated in various ways even as he reflects, and feels the moral weight of what he knows.

This makes a difference for agency and self-knowledge. For, as we have seen, the deontological self, being wholly without character, is incapable of self-knowledge in any morally serious sense. Where the self is unencumbered and essentially dispossessed, no person is left for *self*-reflection to reflect upon. This is why, on the deontological view, deliberation about ends can only be an exercise in arbitrariness. In the absence of constitutive attachments, deliberation issues in 'purely preferential choice,' which means the ends we seek, being mired in contingency, 'are not relevant from a moral standpoint' (Rawls 1975: 537).

When I act out of more or less enduring qualities of character, by contrast, my choice of ends is not arbitrary in the same way. In consulting my preferences, I have not only to weigh their intensity but also to assess their suitability to the person I (already) am. I ask, as I deliberate, not only what I really want but who I really am, and this last question takes me beyond an attention to my desires alone to reflect on my identity itself. While the contours of my identity will in some ways be open and subject to revision, they are not wholly without shape. And the fact that they are not enables me to discriminate among my more immediate wants and desires; some now appear essential, others merely incidental to my defining projects and commitments. Although there may be a certain ultimate contingency in my having wound up the person I am—only theology can say for sure—it makes a moral difference none the less that, being the person I am, I affirm these ends rather than those, turn this way rather than that. While the notion of constitutive attachments may at first seem an obstacle to agency—the self, now encumbered, is no longer strictly prior—some relative fixity of character appears essential to prevent the lapse into arbitrariness which the deontological self is unable to avoid.

The possibility of character in the constitutive sense is also indispensable to a certain kind of friendship, a friendship marked by mutual insight as well as sentiment. By any account, friendship is bound up with certain feelings. We like our friends; we have affection for them, and wish them well. We hope that their desires find satisfaction, that their plans meet with success, and we commit ourselves in various ways to advancing their ends.

But for persons presumed incapable of constitutive attachments, acts of friendship such as these face a powerful constraint. However much I might hope for the good of a friend and stand ready to advance it, only the friend himself can know what that good is. This restricted access to the good of others follows from the limited scope for self-reflection, which betrays in turn the thinness of the deontological self to begin with. Where deliberating about my good means no more than attending to wants and desires given directly to my awareness, I must do it on my own; it neither requires nor admits the participation of others. Every act of friendship thus becomes parasitic on a good identifiable in advance. 'Benevolence and love are second-order notions: they seek to further the good of beloved individuals that is already given' (191). Even the friendliest sentiments must await a moment of introspection itself inaccessible to friendship. To expect more of any friend, or to offer more, can only be a presumption against the ultimate privacy of self-knowledge.

For persons encumbered in part by a history they share with others, by contrast, knowing oneself is a more complicated thing. It is also a less strictly private thing. Where seeking my good is bound up with exploring my identity and interpreting my life history, the knowledge I seek is less transparent to me and less opaque to others. Friendship becomes a way of knowing as well as liking. Uncertain which path to take, I consult a friend who knows me well, and together we deliberate, offering and assessing by turns competing descriptions of the person I am, and of the alternatives I face as they bear on my identity. To take seriously such deliberation is to allow that my friend may grasp something I have missed, may offer a more adequate account of the way my identity is engaged in the alternatives before me. To adopt this new description is to see myself in a new way; my old self-image now seems partial or occluded, and I may say in retrospect that my friend knew me better than I knew myself. To deliberate with friends is to admit this possibility, which presupposes in turn a more richly-constituted self than deontology allows. While there will of course remain times when friendship requires deference to the self-image of a friend, however flawed, this too requires insight; here the need to defer implies the ability to know.

So to see ourselves as deontology would see us is to deprive us of those qualities of character, reflectiveness, and friendship that depend on the possibility of constitutive projects and attachments. And to see ourselves as given to commitments such as these is to admit a deeper commonality than benevolence describes, a commonality of shared self-understanding as well as 'enlarged affections.' As the independent self finds its limits in those aims and attachments from which it cannot stand apart, so justice finds its limits in those forms of community that engage the identity as well as the interests of the participants.

To all of this, deontology might finally reply with a concession and a distinction: it is one thing to allow that 'citizens in their personal affairs . . . have attachments and loves that they believe they would not, or could not, stand apart from,' that they 'regard it as unthinkable . . . to view themselves without certain religious and philosophical convictions and commitments' (Rawls 1980: 545). But with public life it is different. There, no loyalty or allegiance could be similarly essential to our sense of who we are. Unlike our ties to family and friends, no devotion to city or nation, to party or cause, could possibly run deep enough to be defining. By contrast with our private identity, our 'public identity' as moral persons 'is not affected by changes over time' in our conceptions of the good (Rawls 1980: 544–5). While we may be thickly-constituted selves in private, we must be wholly unencumbered selves in public, and it is there that the primacy of justice prevails.

But once we recall the special status of the deontological claim, it is unclear what the grounds for this distinction could be. It might seem at first glance a psychological distinction; detachment comes more easily in public life, where the ties we have are typically less compelling; I can more easily step back from, say, my partisan allegiances than certain personal loyalties and affections. But as we have seen from the start, deontology's claim for the independence of the self must be more than a claim of psychology or sociology. Otherwise, the primacy of justice would hang on the degree of benevolence and fellow-feeling any particular society managed to inspire. The independence of the self does not mean that I can, as a psychological matter, summon in this or that circumstance the detachment required to stand outside my values and ends, rather that I must regard myself as the bearer of a self distinct from my values and ends, whatever they may be. It is above all an epistemo-logical claim, and has little to do with the relative intensity of feeling associated with public or private relations.

Understood as an epistemological claim, however, the deontological conception of the self cannot admit the distinction required. Allowing constitutive possibilities where 'private' ends are at stake would seem unavoidably to allow at least the possibility that 'public' ends could be constitutive as well. Once the bounds of the self are no longer fixed, individuated in advance and given prior to experience, there is no saying in principle what sorts of experiences could shape or reshape them, no guarantee that only 'private' and never 'public' events could conceivably be decisive.

Not egoists but strangers, sometimes benevolent, make for citizens of the deontologi-cal republic; justice finds its occasion because we cannot know each other, or our ends, well enough to govern by the common good alone. This condition is not likely to fade altogether, and so long as it does not, justice will be necessary. But neither is it guaranteed always to predominate, and in so far as it does not, community will be possible, and an unsettling presence for justice.

Liberalism teaches respect for the distance of self and ends, and when this distance is lost, we are submerged in a circumstance that ceases to be ours. But by seeking to secure this distance too completely, liberalism undermines its own insight. By putting the self beyond the reach of politics, it makes human agency an article of faith rather than an object of continuing attention and concern, a premise of politics rather than its precarious achievement. This misses the pathos of politics and also its most inspiring possibilities. It overlooks the danger that when politics goes badly, not only disappointments but also dislocations are likely to result. And it forgets the possibility that when politics goes well, we can know a good in common that we cannot know alone.

SELECTED BIBLIOGRAPHY

Ackerman, B. A., *Social Justice in the Liberal State* (New Haven: Yale University Press, 1980).

Aristotle, *Nicomachean Ethics*. T. Irwin, trans. (Indianapolis: Hackett, 1985).

Aristotle, *Politics*. B. Jowett, trans. (New York: Modern Library, 1943).

Axelrod, Robert M., *The Evolution of Cooperation* (New York: Basic Books, 1984.

Barber, Benjamin, *The Conquest of Politics*. (Oxford: Oxford University Press, 1988).

Barker, Ernest, ed., *The Social Contract: Locke, Hume and Rousseau* (Oxford: Oxford University Press, 1962).

Bean, Philip, *Punishment: A Philosophical and Criminological Inquiry* (1969).

Beccaria, Cesare, *On Crimes and Punishment*. H. Paolucci, trans. (Indianapolis: Bobbs-Merrill, 1963).

Becker, Lawrence. *Reciprocity* (Boston: Routledge and Kegan Paul, 1986).

Bedau, Hugh. "Capital Punishment and Retributive Justice" in Tom Regan, ed., *Matters of Life and Death: New Essays in Moral Philosophy* (New York: Random House Inc., 1980).

Bentham, Jeremy, *Introduction to the Principles of Morals and Legislation* (New York: Hafner, 1948).

Buchanan, James M., *The Limits of Liberty: Between Anarchy and Leviathan* (Chicago: University of Chicago, 1975).

Calhoun, Cheryl H. "Justice, Care, Gender Bias." *Journal of Philosophy* 85(9):451–63.

Camus, Albert, "Reflections on the Guillotine" in *Resistance, Rebellion and Death* (New York: Vintage, 1961).

Cohen, Ronald, *Justice: Views from the Social Sciences* (1986).

Daniels, Norman, *Reading Rawls* (New York: Basic Books, 1975).

Dworkin, Ronald, *Taking Rights Seriously* (Cambridge, MA: Harvard University Press, 1977).

Feinberg, Joel, *Doing and Deserving* (Princeton: Princeton University Press, 1970).

Feinberg, Joel, "The Nature and Value of Rights." *Journal of Value Inquiry* 4:243–57 (1970).

Feinberg, Joel, *Social Philosophy* (Engelwood Cliffs, NJ: Prentice-Hall, 1973).

Fishkin, James, *Justice, Equal Opportunity and the Family* (New Haven: Yale University Press, 1983).

Flew, *The Politics of Procrustes* (Buffalo: Prometheus, 1981).

French, P., T. Uehling, and H. Wettstein, eds. *Social and Political Philosophy (Midwest Studies in Philosophy VII)* (Minneapolis: University of Minnesota Press, 1982).

Galston, William, "Equal Opportunity and Liberal Theory" in F. Lucasch, ed., *Justice and Equality: Here and Now* (Ithaca: Cornell University Press, 1986), pp. 89–107.

Galston, William, *Justice and the Human Good* (Chicago: University of Chicago Press, 1980).

Gauthier, David, *Morals by Agreement* (New York: Oxford University Press. 1986).

Gerstein, Robert S. "Capital Punishment: A Retributivist Response." *Ethics* 85:75–79.

Hampton, Jean, *Hobbes and the Social Contract Tradition* (Cambridge: Cambridge University Press, 1978).

Havelock, Eric, *The Greek Concept of Justice: From its Shadow in Homer to its Substance in Plato* (Cambridge: Cambridge University Press, 1978).

Hegel, G. W. F., *The Phenomenology of Spirit*. A. V. Miller, trans. (Oxford: Oxford University Press, 1977).

Hegel, G. W. F., *Philosophy of Right*. T. M. Knox, trans. (New York: Oxford University Press, 1967).

Heller, Agnes, *Beyond Justice* (Oxford: Oxford University Press, 1981).

Heschel, Abraham, *The Prophets* (New York: Harper and Row, 1962).

Hobbes, Thomas, *Leviathan* (New York: Hafner, 1926).

Homans, George C., *Social Behavior: Its Elementary Forms* (New York: Harcourt Brace Jovanovich, 1961).

Honderich, T., *Punishment: The Supposed Justifications* (Harmondsworth: Penguin, 1976).

Hume, David, *A Treatise of Human Nature*, Selby-Bigge, ed. (Oxford: Oxford University Press, 1978).

Hume, David, *Enquiries Concerning the Principles of Morals*, Selby-Bigge, ed. (Oxford: Oxford University Press, 1978).

Jackson, W. W., *Matters of Justice* (London: Croom Helm, 1986).

Jacoby, Susan, *Wild Justice* (New York: Harper and Row, 1983).

Kant, Immanuel, *The Grounding of the Metaphysics of Morals*. J. Ellington, trans. (Indianapolis: Hackett, 1981).

Kant, Immanuel, *The Metaphysical Elements of Justice*. J. Ladd, trans. (Indianapolis: Bobbs-Merrill, 1965).

Kaufmann, Walter, *Without Guilt and Justice* (New York: Wyden, 1973).

Lerner, Melvin J., *Belief in a Fair World* (New York: Plenum, 1980).

Locke, from *The Second Treatise on Government* (Indianapolis: Hackett, 1983).

Lucasch, Frank, ed., *Justice and Equality* (Ithaca: Cornell University Press, 1986).

MacIntyre, Alasdair, *After Virtue* (Notre Dame: University of Notre Dame Press, 1981).

MacIntyre, Alasdair, *Whose Justice? Which Rationality?* (Notre Dame: University of Notre Dame Press, 1988).

Marx, *Selected Writings*, David McClellan, ed. (New York: Oxford University Press, 1977).

Melden, A., *Rights and Persons* (Berkeley, University of California Press, 1977).

Mencius, *The Mind of Mencius*. D. C. Lau, trans. (New York: Penguin, 1970).

Mill, John Stuart, *On Liberty* (Indianapolis: Hackett, 1978).

Mill, John Stuart, *Utilitarianism* (Indianapolis: Hackett, 1980).

Miller, David, *Social Justice* (Oxford: Oxford University Press, 1976).

Marongiu, Pietro, and Graeme Newman, *Vengeance* (Toronga, NJ: Rowman and Littlefield, 1987).

Martin, Rex, *Rawls and Rights* (Lawrence, KS: Kansas University Press, 1985).

Nagel, Thomas, *Mortal Questions* (Cambridge: Cambridge University Press, 1979).

Nagel, Thomas, *The View from Nowhere* (New York: Oxford University Press, 1986).

Nietzsche, Friedrich, *On the Genealogy of Morals*. W. Kaufmann, trans. (New York: Random House, 1967).

Nozick, Robert, *Anarchy State and Utopia* (New York: Basic Books, 1974).

Nozick, Robert, *Philosophical Explanations* (Cambridge: The Belknap Press of Harvard University Press, 1981).

Nussbaum, M., *The Fragility of Goodness* (New York: Cambridge University Press, 1985).

Parfit, Derek, *Reasons and Persons* (Oxford: Oxford University Press, 1984).

Pincoffs, Edmund, *Quandaries and Virtues* (Lawrence: University Press of Kansas, 1986).

Plato, *The Apology*, in *The Trial and Death of Socrates*. G. M. A. Grube, trans. (Indianapolis: Hackett, 1982).

Plato, *Crito*, in *The Trial and Death of Socrates*. G. M. A. Grube, trans. (Indianapolis: Hackett, 1982).

Plato, *The Republic*. G. M. A. Grube, trans. (Indianapolis: Hackett, 1982).

Posner, Richard, *The Economics of Justice* (Cambridge, MA: Harvard University Press, 1983).

Rapacynski, Andrzej, *Nature and Politics* (Ithaca, NY: Cornell University Press, 1987).

Raphael, D. D., *Moral Philosophy* (Oxford: Opus Books, 1981).

Rawls, John, *A Theory of Justice* (Cambridge: Harvard University Press, 1971).

Rawls, John, "Justice as Fairness." *Philosophical Review* 57 (1958).

Rescher, Nicholas, *Unselfishness* (Pittsburgh: University of Pittsburgh Press, 1975).

Rousseau, Jean-Jacques, *Discourse on the Origins of Inequality*. D. Cress, trans. (Indianapolis: Hackett, 1987).

Rousseau, Jean-Jacques, *Emile*. A. Bloom, trans. (New York: Basic Books, 1979).

Rousseau, Jean-Jacques, *The Social Contract*. D. Cress, trans. (Indianapolis: Hackett, 1987).

Sandel, Michael, *Liberalism and the Limits of Justice* (Cambridge: Cambridge University Press, 1982).

Sen, A. K., *Collective Choice and Social Welfare* (San Francisco: Holden-Day, 1970).

Sidgwick, Henry, *Methods of Ethics*, second ed. (London: Macmillan, 1907).

Singer, Peter, *Practical Ethics* (Cambridge: Cambridge University Press, 1979).

Smith, Adam, *An Inquiry into the Nature and Causes of the Wealth of Nations*, R. H. Campbell and A. S. Skinner, eds. (Oxford: Oxford University Press, 1976).

Smith, Adam, *The Theory of Moral Sentiments* (London: George Bell and Sons, 1880).

Solomon, Robert C., *A Passion for Justice* (New York: Addison-Wesley, 1990).

Solomon, Robert C., *In the Spirit of Hegel* (New York: Oxford University Press, 1983).

Soltan, Karol, *The Causal Theory of Justice* (Berkeley: University of California, 1987).

Sterba, James, *The Demands of Justice* (North Bend, IN: Notre Dame University Press, 1980).

Tawney, R. H., *Equality* (New York: Barnes and Noble, 1964).

Taylor, Charles, "The Nature and Scope of Distributive Justice," in F. Lucasch, ed., *Justice and Equality: Here and Now* (Ithaca: Cornell University Press, 1986), pp. 34–67.

Ten, C. L., *Crime, Guilt and Punishment: A Philosophical Introduction* (Oxford: Oxford University Press, 1987).

Thomas, Laurence, "Ethical Egoism and our Psychological Dispositions." *American Philosophical Quarterly* 17:73–78.

Vlastos, Gregory, "Justice and Equality," in R. Brandt, ed., *Social Justice* (Engelwood Cliffs: Prentice-Hall, 1962).

Van den Haag, Ernst, "Deterrence and Uncertainty." *Journal of Criminal Law, Criminology, and Police Science* 60(2) (1969).

von Hayek, Friedrich, *The Mirage of Social Justice* (Vol. 2 of *Law, Legislation, and Liberty*) (London: Routledge and Kegan-Paul, 1976).

Walzer, Michael, "Justice Here and Now," in F. Lucasch, ed., *Justice and Equality: Here and Now* (Ithaca: Cornell University Press, 1986), pp. 136–50.

Walzer, Michael, *Spheres of Justice* (New York: Harper and Row, 1983).

Westermarck, Edward, *The Origin and Development of the Moral Ideas* (2 vols.) (London: Macmillan, 1912).

Williams, Bernard, *Ethics and the Limits of Philosophy* (Cambridge: Harvard University Press, 1985).

Williams, Bernard, "The Idea of Equality," in *Problems of the Self* (New York: Cambridge University Press, 1973). Originally in *Philosophy, Politics and Society*, P. Laslett and W. C. Runciman, eds. (Oxford: Blackwell, 1964), pp. 110–31.

Williams, Bernard, "Justice as a Virtue," in Amelie Rorty, ed., *Essays on Aristotle's Ethics* (Berkeley: University of California Press, 1980).

Winfield, Richard, *Reason and Justice* (Buffalo: S.U.N.Y. Press, 1988).

Wolff, Robert Paul, *Understanding Rawls* (Princeton: Princeton University Press, 1977).

Wolgast, Elizabeth, *The Grammar of Justice* (Ithaca: Cornell University Press, 1987).